Lecture Notes on Coastal and Estuarine Studies

Managing Editors:
Malcolm J. Bowman   Richard T. Barber
Christopher N. K. Mooers   John A. Raven

19

K. N. Fedorov

Translator
Nadia Demidenko

Technical Editor
Chris Garrett

The Physical Nature and Structure of Oceanic Fronts

Springer-Verlag
Berlin Heidelberg New York London Paris Tokyo

Managing Editors

Malcolm J. Bowman
Marine Sciences Research Center, State University of New York
Stony Brook, N.Y. 11794, USA

Richard T. Barber
Coastal Upwelling Ecosystems Analysis
Duke University, Marine Laboratory
Beaufort, N.C. 28516, USA

Christopher N. Mooers
Dept. of Oceanography, Naval Postgraduate School
Monterey, CA 93940, USA

John A. Raven
Dept. of Biological Sciences, Dundee University
Dundee, DD1 4HN, Scotland

Contributing Editors

Ain Aitsam (Tallinn, USSR) · Larry Atkinson (Savannah, USA)
Robert C. Beardsley (Woods Hole, USA) · Tseng Cheng-Ken (Qingdao, PRC)
Keith R. Dyer (Merseyside, UK) · Jon B. Hinwood (Melbourne, AUS)
Jorg Imberger (Western Australia, AUS) · Hideo Kawai (Kyoto, Japan)
Paul H. Le Blond (Vancouver, Canada) · Akira Okuboi (Stony Brook, USA)
William S. Reebourgh (Fairbanks, USA) · David A. Ross (Woods Hole, USA)
S.- Sethuraman (Raleigh, USA) · John H. Simpson (Gwynedd, UK)
Absornsuda Siripong (Bangkok, Thailand) · Robert L. Smith (Covallis, USA)
Mathis Tomaczak (Sydney, AUS) · Paul Tyler (Swansea, UK)

Title of the Original Russian Edition:
K.N. Fedorov, Fizicheskaya Priroda i Struktura
Pribrezhnogo Okeanicheskogo Fronta
© 1983 by Gidrometeoizdat, Leningrad

ISBN-13: 978-0-387-96445-4   e-ISBN-13: 978-1-4684-6343-9
DOI: 10.1007/978-1-4684-6343-9

Author

K.N. Fedorov, Dr. Sc.
Professor in Physical Oceanology
P.P. Shirshov Institute of Oceanology
Acad. Sci. USSR
23 Krasikova Street
117218 Moscow, USSR

Translator

Nadia Demidenko
Slavonic Languages Section
Translations Bureau
Department of the Secretary of State
Ottawa, Ontario
K1A OM5 Canada

Technical Editor

Chris Garrett
Department of Oceanography
Dalhousie University
Halifax, N.S.
B3M 4J1 Canada

Preface to the English Translation

This translation has been prepared by Nadia Demidenko under the direction of John Kilby, Chief of the Slavonic Languages Section of the Translation Bureau of the Department of the Secretary of State, Government of Canada, and with coordination by John Camp, Deputy Director of Editorial and Publishing Services, Department of Fisheries and Oceans.

Technical editing was carried out in consultation with the author, and camera ready copy prepared by Kathleen Sawler and Lynda Corkum at Dalhousie University, using the TEX computer typesetting system*, with technical assistance from Dr. Bert Buckley, Mr. Phil Green and others. Georges Merinfeld transferred the index.

The author and technical editor are profoundly grateful to all of these individuals and their organizations.

---

* Donald E. Knuth 1983. The TEXbook. Addison Wesley Publishing Company, 483 pp.

Author's Introduction to the English Translation

I consider as a great honour the publication of my monograph on oceanic fronts in the English language, and, in particular, I appreciate Prof. C. Garrett's friendly efforts in arranging for its translation and in taking upon himself the editorial work.

This book, like the previous one ("The Thermohaline Finestructure of the Ocean") was born out of the author's continuing fascination with the role natural boundaries of all sorts and scales play in the ocean's behaviour. In a sense the theme is enlarged here to encompass those lateral discontinuities in the ocean which in most cases account for the evident lack of continuity in the vertical. The author's aim in both instances was to attract attention to the accumulating body of knowledge (and evidence), both observational and theoretical, pointing to the necessity of some drastically new approaches in our attempts to understand the ocean.

While the appearance of this book in Russian some 3 years ago most definitely stimulated interest in oceanic fronts and promoted subsequent new research programmes, I see the present English edition more as a means of bringing together parallel research efforts of different oceanographic communities which sometime are needlessly selfcontained, due mainly to linguistic constraints. I am therefore aware, that, accordingly, judgements on the usefulness of different parts of my book may be different in the two cases.

I shall be very interested to receive critical responses from readers of the English edition. These may prove extremely useful if one day a second revised edition of this book is contemplated.

Moscow   K.N. Fedorov

TABLE OF CONTENTS

Chapter 4.
FRONTS AND THE STRUCTURE OF THE OCEAN

Chapter 5.
PROBLEMS FOR FUTURE RESEARCH AND THE CONCERNS OF
ASSOCIATED DISCIPLINES

INTRODUCTION

During the past several years, research into the frontal divisions of the ocean has been particularly intensive. The significance of this lies not only in the fact that, in the five years from 1976 to 1980, more than 500 papers on this question were published in various journals throughout the world and at least three major international discussions were held (see p. 17). The newness of the discussion of the topic stems from the qualitative reinterpretation of the physical essence of the phenomenon and its role in the ocean, particularly in the processes of mixing and structure formation.

While in the past the conventional view of fronts as boundaries between large-scale water masses of the ocean only led to the recognition of convenient classification limits created by nature itself, there is now a tendency to study oceanic fronts as integral elements of the dynamics of oceanic waters. As we understand it, fronts are being associated more and more with the dynamic and kinematic features which arise when kinetic energy and enstrophy are transmitted through a cascade of scales characterizing various forms of motion of a stratified medium in laterally confined oceanic basins. We are beginning to get a better understanding of the role synoptic-scale oceanic eddies play in the process of frontogenesis in the ocean. We are beginning to perceive the laws that govern the creation and destruction of fronts due to sudden disruption of the normal processes which match the response of the ocean to the fluxes of heat and buoyancy across the ocean/atmosphere boundary. We can no longer ignore the role of the Earth's rotation in the formation, maintenance and evolution of fronts, as well as in the development of instability at these fronts. Finally, we are turning to fronts as the most likely, and possibly the most effective mechanism for mixing and the transfer of heat and salt through the hydrostatically stable pycnocline. We no longer doubt that the link between the vertical finestructure of the ocean and mesoscale horizontal inhomogeneities inevitably involves fronts.

Outdated scientific concepts are not replaced by new views and concepts instantly. For example, it took about 10 years (1965–1975) to revise the conventional ideas regarding the continuous vertical stratification of ocean waters, which had dominated in physical oceanography for a whole century, by introducing the radically different and more realistic concept of a vertical thermohaline finestructure [61]. Just as this concept had to struggle against the conservatism of practising oceanographers and the scepticism of theoreticians ten years ago, so must the new concept of oceanic fronts take the same course towards recognition today. It views the physical structure of ocean waters as being more complex, but at the same time much more comprehensible and logical. The analogy presented above is far from superficial. Indeed, why should small-scale division of a huge continuum (as the ocean appeared to us only yesterday) be considered characteristic only of the vertical direction?! The very presence in the ocean of thermohaline finestructure with a characteristic ratio of vertical to horizontal scales in the neighbourhood of 10^{-3} [61] suggests the existence of horizontal boundaries which are located at comparatively small distances from each other.

However, conventional descriptive oceanography, by analogy with climatology, acknowledged the existence of such boundaries only in the planetary sense. Subtropical convergences, subpolar fronts and polar fronts necessarily separated the main elements of the large-scale general circulation of the oceans in such a way that the generally accepted classification of water masses served to simplify or schematize the observed picture, rather than to describe it realistically. Global or quasiglobal scales were ascribed as required to fronts which separated the main elements of the general circulation. By their positions on maps or the properties ascribed to them, these fronts were often the fruit of crude averaging, and did not correspond very well to actual observations in the ocean.

One of the most important differences between today's ideas on oceanic fronts and conventional concepts is the acceptance of the "multiscale" nature of the phenomenon. Fronts in the ocean are encountered over the whole spectrum of spatial scales from one or more metres up to the dimensions of the ocean itself [82]. The description of oceanic fronts and the study of their dynamics no longer requires the concept of "water mass" as the logical or physical basis. One of the purposes of the present investigation is to present an up-to-date description of the physical nature and structure of frontal zones and divisions of the World Ocean without resorting to a conventional discussion of water masses.

Consequently, we can no longer be satisfied with the frequently cited definition of a front: "a front is the boundary between water masses". Other definitions suggested in the past are also neither universal, nor sufficiently acceptable from a physical point of view. For example, the definitions: "A front will mean a band along the sea surface across which the density changes abruptly. The stable layer separating the water of lower density from the denser water is called the frontal layer" [99] cannot be regarded as suitable, for we are aware of the existence of oceanic boundaries [214] with very large horizontal temperature and salinity gradients which compensate each other so that there is a complete absence of horizontal density gradient. Should we stop calling these boundaries fronts?! On the other hand, definitions such as "a front is a zone where the jump in temperature exceeds $1°C$ in 10 km" [252] or "a front is a band on the surface of the sea where the jump in temperature exceeds $1°C$ in 10 miles. The layer separating the warm water from the cold is called a frontal zone" [176] are not only based on totally arbitrary quantitative criteria, but also describe a front as a purely thermal phenomenon.

However, we know that purely saline fronts, which are visibly and dynamically much more sharply defined than thermal ones, can be encountered in the ocean.

Many researchers studying oceanic fronts over the past few years have come to the pessimistic conclusion that it is hardly possible to formulate a good universal definition of an oceanic front. Although basically of the same opinion, we feel that a discussion of the problem of defining an oceanic front as a phenomenon could prove useful to the reader, and some, albeit incomplete and formal criteria or characteristics of a front should be developed for use as instruments of research. Since oceanic fronts or frontal interfaces are always characterized by an extreme value of the spatial gradient of this or that physical property (temperature, salinity, density, velocity), this characteristic of fronts, which is insufficient for a universal definition, can serve as a basis for a formal quantitative criterion which could to some extent make up for the lack of a physical definition. The same criterion could help us to establish the difference (basically physical) between "frontal interface" and "frontal zone". This difference, which has been noted in nature by numerous researchers (e.g. [6]) but never discussed in the earlier papers of the author [70], will be referred to throughout this book. However, we should once again point out that a formal analysis of a field of spatial gradients from a quantitative point of view is inadequate for understanding frontal divisions in the ocean as an independent physical phenomenon. An important characteristic of frontal interfaces in the ocean is the persistence of spatial gradients to a high degree, despite their tendency to dissipate due to turbulent (and sometimes molecular) diffusion. This property of self-preservation of fronts is undoubtedly determined by the peculiarities of the local dynamics of the zones with relatively high gradients of the main thermodynamic properties of seawater. It is with this property that the various visible manifestations of fronts on the surface of the ocean are most likely associated. In our opinion, it is this characteristic that makes oceanic thermohaline fronts an independent dynamic phenomenon, in contrast to, say, the boundaries of the patches of any passive admixture in a region of active horizontal turbulent diffusion. Though oceanic fronts, together with eddies, can at will be interpreted as a manifestation of horizontal turbulent diffusion (or two-dimensional turbulence) [201, 262] which could result in a significant redistribution of scalar admixtures [178], a deterministic approach to the study of oceanic fronts would in many respects be more promising than a statistical one. These and other similar arguments have made it necessary to include a special section (1.2) on definition, terminology and criteria in chapter 1 of this book, where the above discussion is continued and carried on more specifically. However, it can be said that the above information is already enough to help the reader familiarize himself at the very beginning of the book with a brief historical outline of the development of ideas regarding oceanic fronts (see section 1.1).

The experience accumulated by meteorologists studying fronts in the atmosphere is extremely useful when studying oceanic fronts. Numerous dynamic analogues can be directly borrowed from atmospheric physics. At the same time, there are major differences between the physical structure of the ocean and the structure of the atmosphere. Certain manifestations of oceanic fronts are quite specific, and can be observed in the ocean alone. The differences in the characteristic space and time scales of the given phenomenon in the ocean and atmosphere are also significant. An analysis of all the parallels and differences between the conditions of frontogenesis in these two environments of the Earth should make it much easier to understand the dynamics of oceanic fronts (see section 2.4).

However, we should note that meteorologists studying fronts have experienced difficulties related to the convention of the two- dimensional approach to atmospheric processes in synoptic meteorology (in the horizontal plane XY or on isobaric surfaces). Oceanographers are even greater adherents of the two-dimensional approach, but in the XZ plane, rather than the XY plane. This is associated with the habit of plotting vertical sections along the course of a vessel. Neither of these approaches can satisfy the requirements of the current goals and the present level of understanding of this phenomenon. As in the atmosphere, fronts in the ocean are three-dimensional, and although they can sometimes be approximated by two-dimensional models, the most interesting phenomena at fronts always take place in three dimensions, and can be explained only within the framework of three-dimensional nonstationary problems. Because of this, the introduction of remote (sensing) methods of observation and measurement promises considerable improvement in the quality of our field data, and opens new and promising horizons in research [50, 71].

The author developed a desire to study oceanic fronts naturally, as a result of studying the vertical finestructure of the ocean over many years. In this sense, the present book serves as a logical continuation of the monograph "Thermohaline Finestructure of the Ocean" [61]. It is interesting to note that this development of scientific interests occurred in quite a number of researchers studying the finestructure of oceanic waters, e.g. J. Woods, G. Roden and R. Pingree. The author's discussions with them revealed a natural similarity of views on the relationship between finestructure and fronts on the one hand, and between fronts and other mesoscale oceanic phenomena on the other. Many of the author's points of view on oceanic fronts were greatly influenced by his participation in the Liège International Symposium on Oceanic Turbulence in May 1979, where oceanic fronts were discussed as one of the elements of the various manifestations of turbulence in the ocean in scales from "millimetre to megametre".

The author of this book had predecessors whose goal it was to correlate the various information on oceanic fronts from different perspectives. Among these predecessors, we should first of all mention Ye.I. Baranov, the author of a most interesting chapter on the dynamics and structure of waters at the frontal zone of the Gulf Stream, written for the symposium entitled "Investigations on the Circulation and Transfer of Waters in the Atlantic Ocean" [6] which was published in 1971. In 1975, V.M. Gruzinov published his monograph entitled "Frontal Zones of the World Ocean" [19] which can be regarded as the latest most complete summary of facts that have greatly contributed to the study of large-scale (climatic) oceanic fronts. This book is based mainly on oceanographic data obtained by means of the standard instruments and standard methods prior to 1970. Naturally, these books do not reflect the new results obtained over the past ten years by means of new experimental instruments and new methods, including remote sensing. Furthermore, these authors could not have presented a physical analysis of the important aspects of the problem which became apparent in connection with the discovery of synoptic eddies in the ocean and in connection with the acknowledgement of the close relationship between small-scale turbulence, internal waves and the finestructure of the physical fields in the ocean. Most of the new data obtained in recent years are scattered throughout hundreds of journals. The present monograph is expected to combine the new results, correlate them with the old data, and give the whole ensemble of available information a modern physical interpretation. Its internal structure is the result of a compromise between the desire of the author to present the available material as fully as possible and the unavoidable imperfection of the chosen principle of its organization. Chapter 1 contains the necessary introductory information about the subject and methods of investigation. Chapter 2 presents an up-to-date physical description of the phenomenon, including a brief review of the theory of oceanic frontogenesis. Chapter 3 is devoted to the phenomenological characteristics of certain types of frontal phenomena. It would hardly be possible or reasonable to attempt to cover all the existing types of oceanic frontal interfaces in this chapter. The unavoidable shortcoming related to the arbitrariness of the author's choice is compensated for by the classification of frontal zones and interfaces, presented in section 1.3. Chapter 4 contains information about the structure of oceanic waters in frontal zones and especially near frontal interfaces. The final, fifth chapter outlines the prospects of further research into oceanic fronts for the next $10 - 15$ years.

In his attempt to promote the study of oceanic fronts in every possible way, the author sought the support of Corresponding Member of the USSR Academy of Sciences A.S. Monin and Prof. G.I. Barenblatt, who assisted in organizing a fruitful discussion of numerous aspects of this problem at a theoretical seminar of the P.P. Shirshov Institute of Oceanology of the USSR Academy of Sciences in 1977. Staff-members of the Laboratory of Experimental Ocean Physics, which is headed by the author, helped out with the numerous measurements in field and laboratory conditions. Data obtained by A.G. Zatsepin, G.I. Shapiro, M.N. Koshlyakov and others have been utilized in a number of sections of the book. The author was also greatly assisted by the bibliography of papers on oceanic fronts, compiled in 1976 by the Rosenstiel School of Marine and Atmospheric Sciences of Florida University, and kindly made available to the author by C. Mooers. N.P. Kuz'mina made a firsthand contribution to the writing of section 2.6. The biological part of section 5.2 was written by M.Ye. Vinogradov. The discussion of a number of sections of the book with A.S. Monin and D.G. Seidov was extremely useful. The author expresses his most sincere gratitude to all of them.

There is no doubt that this book, which was written at a time of fundamental changes in views and a breathtakingly rapid accumulation of new facts, contains a great many hypotheses, conjectures and intuitive assumptions. However, the author would be quite satisfied if these hypotheses and assumptions later proved to be correct in the same proportion as in his previous monograph "Thermohaline Finestructure of the Ocean."

Chapter 1

THE SUBJECT AND METHODS OF RESEARCH

1.1 Historical summary of the development of ideas regarding oceanic fronts

The goal of any historical synopsis is not simply to list the facts in chronological order, but rather to attempt to interpret and comprehend this or that peculiarity of historical development. In the present case, where we are dealing with the study of frontal interfaces in the ocean, it would be useful to comprehend why the interest in oceanic fronts grew so tremendously only at the end of the past decade, despite the facts that the phenomenon itself has been known to mariners for more than 350 years and that the scientific concept of an oceanic front has been known to oceanographers for at least half a century.

Indeed, although the earliest references to abrupt changes, in the properties of oceanic waters, of evidently frontal origin are encountered in the navigational directions and journals of 15th to 17th century seamen, the development of ideas relevant to the phenomenon did not begin in earnest until the start of the twentieth century. Only a few researchers and mariners of the past century mentioned facts which described the highly complex local structure of the major boundary currents in the ocean, but their observations could not be properly interpreted until the basic frontal interfaces were understood in the most general way as elements of a large-scale circulation of oceanic waters. Facts had to be accumulated over many years and theoretical hydrodynamic concepts had to be developed in order to comprehend the physical nature of a remarkable phenomenon such as the Gulf Stream, and to combine the concept of "current" and "front" into a single one on the basis of this shining example. The reader will find an excellent synopsis of the early stage of this instructive historical process in Chapter 1 of H. Stommel's book "The Gulf Stream" [237].

The concept "front", which appeared in meteorology at the beginning of the twentieth century, was not originally accepted in oceanography. V. Bjerknes' famous theorem had already found its application in the dynamics of ocean currents, but scientists still did not think that the external manifestations of the different forms of movement in the ocean were sufficient grounds for drawing major physical analogies with the atmosphere. On the whole, it appeared that the ocean was characterized by more gradual changes in its physical and dynamic properties than the atmosphere, and so the boldest analogies in the most substantial oceanic studies went only so far as to state that vast zones of convergence and divergence of currents existed. This point of view somewhat illogically coexisted with the reports that numerous strange and inexplicable phenomena were observed on the surface of the ocean, and that these phenomena were followed by abrupt local changes in the physicochemical properties of the water and its movement. For a long time, the observations and ideas could not be combined into a single concept.

For example, it is interesting to note that in 1935, the German journal "Annalen der Hydrographie und Maritimen Meteorologie" was already publishing papers on the diagnostics and forecasts of "frontal processes" and on "frontal zones" as applied to the atmosphere, but it still presented only descriptive discussions about such phenomena as rough water* zones or rip lines (convergences of currents) on the surface of the ocean. The well-known German marine scientists of the beginning of the twentieth century O. Krümmel, G. Shott and H. Thorade quite cautiously and conditionally associated these phenomena with regions of current boundaries ("Stromgrenzen"), upwellings of abyssal waters and regions of convergence and divergence. On the other hand, E. Witte [258] assumed that there was a more definite relationship between the "flow antagonism" of currents (literal translation of the German word "Stromkabbelung") and frontal phenomena, particularly increased density, during mixing (see section 4.5). E. Römer [216, 217], A. Schumacher [226] and others have mentioned places where rough water and rip zones are most commonly encountered by seamen, e.g. Pentland Firth off the northern tip of Great Britain, the Strait of Gibraltar, the waters off the eastern coast of North America and the western coast of Mexico, the region of the Agulhas Current, the seas of the Indonesian Archipelago, the region of the Guinea Current and the region adjacent to the west coast of Africa. The same authors have statistically analyzed and summarized the log records of commercial vessels passing close to the African shore in order to establish the places where rough water and rip zones are most commonly encountered and how the frequency of their occurrence varies with the seasons. For example, it was found that the zones were observed with the greatest frequency (27 times within a month) in October in the area south of the Cape Verde Islands at around $9°42'N$ and $15°54'W$ [216], whereas they were the most frequent in August

---

* Tolcheya (Russ.) – particularly steep and short, frequently random waves. Equivalents in other languages: "Stromkabbelung" (Ger.) (apparently used to denote rip as well); "rip" or "current rip" (Eng.); "clapotis" or "clapotage" (Fr.); "shiome" (Jap.) (also denotes rip); "Stromrafeling" (Dut.).

(in 56% of all the observations on currents) in the region of the Guinea Current at around $5°N$ and slightly farther south [226]. We can now unequivocally associate the frequency of occurrence of rough water and rip bands in these areas with the seasonal cycle of upwelling in the Gulf of Guinea where, according to the characteristics of equatorial circulation, the upwelling of waters reaches its maximum intensity in July – October. However, the data on upwelling $40 - 50$ years ago were still of a highly fragmentary nature and the seasonal cycle was still unknown. The maps plotted by A. Schumacher [226] for five months of the year, showing all the points at which rough water and rip bands were observed in the area west of the African coast from 1855 to 1900, are uniquely informative (Fig. 1.1, a, b).

The investigations of German scientists did not remain unnoticed. The Japanese researcher M. Uda [249] analyzed them and augmented them with his own oceanic data and the results of laboratory experiments. He did not stop at one-time observations of phenomena on the surface of the ocean, but went much farther than his German predecessors by conducting measurements in the ocean in order to get a spatial idea (in horizontal and vertical planes) of the nature of the distribution of the physical and dynamic characteristics of the waters near frontal interfaces. The strange thing is that Uda, like the German authors, was not yet using the term "front".

Basically, we have good reason to conclude that detailed research into oceanic fronts began in 1938 with Uda's "Researches on "siome" or current rip in the seas and oceans" [249]. However, it is remarkable that even this investigation did not start a wide-scale scientific advance on the problem of oceanic frontogenesis, even despite the important dynamic premises presented in A. Defant's earlier papers [104, 105], which are without a doubt known to a wide circle of oceanographers. If we recall, the same fate befell the first endeavours of the German hydrophysicist Kalle (see [61]) in the area of thermohaline finestructure, which did not result in immediate awareness of the importance of the detected phenomenon and were practically forgotten for three whole decades up to the onset of "better times". Of course, World War II put a stop to the development of the ocean sciences. However, there were also purely scientific reasons, these being that the scientific climate was not yet suitable for such overly complex ideas about the structure of oceanic waters. Even the most general physicogeographic description of the ocean was not yet complete in those years, and scientists strove mainly to comprehend the ocean as a whole by means of simplification through systematization of available data. The time was not yet right for the natural complication of views on the individual, be it even the major, aspects of the question.

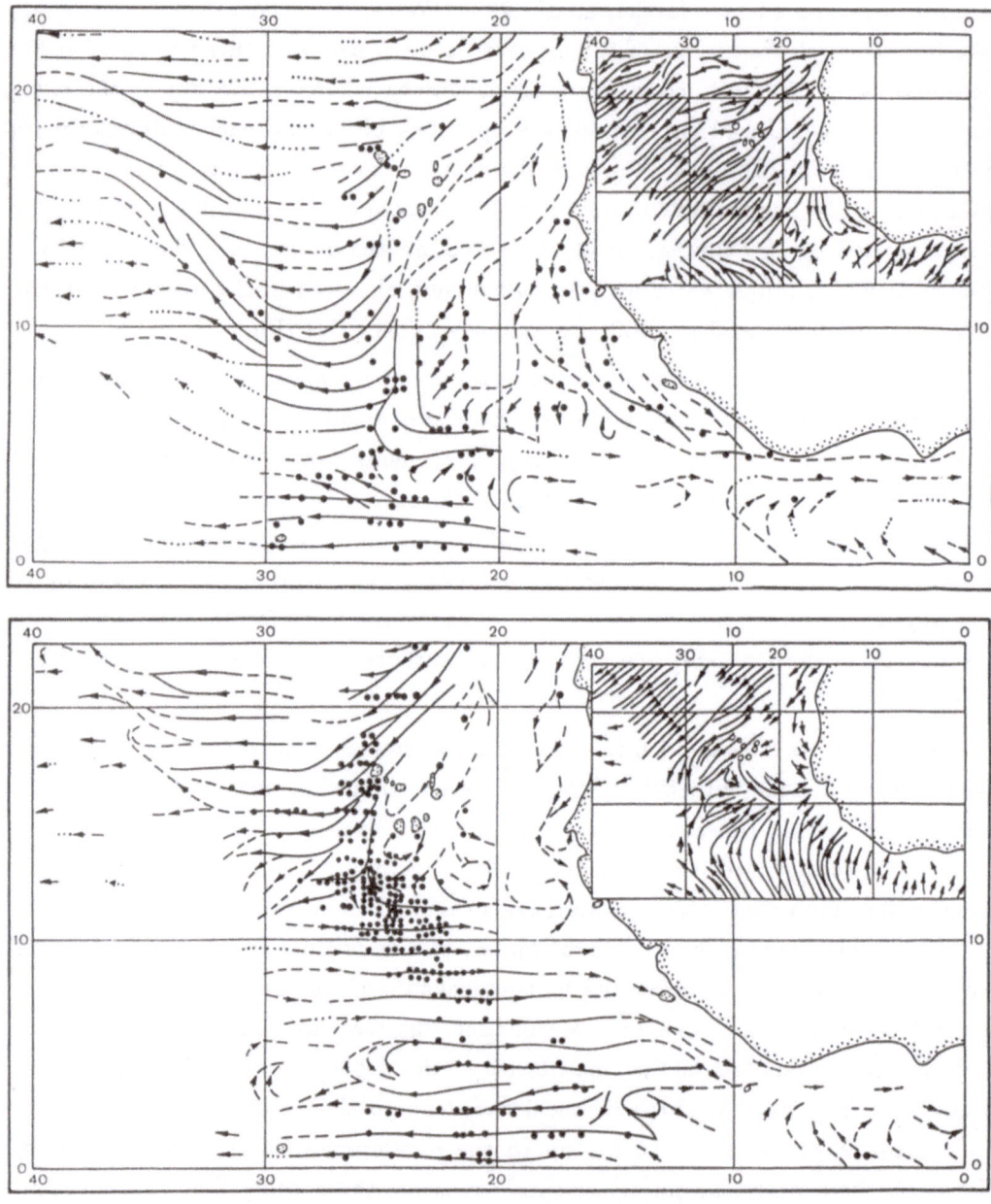

Fig. 1.1. Points at which mariners observed rough water or rip bands (black dots) in the area west of the African coast in January (a) and in August (b) from 1855 to 1900 (according to the data of A. Schumacher [226]). Unbroken arrows – wind direction; broken arrows – direction of currents on the surface of the ocean.

However, the situation was not entirely the same in all countries. The necessity of studying oceanic fronts in detail in Japan was dictated primarily by its interest in the coastal fishery which provided the population with essential protein food. It is therefore not surprising that 40 – 50 years ago, Japanese oceanographers could plot fronts with simultaneous participation of several research vessels, which is sometimes impossible to organize even today. For a long time, the scientists of other countries approached the study of oceanic fronts mainly from the descriptive geographic- climatological point of view. In this case, interest was stimulated mainly by the desire to systematize the ever-growing ideas regarding the geographical distribution of the physical characteristics of the general circulation and the water masses of the ocean, as well as the application of some of the most general and formal analogies from atmospheric research to oceanic research. From this stems the colossal difference in the approach to this particular question in the most important prewar oceanographic investigations. In the above-mentioned basic phenomenological study by M. Uda, we find the first detailed descriptions of various manifestations of fronts on the surface of the ocean, and of the thermohaline finestructure typical of frontal zones. Uda proposed a classification of fronts based on their characteristic features, as well as the first physical hypotheses concerning the processes of frontogenesis in the ocean. At the same time, N.N. Zubov's classic monograph entitled "Seawater and Ice" (1938) does not yet mention oceanic fronts, while the well-known treatise "Oceans" by Sverdrup, Johnson and Fleming (1942) describes the main convergence zones in the World Ocean only in general outline, and mentions in passing, with reference to [249], the presence of numerous eddies and visible "demarcation lines" on the surface between the individual branches of the Kuroshio Current.

The fact that the observations of the German scientists and the pioneer work of M. Uda were not given proper attention in the most recent major treatises resulted in a situation where questions to which Uda had found answers more than 40 years ago were recently again being referred to as "unresearched".

Because of this, let us sum up the most important results of this author as we see them today in the light of the most recent data.

1. Uda distinguished the phenomenon of pronounced active fronts on the surface of the ocean ("siome") against a background of an overall increase in the horizontal gradients of temperature, salinity, density and current velocity. For instance, in the transitional zone between the Kurile (Oyashio) and Kuroshio currents, all of which was regarded as being a frontal zone or a part of the subpolar front by

researchers of later years, Uda localized areas where "siome" were observed during different months (Fig. 1.2 a, b). In this we can perceive the physical bases for acknowledging the existence of a whole hierarchy of scales of the phenomenon, as well as the possibility of intensive local interaction (local dynamics of fronts) against a background of a general climatic inhomogeneity in the spatial distribution of the physical characteristics of oceanic waters. Here, we also see the phenomenological bases for discriminating between the concepts "frontal zone" and "frontal interface" (or "front").

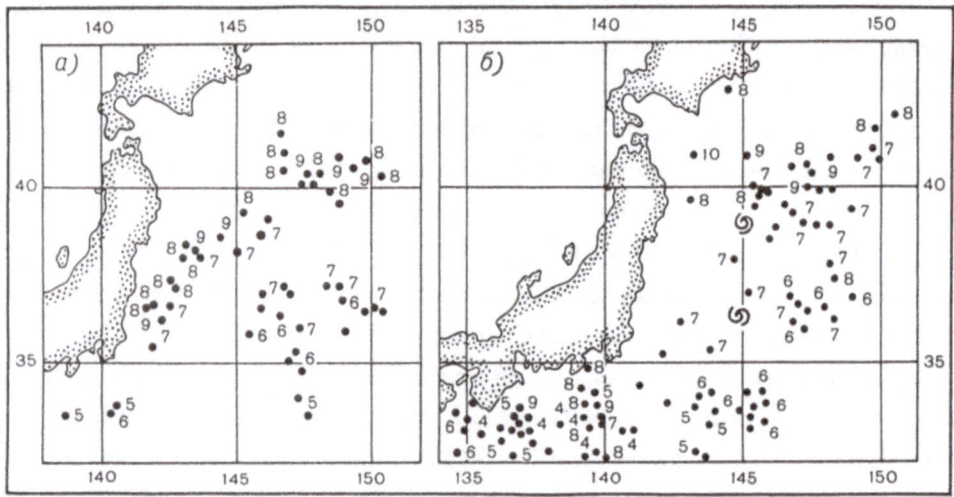

Fig. 1.2. Points at which "siome" were observed by fishing vessels in the Kuroshio frontal zone east of Japan in 1936 (a) and 1937 (b), based on the data in ref. [249]. Figures near dots denote month; spirals mark the location of eddies observed on 6 June 1980, based on satellite data (see section 3.1, fig. 3.7).

2. Uda gave a detailed description of the various visible and sound manifestations of "siome", i.e. the abrupt increase in the steepness of the waves and ripples (the appearance of a rip current, the decay of which is accompanied by a characteristic hissing sound); the accumulation of floating objects, refuse, algae and plankton typical of rip lines; the abundance of life (fish, cetaceans, birds) directly near the front; a change in the direction of currents across a front, etc.

3. Uda studied the dependence of the locations of "siome" on the distribution of the relative vorticity of motion, and the regions of current convergence and divergence which were always close to each other. On the basis of abundant experimental material collected in the ocean, he came to the conclusion that the locations of "siome" coincided in the majority of cases with the lines of convergence to which

the c y c l o n i c (my spacing, K.F.) vorticity of near-frontal currents corresponded. This conclusion is in complete accordance with the assumptions of Defant [104, 105] and the later discussion of the dynamics of oceanic fronts by Petterssen and Austin [206].

4. Uda noticed that "siome" were usually observed near the boundaries of water masses and c u r r e n t s (my spacing, K.F.). After that, many researchers took a step backward and began to regard the boundaries between "water masses" as fronts, forgetting entirely about the dynamic aspect of the problem.

5. On the basis of his own observations, Uda developed the first classification of fronts, treating the origin of "siome" from the point of view of the characteristics of a large-scale field of motion in the far field of the given phenomenon. At the same time, he actually associated fronts with the deformation fields in the ocean, still without using the actual term "deformation field", though the latter was already being widely used at that time in dynamic meteorology.

6. Uda associated the frequent appearance of "siome" in some areas with the peculiarities of the bottom topography, particularly in the coastal region.

7. Uda described the appearance of a characteristic thermohaline finestructure, i.e. "dichothermal" and "mesothermal" layers, in vertical profiles and sections in the frontal zones. Though he did not use the term "intrusion" which appeared later, the process he described at the time was an intrusive process typical of frontal zones (see section 4.2).

8. Uda described a number of laboratory experiments carried out by his contemporaries, and himself conducted a series of laboratory experiments which, despite a certain degree of simplicity, demonstrated quite well the instability of frontal interfaces and the tendency toward the formation of wave-eddy- type turbulence on them.

The later doubts of a number of researchers contrast sharply with the highly concrete results of Uda. For example, "... there have been considerations that oceanic fronts should be distinguished on the basis of not only convergence, but also divergence. Furthermore, it was suggested that oceanic fronts were related to the maximum vertical velocities of the water. This is not altogether clear even now, as special investigations of this question are very limited" [53]. There have been even more extreme points of view, namely that fronts should not even be identified with convergences, but that they should be extended along the axes of the main branches of the general circulation, i.e. along the zero values of the vorticity of the horizontal component of velocity of such major currents as the Gulf Stream.

Uda's rejection of the phenomenological approach [249] in favour of the theoretical approach contributed quite significantly to this difference of opinions. For example, as a result of the second approach, Yu.A. Ivanov and V.G. Neiman formed two different concepts of "physical front" and "dynamic front" [28], in addition to confusing the concept "front" with "frontal zone".

Generally speaking, few field investigations specially devoted to oceanic fronts as a phenomenon were carried out during the first postwar decades. Furthermore, these investigations were so irregular that the points of view cited above should hardly surprise anyone. Apparently, the simplified flow diagrams for the principal fronts with which certain arbitrary oceanic boundaries of an idealized form were associated (e.g. see [52]) constantly contradicted the data which were nevertheless accumulated thanks to the efforts of the most inquisitive researchers [4, 5, 117, 120, 122]. At that time, the theoretical side of the topic was not even close to being adequately developed. The idea of eddy formation at fronts was still at the level of conjectures, rather than hypotheses. Defending the exclusive right of convergences to be called fronts, V.N. Stepanov wrote the following in 1960: "In the water convergence zone, as in the region of meteorologic fronts, we observe extremely important phenomena such as cyclonic eddies, the formation of which is completely excluded in the divergence zone. ... and the greater the difference in the properties of the water masses on both sides of the front, the more intensive the cyclonic eddies. It is quite probable that a series of cyclones forms during significant disturbances of the equilibrium of forces at the front, as is the case in the atmosphere. Moving along the front, they gradually die down as they move farther away from it. Judging by information from the Gulf Stream, cyclones, "breaking away" from the front, move into the region of warm waters where they occlude" [52]. For the time at which it was made, this analogy with the atmosphere appears quite daring. However, the use of even simple eddy analogies (see the discussion of Barkley's treatise in V.M. Gruzinov's monograph [19] on p. 135-137) to explain the spatial structure of frontal oceanic zones was also very rare in later years, due to both the unusual complexity of observed spatial distributions of temperature and salinity in these zones, and also because of the highly idealized nature of the analogies themselves. It should be said that the mechanistic use of these apparent analogies with the atmosphere in relation to the ocean sooner interfered with, rather than promoted, a correct understanding of the laws underlying oceanic frontogenesis. At the same time, the profound physical significance of the comparisons made back in the 1950's between the structure and dynamics of currents in the atmosphere and the ocean

[193, 219, 220] did not result in an immediate reinterpretation of the concepts on oceanic fronts given in descriptive oceanography.

It can be said that the decade from 1950 to 1960 was a time when the modern theory of oceanic currents was only beginning to develop, and experience gained in relating theoretical results and observational data was still quite disappointing. This period was best characterized by H. Stommel [236] who wrote, "In the past there were very few points of contact between the ocean as visualized by conventional analysis of serial observations on the one hand and the ocean as portrayed by simplified laminar theoretical models on the other. I think the reason is not hard to find: neither model had been developed to a level of sophistication corresponding to the essential complexity of the oceanic phenomenon it was trying to describe. From the scattered pieces of evidence that are at present available it appears that the dynamics of the oceanic circulation, and the transport of various properties in the sea, may actually be dominated by the large-scale, transient, turbulent processes which hitherto have been ignored by observers, and which theoretical workers had hoped to bypass".

At the time, there was a very great need for prominent guiding hypotheses which would determine the strategy and tactics for organizing observations. It is this problem that H. Stommel dealt with in his paper "Varieties of oceanographic experiences" [236].

Despite the difficulties characteristic of any period of development, a series of very important field investigations was still carried out during the first postwar decades in the frontal zone of the Gulf Stream. Their purpose was to examine the structure and nature of the meandering of this powerful warm current of the Atlantic. In June 1950, the Woods Hole Oceanographic Institute of the USA conducted an expedition in the Gulf Stream area; this expedition, which was named "Operation Cabot" [120, 122], involved seven vessels which for the first time mapped a major meander of the Gulf Stream in all its details (see fig. 3.1 in section 3.1) in the process of its transformation into a gigantic separate cyclonic eddy.

The results of a detailed temperature survey obtained by the "Cabot" expedition confirmed Church's earlier assumptions [237] that the main stream of the Gulf Stream (and consequently the front associated with it) east of Cape Hatteras did not retain a fixed position, but migrated with an undulating meandering motion.

These results were later augmented by the investigations of Ye.I. Baranov [4-8] who became especially interested in the spatial finestructure of the frontal zone of the Gulf Stream, and the relationship between this structure and the meandering of the current. He greatly outstripped other researchers in his analysis of the processes of eddy formation in this zone.

It is interesting to note that neither the results of the "Cabot" expedition, nor the results of almost ten years of research by Ye.I. Baranov and his colleagues led to any universal physical interpretation of the eddy-forming role of the Gulf Stream and other western boundary currents. P. Saunders' aerial observations of the generation of an anticyclonic eddy from a meander north of the Gulf Stream [224] and F. Fuglister's observations of cyclonic rings [121] were required to finally develop and firmly establish a clear concept of frontal eddies or "rings", as we understand it today on the basis of the examples of Gulf Stream or Kuroshio "rings" [164].

At the beginning of the 1960's, we gradually began to realize that a vortex motion could be a much more widespread form of motion of oceanic waters, even far from such powerful currents as the Gulf Stream. In the paper cited above [236], Stommel wrote the following about this: "We are not interested in describing these eddies in isolation; we are concerned with discovering whether they play a significant role in driving the large-scale circulation. Is there interaction of eddies and large-scale circulation in the ocean as there is in the atmosphere? The problem is basic to a theoretical understanding of the general circulation of the ocean".

We can add that the same problem was also fundamentally relevant to our understanding of oceanic frontogenesis and the role of fronts as integral elements of the "large-scale, transient, turbulent processes" referred to by Stommel in another part of the same paper.

In our opinion, this interpretation became possible only after oceanology had passed through the following three important stages during the 1970's:

1) the general dynamic similarity between the ocean and the atmosphere began to be interpreted more thoroughly in addition to the comprehension of a number of concrete and significant physical differences;

2) vortex motions differing in nature and scale, including synoptic-scale eddies* were detected in the ocean and studied in a first approximation;

3) a thermohaline finestructure was detected in the ocean, and the main structure-forming processes were studied in a first approximation.

On the other hand, by the end of the 1970's, the above-mentioned important stages created a scientific atmosphere which greatly stimulated an interest in the further study of oceanic fronts. Naturally, the study of the relationship between

---

* See the recent publication "Synoptic Eddies in the Ocean" by V.M. Kamenkovich, M.N. Koshlyakov and A.S. Monin (Leningrad, Gidrometeoizdat, 1982).

fronts and synoptic-scale features gradually began to take priority. It is interesting
to note that the first direct detection and detailed description of this type of feature
during the Soviet experiment "Polygon-70" [33] was accompanied by the discovery
and description of a thermohaline front [72] which appeared in connection with the
observed anticyclonic eddy.

The following interesting international scientific discussions organized in recent
years also played an important role in the further development of research into
oceanic fronts:

1) the Chapman Conference on Oceanic Fronts in New Orleans, USA in October
   1977;

2) the Seminar on Oceanic Fronts in Coastal Waters at Stony Brook, USA in May
   1977;

3) the Interdisciplinary Discussion on Oceanic Fronts during the 14th General
   Meeting of SCOR in Brest, France in November 1978;

4) the Second Symposium on Turbulence in the Ocean organized by SCOR and
   IOC in Liège, Belgium in May 1979 within the framework of the International
   Decade of Ocean Exploration (IDOE).

The first three of these discussions played a particularly important role in that
they were centered around the undeservedly forgotten and often unknown investigations of M. Uda [249], and attracted universal attention to them. The report
delivered by Prof. Uda at the opening of the Interdisciplinary Discussion in Brest
in 1978, which was followed by a display of numerous photographs and slides taken
by the author himself, proved to be highly relevant and useful in that it again emphasized the physical tangibility and originality of the phenomenon being studied.
The superb photographs of the lines of rips and foam, rough water zones, accumulations of floating objects and algae, marked changes in water colour and other
diverse surface manifestations of oceanic fronts all put this discussion on a certain
level, for the participants were dealing not with abstract demarcation boundaries
between equally abstract water masses, or with mere tightening of isotherms, but
with a real, complex and varied dynamic phenomenon, the physical aspects of which
can and should be studied only on the basis of carefully organized observations and
measurements in the ocean.

The success of the new approach to the problem of oceanic fronts proved to
be closely related to the recently discovered possibilities of utilizing remote sensing
methods of observation and measurement. Many of the reports presented at the
Chapman Conference and pratically all the major reports delivered in Brest were

based on satellite information and primarily on images of the ocean surface in the IR spectrum. The thermal contrasts of fronts on the surface of the ocean proved to be clearly distinguishable on the images obtained even with geostationary satellites located at a height of about $36,000$ km above the Earth. Researchers studying oceanic fronts were among the first to benefit tangibly from the development of new space technology.

Subsequent developments no longer pertain to the history of this topic, but comprise the essence of modern study. Therefore, the reader will learn about the most recent investigations as material is presented in later chapters.

1.2 Definitions, terminology and criteria

A small group of specialists interested in oceanographic fronts, which included John Woods, Gunnar Roden, Christopher Mooers, A.S. Monin and the author of this book met in September 1976 during the Joint Oceanographic Assembly in Edinburgh. The animated discussion that took place was specially devoted to defining what should be called a front in the ocean. All agreed that due to the diversity of the physical nature, external manifestations and scales characterizing oceanic fronts, it would be not only impossible, but inappropriate to establish a brief, unequivocal and universal definition for them. This opinion reflects the dissatisfaction with the results of the numerous attempts to present a suitable physical definition of the terms "front", "frontal interface" and "frontal zone". Such dissatisfaction was expressed many times in the literature (e.g. [48]), but this never prevented the critics themselves from trying their skills in this difficult matter and attempting to correct this unsatisfactory state of affairs. For instance, the Encyclopedic Dictionary of Geographic Terms [73] recently added the following definitions by A.S. Polosin:

"Frontal zone: 1) in oceanology – a region in which the position of an oceanic front undergoes annual, diurnal or long- term changes. Within any relatively wide frontal zone, there is always a narrow frontal strip; 2) in meteorology – a transitional zone between two air masses, usually more diffuse and wider than a front".

"Oceanic front – an inclined interface between two different masses of water in the ocean. Here we observe the maximum horizontal gradients of all the oceanographic characteristics (temperature, salinity, etc.)".

In our opinion, these definitions are neither worse, nor better than other definitions, with which they share a common property, i.e. although correct to a certain extent, they still do not capture and do not reflect the peculiarities of the local dynamics which in the real ocean transform a mere tightening of isolines into a fairly definite physical phenomenon – a frontal interface (or front).

Special mention should be made of the attempt to define an oceanic front from a diametrically opposite position. For example, Owen [201] recently presented the following definition on the basis of the properties of frontal interfaces which are important to the fishing industry and marine biology: "A front is a line or linear zone that defines an axis of laterally convergent flow, below or above which vertical flow is induced". Unlike the definitions mentioned earlier, this one does not apply in any way to the horizontal gradients of the physical characteristics with which the "convergent flow" is associated. Consequently, we are faced with another extreme which makes it difficult to use this definition.

In other words, the available definitions do not help us to understand how, on the basis of the initial intensification of the horizontal gradients of physical characteristics, a marked frontal interface or several fault-type interfaces can appear and exist over a fairly broad area in the frontal zone. This is not surprising, as the mechanisms of the local dynamics, which help to keep the gradients of the physical properties at the frontal interface at a high level for a long time, have not yet been studied well enough. Hypothetically, one can indicate the following factors and processes which may be conducive to the intensification of gradients at the boundaries of a frontal interface:

 – the Earth's rotation and the appearance of internal Ekman layers;
 – turbulent entrainment;
 – inertial (nonlinear) effects;
 – increase of density during mixing.

Apparently, there is no lack of either quasi-stationary, or transient deformation fields in the ocean, which is constantly under the dynamic influence of the atmosphere and is always filled with eddies of various scales (see section 2.5). Consequently, there should always be both quasi-stationary and transient regions of high (compared with the mean values) spatial gradients of physical properties, i.e. frontal zones. However, the key question is to what extent must the external frontogenetic factors intensify the spatial gradients of the main thermodynamic characteristics in any part of the frontal zone to get one of the above-mentioned factors or processes to participate in the local dynamics of self-preservation of the resulting frontal interface? The other question is what specifically is the local dynamics? There is still no convincing and well-defined answer to these questions. It is in these areas that researchers should concentrate their efforts.

Due to the absence of clear answers to the key questions concerning the dynamics of oceanic fronts, we must fall back on arbitrary formal criteria which can help us reach some agreement as to what can be regarded as a "frontal zone", "frontal interface" or "front". In this respect, it would be more convenient to continue this treatise using more clearly defined criteria as compared with the ones in the above definitions [73] or the criteria formulated earlier by other researchers [6, 70]. Therefore, let us regard an oceanic frontal zone as a zone in which the spatial gradients of the main thermodynamic characteristics are very high* in comparison with the mean uniform distribution between the stable climatic or other extremes. In turn, the frontal interface is a surface within the frontal zone, which coincides with the surface of the maximum gradient of one or several characteristics (temperature, salinity, density, velocity, etc.). Then, strictly a "front" can be regarded as the result of the intersection of the frontal interface with any given surface, particularly with the free surface of the ocean or with an isopycnal surface. With this approach, an oceanic thermocline or pycnocline can also be regarded as a frontal zone. However, it is not usually associated with any real surface front, except in cases where the thermocline (or pycnocline) outcrops at a free surface, as is the case in upwelling zones where abyssal waters rise to the surface.

This point of view is not inconsistent with the frequently observed phenomenon of "multifrontal" frontal zones (see sections 2.1 and 4.1). Moreover, the presence of several frontal interfaces within the same frontal zone implies the presence of several regions of divergence which, according to the observations of Uda [249], are always located at some distance from the frontal interfaces (convergences). The alternation of convergences and divergences within the frontal zone may be associated with eddy formation.

We should note that, unlike meteorologists, oceanographers do not have a universally accepted division of fronts into "warm", "cold" and "occluded fronts". Perhaps it is time to develop a physical or phenomenological classification of oceanic fronts. This question will be discussed in the following section.

It is also worth mentioning that even researchers who are aware of the necessity of acknowledging the essential differences between the concepts "frontal zone", "frontal interface" and "front" often do not adhere to this terminology themselves, and prefer to use the word "front" in all the cases for the sake of brevity. This also applies to the majority of the author's earlier works. However, we should stress that it is quite acceptable to replace the expression "frontal interface" with the shorter

---

* The quantitative aspect of the criterion can always be determined arbitrarily for each specific case.

term "front", or to use them as synonyms. On the other hand, it is absolutely necessary to differentiate between "frontal zone" and "frontal interface" ("front").

Of all the researchers who have attempted to find a universal definition for oceanic fronts, Ye.I. Baranov [6] expressed views which were the closest to our own; like meteorologists, he drew a distinction (basically in scale) between a frontal zone and a front, and acknowledged that several fronts could exist in the same frontal zone. According to Ye.I. Baranov, "a frontal zone is a wide, transitional and relatively time- and space-stationary zone between water masses with different quantitative characteristics of physical and chemical properties." The drawback of this definition is related to the fact that it is based on the concept "water mass", which in turn is very difficult to define, and to the fact that the claim of relative stationarity in time and in space ties it down exclusively to phenomena of a climatic nature and scale.

That which Ye.I. Baranov defines as "front" is by our definition a "frontal interface", an expression which has of late become quite common. In turn, "front" (as interpreted on p.22) is called "line of the front" by Ye.I. Baranov; this term is hardly ever used by Baranov or any other author. With all their terminology, the latter differences between the two approaches are insignificant in comparison with the discrepancies that arise because of the association of the concept "frontal zone" with the concepts "water mass" and "climatic zonation" in Ye.I. Baranov's approach [6].

We would further like to discuss some of the aspects of the practical use of the criteria which determine the usage of this or that term, and also to express some ideas on how such criteria can be used as instruments of research.

From the author's point of view, one of the advantages of the criteria formulated above is that they allow us to correctly take account of the correlations of the spatial scales in a clearly multiscale situation.

Let us examine this property of the suggested criteria for the main frontal zones in the sea surface temperature (SST) field, say for the North Atlantic or northern part of the Pacific Ocean. As established, the horizontal SST gradient across the main frontal zones should be quite high in comparison with the mean meridional climatic gradient which for the indicated parts of these oceans is equal to approximately $0.003°C/km$. On the other hand, observations have shown [214, 251] that these particular frontal zones are characterized by mean horizontal SST gradients of 0.03 to $0.15°C/km$. Therefore, the introduction of an arbitrary numerical criterion

$$G_{fz} \geq 10\, \overline{G}_c, \tag{1.1}$$

where G_{fz} denotes the horizontal SST gradient in the frontal zone and $\overline{G}_c$ denotes the mean climatic SST gradient in the given region, allows us to determine the position and boundaries of all the main frontal zones on the basis of these measurements with a fairly rough spatial resolution ($\sim 10\ km$). The whole subtropical convergence in this case will appear (within the limits of the chosen criterion) as a single zone 100 to 300 km in width.

If, in turn, we wish to find the positions of all the marked thermal fronts within the zone of subtropical convergence, we can express the quantitative criterion for their determination as

$$G_f \geq 10\, \overline{G}_{sc}, \tag{1.2}$$

where G_f denotes the horizontal SST gradient at the fronts, and $\overline{G}_{sc}$ denotes the mean horizontal SST gradient across the subtropical convergence zone, known to us from observations.

Voorhis's observations [251] in the subtropical convergence zone of the Atlantic Ocean in the winter and spring have shown that $\overline{G}_{sc} \approx 0.03°C/km$, which accords fairly well with criterion (1.1) which is applied at the initial stage of the problem. At the second stage, we shall be determining the fronts with gradients of $G_f \geq 0.3°C/km$. However, for this we shall require measurement data with a significantly higher degree of spatial resolution (1 km or better).

The apparent arbitrariness of the proposed criteria should not confuse anyone. Very rarely does the researcher approach a problem blindly. His prior knowledge and the experience of other researchers should suggest the values of the parameters that determine the quantitative aspect of these criteria. In some cases, additional research may be required for this. The known correlations between the mean horizontal SST gradients in the frontal zones of the Kuroshio and Gulf Stream (which reach $0.1-0.15°C/km$ [214]), and the local SST gradients at individual fronts within these zones ($1-1.5°C/km$ [6, 17, 66, 174]) confirm the above reasoning. A study on the frequency of occurrence of thermal oceanic fronts, based on the use of this type of criterion [31] (see also section 2.1), has shown that the given approach is "workable".

In the opinion of G.I. Barenblatt, this approach can also prove useful during the mathematical interpretation of various problems related to fronts, since the $1-3$-order difference between background and the phenomenon being studied makes it possible to simplify and solve the problems by means of a small-parameter series expansion.

On the other hand, one can say that during mathematical interpretation of problems related to fronts and the exchange of mass, heat and impulse through them, the use of criteria such as (1.1) and (1.2) will automatically help to establish the regions (small and large) in which the dynamic balance of the properties, dictated by the laws of preservation, is likely to be maintained.

These criteria are also consistent with the concept of Woods [266], who by analogy with the analysis of atmospheric frontogenesis, deals with the frontogenetic effect of adiabatic movements in the ocean as well. Consequently, he distinguishes frontal interfaces by the maximum gradients of thermodynamic characteristics on isopycnal surfaces. According to his terminology, a front is "thermoclinic" when the isothermal surfaces are inclined relative to the isopycnal surfaces.

We should also note that the physical content of the term "frontal zone" as interpreted by us does not require that all the main thermodynamic characteristics (temperature, salinity, density and velocity) undergo a marked change in the frontal zone simultaneously. Since our interpretation of the term "frontal interface" or "front" is derived from the concept of frontal zone, the researcher can in accordance with what he observes in nature speak of thermal or salinity fronts, just as he speaks of current convergence lines (e.g. tidal), without the fear of violating the proposed criteria. Furthermore, the existence of several frontal interfaces and several fronts, some of which may be thermal and some haline (if such a thing is possible*), in the same frontal zone is not inconsistent with our criteria either. Considering that the observations carried out in the ocean are very often incomplete (with measurements of current velocity or even salinity not available), we should welcome the use of the terms "thermal front", "temperature front", "front in the temperature field" or "salinity front", "front in the salinity field", etc. in accordance with what was observed in reality. Explanatory definitions like "weak (front)" or "marked" are best followed up with quantitative values.

1.3 Classification of frontal zones and fronts of the World Ocean

One can hardly expect to give in a single book a detailed description of all the available data on all the various types of frontal interfaces ever encountered

---

* The data from observations [101] in the frontal zone of the Oregon upwelling indicate that this type of situation is also encountered in nature.

in the ocean. The more correct approach would be to describe all that which is fundamental to all oceanic fronts, and to find out the as yet unknown system of laws which govern all the existing or conceivable modifications. This cannot be done without at least an elementary classification of the phenomena being studied.

The literature describes several attempts to classify the fronts encountered in the World Ocean on the basis of certain characteristic sets of features [70, 82, 187, 249]. All of these classifications are quite logical, but more or less inadequate because of their excessive diversity of forms and scales of the given phenomenon.

M. Uda [249] was the first to suggest that all the manifestations of fronts ("siome") observed in the ocean be grouped into the following three categories:

a) "siome" of the first type, which is formed as a result of pure convergence of surface currents;

b) "siome" of the second type, which is formed in the convergence zone that develops adjacent to the region of upwelling and is therefore the result of the divergence that generated this upwelling;

c) "siome" of the third type, which is also observed on the surface of the ocean, but develops as a result of the interface between water masses (and currents) in the ocean at the level of the main thermocline*.

Uda's classification is a kind of genetic classification that is based on the deformational properties of the field of motion on a spatial scale which is significantly greater than the observed "siome" itself. However, this classification is a local one, in the sense that it systematizes the dynamic conditions of frontogenesis typical of the area adjacent to Japan and the Kurile islands, which is extremely active frontogenetically, but does not include all the situations typical of the World Ocean as a whole.

Fedorov and Kuz'mina [70] have proposed the following classification of fronts based on scales:

1) large-scale quasi-stationary fronts of climatic origin;

2) mesoscale fronts or synoptic-type fronts;

3) small-scale fronts of local origin.

This classification does not take into account the author's recent suggestions regarding the terminological differences between "frontal zones" and "fronts" (see Introduction and section 1.2). Therefore, the examples given in ref. [70] include

---

* Later (see tables 1.2 and 1.3 below, and subsections 3.1.1 and 3.1.3), we shall see that many of the active subsurface fronts in the open ocean belong to this category, and are associated with the synoptic-scale eddies formed in the pycnocline.

both the mean characteristics for very extensive frontal zones of a climatic nature, as well as the local characteristics of strict fronts or frontal interfaces. If we are to use this classification today without altering the term "front", we must always be aware of its ranges of variability with respect to the main frontal characteristics. These ranges (based on published data) are given in table 1.1.

Table 1.1.

Typical scales characteristic of
the fronts of the World Ocean (observational data)

Width of zone of maximum horizontal gradient of the main property	$10\ m - 10\ km$
Temperature difference across this zone	$1 - 6°C$
Salinity difference across this zone	$0.2 - 10\ ppt*$
Density difference across this zone	$10^{-1} - 10\ kg/m^3 (10^{-1} - 10\sigma_t)$
Horizontal temperature gradient (per km).....	$0.1 - 30.0°C/km$
Horizontal salinity gradient (per km)	$0.1 - 10\ ppt/km*$
Horizontal density gradient (per km)..........	$10^{-1} - 10\ kg/m^3 (10^{-1} - 10\sigma_t/km*)$
Slopes of frontal interfaces	$0.001 - 3.00$

---

* In exceptional cases (e.g. estuarine and river discharge fronts) more than the indicated maximum (see section 3.3).

The figures given in the table characterize only the fronts proper (or frontal interfaces), and do not apply to frontal zones which may be significantly wider (100 km and more) and have much lower mean horizontal gradients of the main characteristics (in the neighbourhood of $3 - 8°C$ per 100 km, e.g. [214, 215]).

In essence, however, this classification was based on such dynamic effects for different spatial scales as the effect of the Earth's rotation (geostrophic equilibrium) for fronts of the 1st and 2nd category, the dominant role of friction for fronts of the 3rd category, and on the difference in the typical period of frontogenesis, which also stems from the differences in spatial scales. The classification also took into account the significant difference in the physical nature and scales of the initial deformation fields. Basically, this classification has the potential for further development as facts on oceanic fronts are accumulated. As shown below, such detail is not only possible, but also logical.

It would be reasonable to attempt to work out a classification which would take into account the differences of the concepts "frontal zone" and "frontal interface" or "front". Frontal interfaces, as the surfaces of the maximum gradients of this or that property, can apparently form within one and the same frontal zone under the effect of various frontogenetic mechanisms. For example, in the frontal zone of the Gulf Stream, we can distinguish the main frontal interface of planetary scale and a large number of secondary ones, some of which are associated with the mesoscale meandering of the main front and with the separation of Gulf Stream "rings", while others may be associated with the intrusions or advective entrainment of cold and fresh slope or even shelf waters into the frontal zone of the Gulf Stream [166, 250], which can be attributed to various causes, including eddy formation on a much smaller scale than that of the "rings" [50, 173, 250]. Basically, one can easily visualize local frontogenesis being caused by any of the conceivable local mechanisms within a frontal zone created by a deformation field of planetary scale.

On the other hand, it is useful to differentiate the processes and phenomena which are relevant to the climate of the ocean (quasi-permanent in nature, with climatic scales of variability) from those which characterize the "weather" of the ocean (relatively short-lived, with synoptic scales of variability) (see section 2.4). Such an approach will enable the classification system to include situations where synoptic frontogenesis occurs within a climatic frontal zone.

Finally, the spatial scales of disturbances in the field of motion of ocean waters (and we can regard fronts as disturbances) determine the relative contributions of such factors as the Earth's rotation and viscous friction to the dynamics of the disturbances. At one end of the scale, we have the inevitable adaptation of motion to geostrophic equilibrium and the "β-effect" for fairly large scales (low Kibel-Rossby numbers, Ki; see [30, p.300]), and at the other end we have a viscous ageostrophic regime of motion for small-scale disturbances (large values of Ki). This applies more to frontal interfaces than to frontal zones. The latter, on the basis of their substantial lateral dimensions, should always be characterized by low Ki values. Proceeding from this, it can be said that spatial scales can serve as a basis for the systematization of frontal interfaces according to their most significant dynamic differences. It is hardly worth attempting in this case to determine that certain boundary value of the Ki number, which separates the geostrophic and ageostrophic regimes. On the one hand, the frontal Kibel-Rossby number $Ki_f = U_f/(fL_f)$, where U_f is the characteristic velocity, L_f the characteristic spatial scale and f the Coriolis parameter, can have different values depending on how we determine U_f

and L_f (e.g. see section 2.8). On the other hand, there exists a broad and continuous spectrum of scales of frontal phenomena, for which Ki_f can vary anywhere from $O(10^{-1})$ to $O(10^1)$. Only at the diametrically opposite ends of this spectrum can the regimes be regarded as geostrophic (on the left) and ageostrophic (on the right). Woods [262] rightly believes that practically all oceanic frontal interfaces are a "semigeostrophic" phenomenon, i.e. one in which a certain essential part (e.g. the jet stream along the front) can be described quite well by a geostrophic balance, while the other important part of it (e.g. circulation in the plane normal to the front) requires a nonlinear or frictional model for its description.

For frontal zones (in contrast to interfaces), it is apparently more convenient to select the principle of division by time scales, which allows us to separate the quasi-permanent (climatic) regions of condensation of isolines of the main thermodynamic characteristics from the short-lived (synoptic) increases of the spatial gradients. The first are often of a planetary nature, whereas the second are predominantly local.

Adherence to the above principles has enabled us to modify and detail the classification proposed earlier in [70] so as to create a real basis on which we can determine quite clearly the nature of any of the characteristic types of oceanic frontal interfaces from all the ones discussed in this book. The proposed classification (see tables 1.2 and 1.3) can have its own significance as well. However, its future use depends greatly on the extent to which other researchers are ready to accept the author's points of view and his basic concepts.

Table 1.2

Frontal zones of the World Ocean

Class	Brief Characteristics	Example(s), references
1. Climatic	1. Associated with the global distribution of solar radiation, evaporation and precipitation, and the deformation fields generated by the general circulation of the ocean and atmosphere and other constant climatic factors	
1.1. Of planetary scale	1.1 The main large-scale elements of the general circulation of oceanic waters (T–S, T)	1.1 Frontal zones of the Gulf Stream [6, 250] and Kuroshio [11], subtropical convergences [251], the Antarctic circumpolar frontal zone [133] (see also sections 2.1, 4.1, 4.3)
1.2. Of local interaction	1.2. The zones of interaction of the secondary branches of the general circulation of the ocean or the waters of various basins, areas and climatic zones (T–S, T)	1.2. Frontal zone between the shelf and slope waters in the N. Atlantic [139]; zones of interaction of the river plumes with the waters of seas and oceans; haline frontal zones of the doldrums [214]
1.3. Topographic	1.3. The result of the interaction of elements of the general circulation of oceans and seas with large-scale features of the bottom topography and shoreline (T–S, T)	Frontal zone over the continental slope of Europe [107]; frontal zone in the Barents Sea [153]; the Maltese frontal zone [94]
1.4. Near-bottom benthic	1.4. A special case of the combined manifestation of 1.2 and 1.3 in the near-bottom frictional layer in areas of intensive water exchange between various basins and areas	1.4. A vast frontal zone at depths of 600-1400 m along the northern continental slope of the Gulf of Cadiz, determined by the outflow from the Mediterranean Sea [80]

2. Synoptic	2. Associated with synoptic-scale processes in the ocean and atmosphere	
2.1. Of the open ocean	2.1. Formed in the deformation fields of synoptic-scale eddies, or appearing as a result of intensive local interaction of the ocean and atmosphere. Migrating (T–S, T)	2.1. Frontal zones on the peripheries of synoptic eddies of a different scale [21, 72, 86, 253] Frontal zones of warm and cold rings of the Gulf Stream [66]; frontal zones which bound the cold trail of typhoons and cyclones in the ocean [63] (see also subsections 3.1.1 and 3.1.3)
2.2. Local	2.2. Formed in the same areas under favourable synoptic conditions of ocean-atmosphere interaction (T, T– S)	2.2. Frontal zones of the areas of coastal [186] or equatorial upwelling (see also section 3.2)

Table 1.3

Frontal interfaces and fronts of the World Ocean

Category	Brief characteristics	Example(s), references
1. Geostrophic and "semi-geostrophic"	1. Low values of Ki_f	
1.1. Climatic	1.1. Associated with constant factors of a climatic nature	
1.1.1. Planetary	1.1.1. Of planetary scale, or elements of planetary scale (T–S, T)	1.1.1. The main front of the Gulf Stream [50, 237, 250], the Antarctic polar front [133], the fronts of subtropical convergence [159, 251]
1.1.2. Circulatory-intrusive	1.1.2. The elements of the frontal zone of interaction of the secondary branches of the general circulation or the waters of various basins, areas and climatic zones (T–S, S, T)	1.1.2. The front between the slope and shelf waters in the NW Atlantic [139]; the salinity fronts at the boundaries between the intrusions of shelf waters and the waters of the Gulf Stream [166] (see also subsection 3.1.3). The thermal front in the Kunashir Strait [249, p. 353]. The thermohaline front in the Strait of Sicily (see section 4.2 and fig. 4.7)
1.1.3. Circulatory-topographic	1.1.3. The elements of a frontal zone of interaction of steady currents with large features of the bottom topography and shoreline (T–S, T)	1.1.3. Thermal fronts of the topographic frontal zone on the continental shelf of W. Europe [107]; the same in the Barents Sea [153] and east of Malta in the Mediterranean Sea [94]

1.1.4. Run-off	1.1.4. A special case of 1.1.2, where one of the branches of the oceanic circulation is freshened by river run- off, or passes along the edge of an extensive run-off lens of a large river (S)	1.1.4. The fronts at the edges of the run-off lenses of the Orinoco, Amazon [123, 134] and other large rivers [151] (see also section 2.1)
1.1.5. Estuarine	1.1.5. A special case of 1.1.2, where sea and river waters interact in a wide and deep estuary (S, S–T)	1.1.5. The salinity front in the Gulf of St. Lawrence [243]
1.2. Synoptic	1.2. Associated with the interaction of the ocean and atmosphere on synoptic scales	
1.2.1. Advective-eddy	1.2.1. Arising in the deformation fields of synoptic-scale eddies in th open ocean, including in frontal zones of a climatic nature. "Cold" and "warm" zones are distinguished. Migrating (T–S, T)	1.2.1. The fronts on the peripheries of synoptic eddies of a different scale [21, 72, 86, 253] and the rings of the Gulf Stream [66] (see also subsection 3.1.1)
1.2.2. Storm	1.2.2. Bound the regions of intensive atmospheric interaction (the trails of prolonged storms, typhoons and tropical cyclones). Migrating (T, T–S)	1.2.2. The weak thermal fronts (not more than $0.2°C/km$) at the boundaries of the cold trail of the tropical cyclone "Ella" in 1978 [63]. The marked thermohaline fronts ($1°C/km$ and $0.6\ ppt/km$) at the boundaries of the trail of the tropical cyclone "Clara" in 1977 [34, 35]
1.2.3. Upwelling	1.2.3. Bound the patches of intensive local upwelling (T, T–S)	1.2.3. The thermal fronts of the South African upwelling [83]; the thermohaline fronts of the Oregon upwelling [186, 195] (see also section 3.2); the intensive "cold" front recorded by Knauss [163] at $3°08'N$ and $120°36'W$

2. **Ageostrophic**	2. High values of Ki_f	
2.1. Climatic	2.1. Associated with constant climatic factors such as:	
2.1.1. Run-off	2.1.1. and 2.1.2. River run-off into the oceans and seas (S, S–T)	2.1.1. The run-off fronts of the Connecticut R. [126]; (see also subsection 3.3.1)
2.1.2. Estuarine		2.1.2. The haline estuarine fronts in Delaware Bay [167] (see also subsection 3.3.2)
2.1.3. Shelf, seasonal	2.1.3. Seasonal warming and tidal mixing on the shelf (T, T–S)	2.1.3. The tidal fronts during the summer on the shelf of the British Isles and Europe [210] (see also section 3.4)
2.1.4. Tidal	2.1.4. Tidal convergences in narrow waters, bays and straits under the conditions of significant local horizontal changes in temperature and (or) salinity of a climatic nature, i.e. in all the frontal zones of categories 1.2 and 1.3; sometimes migrating (T, S, T–S)	2.1.4. Observations of M. Uda [249, p. 365-366] on the migrating fronts in Tateyama Bay. The tidal fronts in Cook Strait [92]
2.1.5. Circulatory-topographic	2.1.5. The same as 2.1.4, but associated with the elements of permanent currents (T, S, T– S)	2.1.5. Author knows of no specific examples
2.1.6. Near-bottom (benthic)	2.1.6. Intensive mixing in the near- bottom friction layer above sills, in straits and in confined waters, associated with the factors mentioned in 2.1.4 and 2.1.5 (T, S, T–S)	2.1.6. The near-bottom fronts above sills, in straits and underwater canyons [98, 177]
2.2. Synoptic	2.2. Associated with the appearance of synoptic deformation fields as a result of favourable synoptic conditions in shallow areas and straits	

2.2.1. Shallow, storm fronts	2.2.1. May arise at the boundaries of regions mixed by storms in the shallow areas of seas and oceans (T, T–S), above banks and in straits	2.2.1. Author knows of no specific examples
2.2.2. Circulatory-topographic	2.2.2. Arise during the interaction of anomalous tidal currents with marked features of the bottom topography and shoreline (T, S, T–S)	2.2.2. Fronts near the extremities of capes, banks and islands [209]

1.4. Modern methods of frontal research

Oceanic fronts are one of the most variable physical phenomena. Even within the quasi-stationary climatic frontal zones, we observe rapid changes in structure, rapid and unpredictable migrations of frontal interfaces, meandering and the formation of eddies. The frontal zones themselves keep shifting, fluctuating around some mean climatic position. Synoptic-type frontal zones and fronts are even more variable; they can "travel" over large distances, become sharper or undergo frontolysis, dissipate or arise again. This is precisely why frontal zones and fronts are so difficult to describe and study, and why they definitely require a synoptic approach, flexible effective programs of field work and systems for processing data in real time or quasi-real time for correction of any experimental plans. We became aware of the necessity of adhering to these requirements only at the end of the last decade (1976-1979). From that time, we observed the rapid development of a new phase in the study of oceanic fronts; it was accompanied by an avalanche-like accumulation of new facts and data which theoretical analysts still cannot keep up with.

In the context of this book, this small section on the modern methods of studying oceanic fronts is of a supplementary nature. Its purpose is to help readers understand why many of the observations on oceanic fronts, which at first glance appear to be quite necessary, have not yet been carried out and also to convince them that specialized research, and not incidental investigations, is required for the study of oceanic fronts. Considering the fact that oceanic fronts will apparently be studied quite actively throughout the next decade, field researchers should definitely be supplied with a truly helpful methodological handbook for conducting such investigations. The present book cannot serve as such, due to the brevity of this section and the minimum number of references to the literature. Those wishing to obtain these references can turn to one of the synopses, e.g. [71].

The discovery of synoptic-scale eddies in the ocean by Soviet researchers has greatly contributed to the development of new methods of research. A whole series of Soviet, Soviet-American and other polygon hydrophysical experiments in the ocean, such as "Polygon-70", MODE and POLYMODE, has resulted from this discovery. These experiments have made it necessary to make some fundamental changes in the methods of classical oceanography and to develop new synoptic and quasi-synoptic methods of oceanic research. Features which earlier were not picked up by the sparse network of stations with $30 - 60$ and even 100 miles between them began to be detected daily by means of focussed experimental research programs based on the principle of narrow-band receiving antennae directed at phenomena of a definite space-time scale.

Therefore, it became necessary to use all types of towed equipment to record continuously, in transit, primarily and mostly temperatures at one or several levels (e.g. thermotrawls). It also became popular to use temperature and salinity gauges installed in water intakes right on the vessel, or towed in various ways [95, 106]. The standard hydrophysical STD probes have been successfully adapted for continuous in-transit recording of temperature and salinity in flow systems based on intensive pumping of ocean water through a small tank, on board the vessel, in which a probe has been placed [68]. Oceanographic stations began to be executed more frequently, i.e. every $5 - 10$ miles, and sometimes even $2.5 - 3$ miles, as was the case in the microsurveys carried out by the author in 1970 [72]. The introduction of the Expendable Bathythermograph (XBT) into oceanographic research in the USA at the end of the 1960's made it possible to obtain accurate vertical temperature profiles in a $500 - 700 \, m$ layer in transit over the shortest possible distances. By the end of the 1970's, American researchers also began to use the Aerial Expendable Bathythermograph (AXBT) quite extensively. The methods of spacing oceanographic stations began to be efficiently adapted to the reconnaissance data obtained by means of towed sensors, or by XBT and AXBT. Basically, it became possible to conduct quasi-synoptic surveys of limited parts of oceanic frontal zones with the participation of only one vessel.

The rapid development of research into oceanic fronts over the past years has also been greatly helped by advances in the technology of remote sensing (mainly infrared) measurement of the thermal field of the ocean surface from airplanes and artificial Earth satellites. We can probably say without exaggeration that remote sensing methods of studying the ocean are especially effective in research on oceanic fronts. Analysis of the thermal contrasts on IR images obtained from satellites with the help of high-resolution scanning radiometers (fig. 1.3) makes it possible, even in the absence of high absolute accuracy in the measurements of the radiative temperature of the ocean surface, to observe the position of thermal fronts, their emergence, intensification, migration and dissipation, to obtain statistical and other information about their variability, and to observe the formation of meanders at the fronts, their separation and the formation of eddies in the ocean. There have been some attempts to determine the rates of frontogenesis and frontolysis on the basis of satellite data. Even a simple visual study of the published IR images of the frontal zones of major boundary currents or the front-rich coastal shelf areas of the ocean shows how complex and variable the spatial structure of frontal zones is (see sections 2.1 and 4.1, and fig. 2.9).

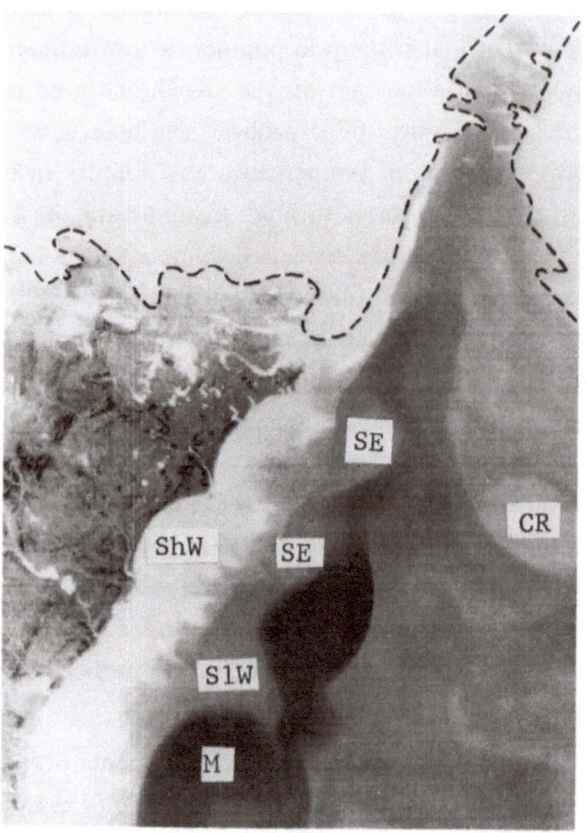

Fig. 1.3. IR image of the frontal zone of the Gulf Stream from the Florida peninsula to Cape Lookout, obtained on 1 April 1974 with the help of ultra-high-resolution (1 km) IR radiometry from the American satellite NOAA-3. The three main gradations of tonalities correspond to the shelf waters (ShW - the lightest, $8 - 10°C$), the slope waters (SlW - of intermediate tone, $12 - 18°C$) and the waters of the Gulf Stream (the darkest, $20 - 22°C$). At this time of the year, the waters of the Sargasso Sea (on the right) are considerably colder than the waters of the Gulf Stream, and are characterized by intermediate tonality ($18 - 19°C$). The abundance and complex distribution of the frontal temperature contrasts between these gradations are noted. We can clearly see spiral eddies (SE) on the left side of the Gulf Stream (see subsection 3.1.2), a meander (M) and a cold cyclonic ring (CR) on the right (see subsection 3.1.3). The dotted line indicates the approximate position of the boundary of stratus cloud. The above image was kindly made available to the author by R. Legeckis (NOAA).

The use of IR radiometry from geostationary satellites (American satellites SMS/GOES) makes it possible to filter away the cloud cover which is far more variable than oceanic fronts, and, by means of digital treatment of IR images, to use objective criteria for locating fronts and statistical methods of analysis. Regularity in obtaining images of the same area makes it possible to put together a film of successive images with any time interval to study the variability of fronts.

Long-term systematic observations of the ocean surface with the help of scanning IR radiometers and analysis of the observations on weather vessels give us every reason to assert that the manifestation of thermohaline fronts on the surface of the ocean is closely associated with the latitudinal seasonal variability of the amount of solar radiation and with the type of interaction between the ocean and the atmosphere. At temperate latitudes during the winter, the convection and vertical wind mixing which penetrate right down to the main thermocline result in a homogeneous layer of water which does not prevent marked horizontal temperature inhomogeneities related, for example, to jet streams or mesoscale eddies from appearing on the surface of the ocean. However, during the summer months, the warm layer near the surface and the density gradient associated with it screen the processes taking place in the main thermocline. Observations of the ocean surface with the help of IR radiometers have also shown that surface manifestations of thermal fronts are not observed in tropical areas where a subsurface warm layer with a density gradient exists practically at all times. In the $25 - 35°C$ latitude range, the emergence of fronts on the surface is observed from artificial Earth satellites during the autumn, winter and spring. At latitudes higher than 35°, fronts can be observed on the ocean surface regardless of the time of year. Thus, satellite methods have made it possible to give a detailed description of the seasonal cycle in the distribution of sea surface temperature (SST) in the most interesting and oceanographically complex areas. We have also learnt recently that the signatures of internal waves on the ocean surface, recorded from artificial Earth satellites, can provide us with information about the enthalpy of the upper mixed layer of the ocean [184].

The methods of such investigations will be improved and will definitely prove useful in determining the nature of oceanic fronts from outer space.

The comparative simplicity of detection of oceanic fronts on satellite IR images enables us to use satellite information about fronts in real time, and with its help, to direct research vessels to the objects of research (frontal zones, meanders and eddies which migrate continuously). This procedure has already been applied successfully in a number of expeditions, e.g. the 27th cruise of the research vessel "Akademik Kurchatov" of the USSR Academy of Sciences Institute of Oceanology in the POLYMODE program [64].

Useful information about oceanic fronts is not only provided by IR radiometry. Under certain conditions, fronts and eddies, for example, can be distinguished quite well on images of the ocean surface in the visible spectrum [29]. The high spatial resolution of such images makes it possible to study numerous structural details of the manifestations of eddies and fronts on the surface of the ocean. The appearance of contrast in this case is promoted by the modulation of surface roughness by eddy and frontal currents. One such image is presented in subsection 3.1.2 (fig. 3.7).

The topography of the ocean surface measured from an artificial Earth satellite with the help of a radar altimeter with about ± 10 *cm* resolution relative to the gravitational geoid can provide valuable information about the positions of current boundaries, meanders, rings and eddies. The experimental investigations [148] carried out on the basis of the radar altimeter data of the American artificial Earth satellite "GEOS-3" have demonstrated the potential of this method and the difficulties related to its application. The main difficulty is that the exact form of the present-day gravimetric geoid from which the elevations of the surface of the ocean should be measured has not yet been determined. This is precisely why the use of this method is still at the experimental stage, and is limited to areas of the ocean where the average dynamic topography of the ocean surface over many years is well-known.

Another promising new method of research lies in the observation from satellites of freely drifting buoys equipped with radiotransponders. The coordinates of each buoy are regularly calculated and retransmitted to land-based centres by the satellite. The trajectories of buoys placed in frontal zones where intensive eddy formation takes place are of special interest.

The most promising approach to the study of oceanic fronts apparently consists of a skilful combination of remote sensing (aerospatial) and in-situ methods, the results of which would supplement each other [64]. An example of this type of multiple approach to the study of the three-dimensional structure of the frontal zone of the Gulf Stream on the basis of synchronous satellite and vessel data is given in the recently published paper by the author and V.Ye. Sklyarov [50] (see also section 4.3).

Attempts to study the three-dimensional pattern of the structure of frontal zones in the finest detail are linked mainly to the desire to study the processes of intrusive stratification at frontal interfaces, and to determine the quantitative effects of the cross-frontal heat and mass transfer generated by this stratification. In order to accomplish this, acoustic floats were placed in the intrusive layers, and their drift with the layers was traced simultaneously with the mapping of the general structure of the frontal zone [254].

This type of method is a highly promising one, and will probably be used many times to study the "calving" of intrusive masses of water during their penetration through the frontal interface from one mass of water into another (see sections 4.2 and 4.4). It should be said that the methodology of studying frontal processes and phenomena under natural conditions is in many cases still descriptive, and will remain as such for a long time to come. It is often still based on conventional methods of data processing and analysis. For example, the method of comparing the vertical temperature, salinity and density profiles on both sides of a front, or the conventional T, S-analysis of these profiles is quite common. Detailed statistical analysis of the thermohaline finestructure of the vertical profiles around fronts is still comparatively rare.

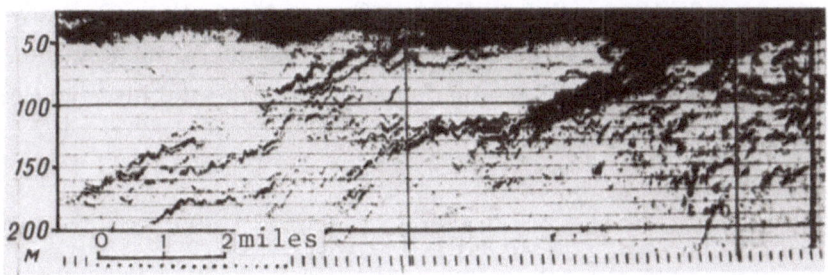

Fig. 1.4. Echogram of the frontal interface recorded by V.P. Shevtsov underway (at about 10 knots) in the coastal region of the Sea of Okhotsk. Slope of interface 1 : 100.

Analysis of the vertical profiles of the optical properties of seawater (e.g. attentuation and scattering) in frontal zones can serve as an important addition to conventional methods. Optical methods can also be used underway to detect frontal interfaces by the change in the spectral composition of light dispersed by the water. Acoustic methods of determining the positions of oceanic fronts are also promising ones. Side-scan sonars and improved echo sounders can and are now being successfully used for underway mapping of the density structure associated with fronts, large intrusive formations and internal waves [75]. Fig. 1.4 depicts a clearly defined front with a significant density gradient and a slope of approximately 1 : 100, recorded by V.P. Shevtsov. An echo sounder operating at a frequency of 30 kHz served as the recording device.

Simultaneous measurements of the temperature, salinity and velocity fields of currents in frontal zones are especially important and necessary in order to comprehend the physical nature of oceanic fronts. The main difficulty of organizing such measurements lies in the practical impossibility of ensuring widely spaced synchronous measurements and the difficulty of obtaining the necessary spatial resolution [188]. It is especially difficult to organize systematic and exact measurements near frontal interfaces. On the other hand, theory indicates that by analyzing the dispersion of freely drifting floats, one can study the spatial characteristics of the velocity field in frontal zones. This can apparently be accomplished by using acoustic current- velocity Doppler profilers from a vessel or vessels in combination with precision satellite navigation controlled by means of a marker radar buoy. The use of fluorescent dyes and other tracer methods could also provide useful results.

In the next few years we can expect the further, rapid development of specialized methods of field research into oceanic fronts, due to the universally growing interest in this problem which cannot be satisfied because of the inadequacy of the present methods and means of conducting truly experimental investigations in the ocean.

Chapter 2

GENERAL PHYSICAL DESCRIPTION OF THE PHENOMENON

2.1 How frequently are fronts encountered in the ocean?

In the light of the discussions in Chapter 1 on the definition of fronts and the criteria for distinguishing oceanic fronts, it would be logical to establish the actual extent of the phenomenon in the ocean. In attempting to find the answer to this question, the reader will find that a research vessel is not ideal for obtaining the necessary basic information. The advantages of remote sensing will become especially apparent here. On the other hand, reflection on the above question will require some knowledge of the scales involved and we will find that the final result greatly depends on what criterion is used to distinguish fronts.

It goes without saying that the answer to this question is required not only for the sake of confirming the correctness of the new interpretation of oceanic fronts. It may also be required to assess the role of fronts in the generation of finestructure in the ocean and in the vertical transport of heat and salt across the oceanic thermocline. We would also like to know whether fronts are encountered with the same frequency in coastal areas as in the open ocean. Finally, the resulting quantitative evaluations may prove useful for checking theoretical assumptions or modelling results.

The first attempt to give a direct answer to the question posed in the heading on the basis of field data was made by E.I. Karabasheva, V.T. Paka and K.N. Fedorov [31] even before satellite IR images became widely accessible for this type of investigation. However, some useful information could also be obtained from earlier investigations which pursued other goals. For example, proceeding from purely theoretical assumptions regarding the transfer of variability over a cascade of scales in the seasonal thermocline, with the involvement of synoptic-scale eddies, Woods [261] came to the conclusion that fronts (without specifying what should be regarded as a "front") are likely to be encountered in the ocean every 100 km on the average. The author's results based on continuous towings of a temperature sensor during 8 cruises of the "Akademik Kurchatov" in the Atlantic Ocean in 1967-1972 showed that the average distance between all the zones with a temperature jump of $\Delta T > 1°C$ was about 500 km on the basis of a sampling of all these zones. The sample of R.V. Abramov and coauthors [1], on the basis of only the 11th cruise of the "Akademik Kurchatov", showed an average separation of approximately 160 km

for the zones with temperature jumps $\Delta T > 1°C$, including the so-called "vergent" *
zones. If we select from these zones only the ones in which the absolute value of
the horizontal temperature gradient is $|G_f(T)| > 0.2°C/km$, the average distance
between them will be equal to 1000 km. From this we see that the end result in this
type of assessment greatly depends on the criterion used for distinguishing fronts.

Let us assume that, in accordance with the criterion proposed in section 1.2,
a thermal front is a narrow extended zone in the ocean with a high temperature
gradient in the transverse direction, and that this horizontal temperature gradient
is at least two orders of magnitude greater than the mean horizontal temperature
gradient for the given region determined from climatic conditions. The mean climatic temperature gradient in the meridional direction in vast regions of temperate
and tropical latitudes varies from 2×10^{-3} to $4 \times 10^{-3}°C/km$. Keeping within the
framework of the above formal definition, we shall in essence be studying the spa-
tial frequency of discrete values of $|G_f(T)| \geq 0.2°C/km$, without knowing anything
about the spatial extent or orientation of the traversed zones with high gradients.
This will be discussed briefly at the end of this section when we will be comparing
processed data from research vessels with those from satellite IR images.

When processing analog data, we must select the interval of spatial sampling
sensibly before forming numerical series. This interval should not exceed the minimum frontal width observed in the study area. However, the interval should not be
too small either, as in this case it would be impossible during statistical processing
to differentiate the gradients associated with fronts from the gradients characterizing short-lived small-scale patchiness of the temperature field which frequently
appears during the day [69] (see section 3.5). This situation should definitely be
kept in mind when comparing the results obtained by different authors.

We shall now analyze the results of temperature measurements in the surface
layer, taken with a towed temperature sensor along the routes of the 11th cruise
of the research vessel "Dmitriy Mendeleyev" and the 22nd cruise of the "Akademik
Kurchatov" (figs. 2.1 and 2.2).

---

* The "vergent" zones or "vergences" in this case (V.N. Stepanov's terminology
[52]) are the warm ("convergent") and cold ("divergent") bands of water of different
widths, intersected by the vessel.

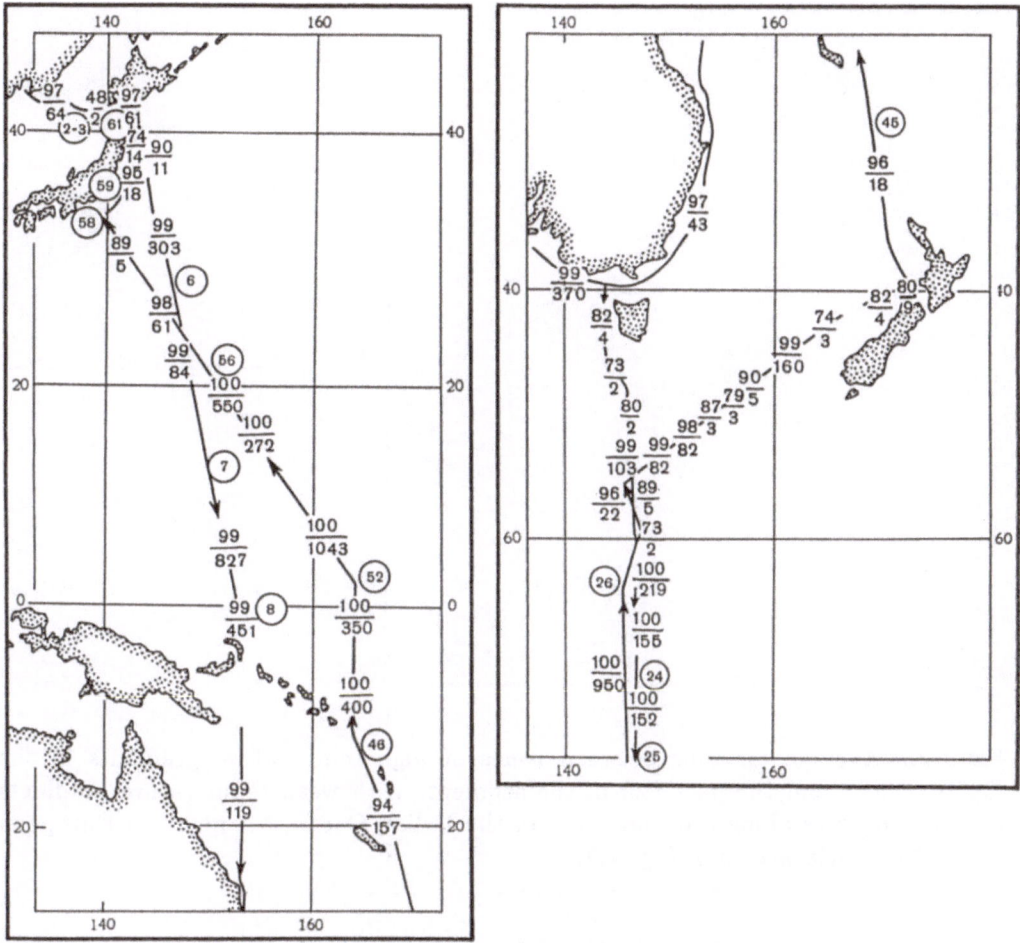

Fig. 2.1. Average distance between the zones of high temperature gradients in the Pacific Ocean (denominator) and the part (in %) of the segments with weak temperature gradients in different sections along the route (numerator). The circled figures denote the numbers and position of the parts of the route which are referred to in the text.

44

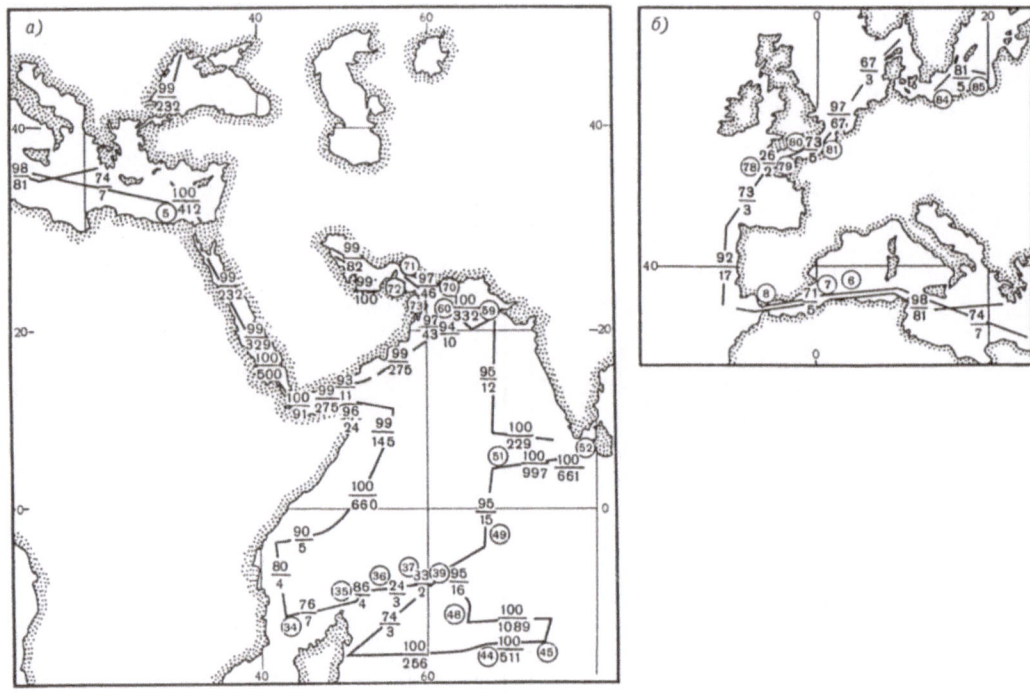

Fig. 2.2 Average distance between zones of high temperature gradients in the Pacific Ocean and the part (%) of the segments with weak temperature gradients in different areas along the route. a – in the Indian Ocean, b – near the European coast. Nomenclature as in Fig. 2.1.

The temperature sensor consisted of an MMT-1 thermistor with a sensitivity of 30 $Ohm/°C$ and a time constant of 3 seconds. The thermistor was towed in the ship's wake at a depth of about 1 m. The signals from the sensor were picked up by an EPP-09 potentiometer with a tape speed that remained at a constant 240 mm/hr throughout the measurements. Depending on the range of the temperatures measured, the sensitivity of the measurement system was regulated so that the entire scale of the potentiometer corresponded to 5 or $10°C$.

The analog record was processed by hand. Temperatures with an accuracy of up to $0.025°C$ on the $5°C$ scale and up to $0.05°C$ on the $10°C$ scale were taken from the strip chart. The digitizing interval for the temperature readings was equal to 1.25 min everywhere (2 points in 1 cm of the strip chart), i.e. the spacing of the readings along the route of the vessel varied from 180 m at a speed of 5 knots to 640 m at a speed of 16 knots.

The discrete temperature values were used to compile punched-tape files of data which by a special program were produced by an electronic computer and recorded on an H-327 automatic recorder on a scale that made it possible to visualize the general picture of the changes in temperature along the routes of the vessels. The files were divided into parts, each of which was then plotted graphically with its own scale of temperature and distance.

The most typical examples of temperature changes along individual segments of both cruises are shown in figs. 2.3 and 2.4.

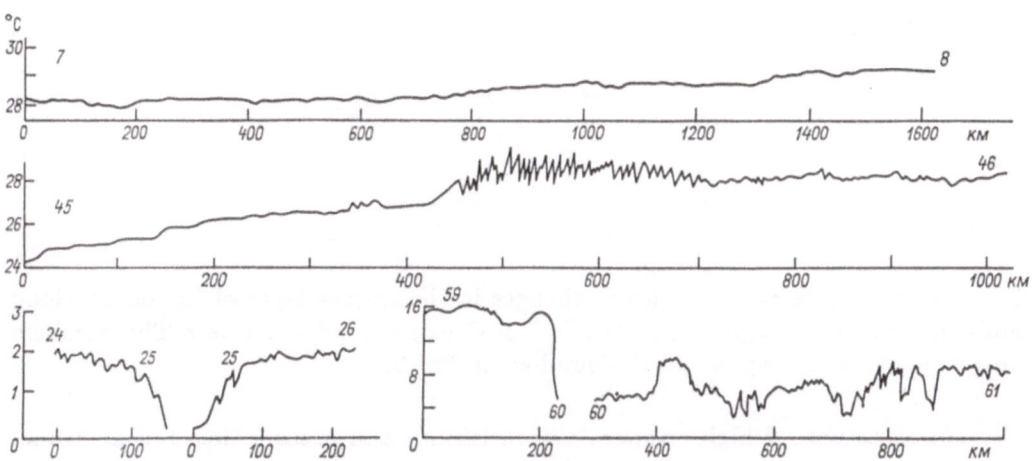

Fig. 2.3. Examples of temperature changes in the surface layer of the ocean, along individual recorded segments in the Pacific Ocean. The numbers near the curves correspond to the numbers in fig. 2.1.

For all the continuously recorded segments, the horizontal temperature gradients $G(T)$ (in $°C/km$) were calculated by the formula $G(T) = \frac{T_i - T_{i-1}}{\Delta \ell}$, where T_i denotes the temperature reading at a point i of the segment, T_{i-1} – the previous temperature reading, and $\Delta\ell$ – the distance between adjacent points in *km*.

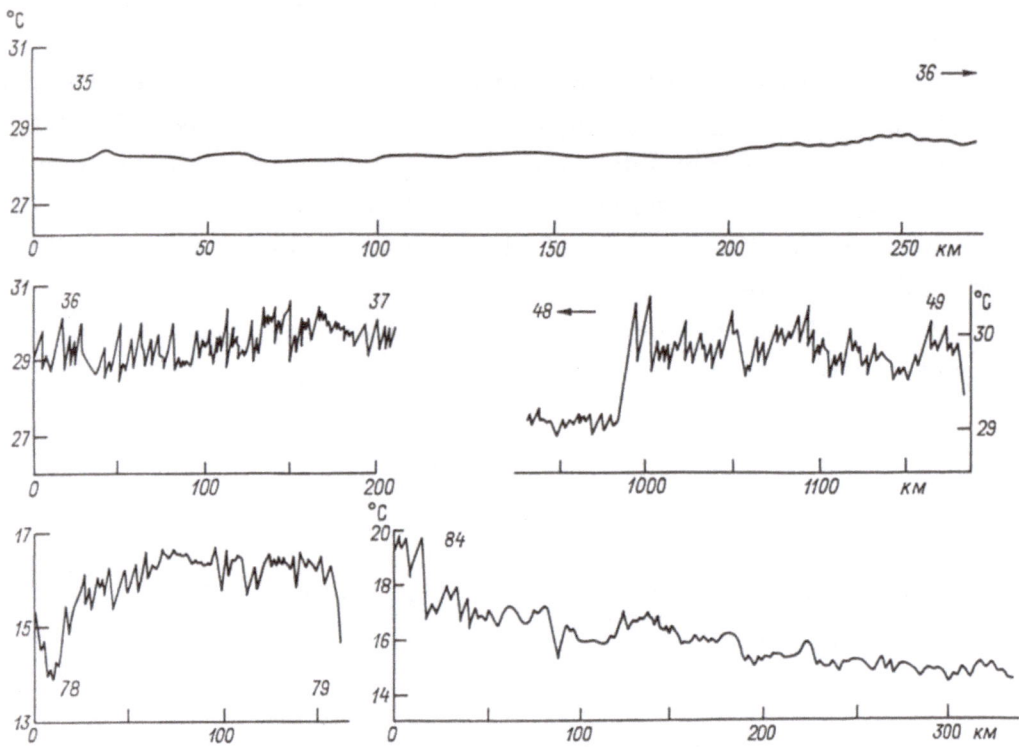

Fig. 2.4. Examples of temperature changes in the surface layer of the ocean along individual recorded segments in the Indian Ocean and adjacent seas The numbers near the curves correspond to the numbers in fig. 2.2.

For areas with a high temperature variability and areas of large- scale fronts the gradient was calculated between the closest adjacent points; for areas with a low variability, it was calculated between points located $4 - 6$ *km* from each other.

Histograms of gradients within the $0 - 0.7°C/$ *km* limits with a 0.05 and $0.02°C/km$ step were calculated and plotted for series of the absolute values of the gradients. Examples of the most typical histograms for individual segments of both routes are shown in figs. 2.5 and 2.6.

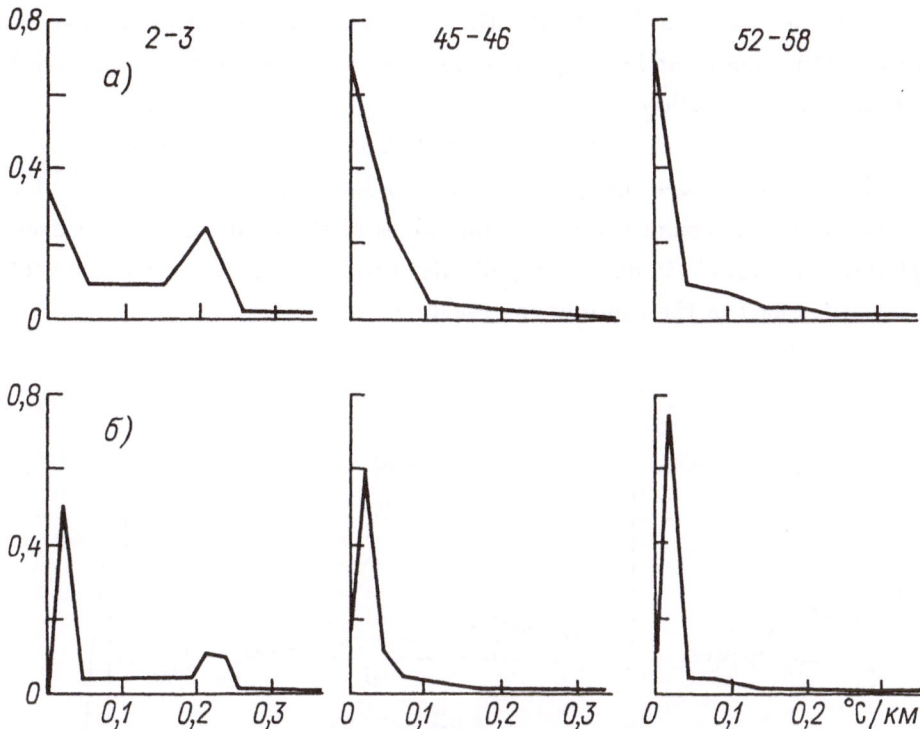

Fig. 2.5. Histograms of horizontal temperature gradients for individual recorded tracks in the Pacific Ocean. a – with a step of $0.05°C/km$, b – with a step of $0.02°C/km$ (with the same numbering of segments).

As we can see from figs. 2.3 and 2.4, the variability in temperature was of a completely different nature in the different areas of the ocean. In the Pacific Ocean, the lowest variability in temperature was observed on segments $6-7$, $7-8$ and $52-56$, located within the anticyclonic circulation formed by the main currents of the northwestern part of the Pacific Ocean (see fig. 2.1). Segment $59-61$, where the waters of the Kuroshio Current border on the cold coastal waters, abounds in very sharp fronts (up to $9°C/km$).

In the middle of segment $45-46$ we observe an unusually variable behaviour of temperature, with frequent alternating high values of the horizontal gradient of opposite signs. The absolute values of these gradients reach $0.2-0.3°C/km$, while the horizontal scales of inhomogeneities are such (with zone widths from 1 to 10 km) that they exclude the possibility of attributing this variability to the temperature sensor being occasionally thrown out of the water during a swell. The zone of high temperature variability occupies approximately 300 km of the total length of the 1000 km stretch with a highly monotonic temperature trend. The observed high variability was recorded during the day in practically calm weather under conditions of strong solar heating. The physical nature of this type of variability will be discussed in section 3.5.

As one would expect, the sections with a large zonal component are characterized by temperature changes with weaker horizontal gradients than for the meridional ones. Although the maximum values of the gradients greatly exceeded $1°C/km$ on individual segments, they did not exceed the limits of $0.6 - 0.8°C/km$ on the greater part of the route in the Pacific Ocean.

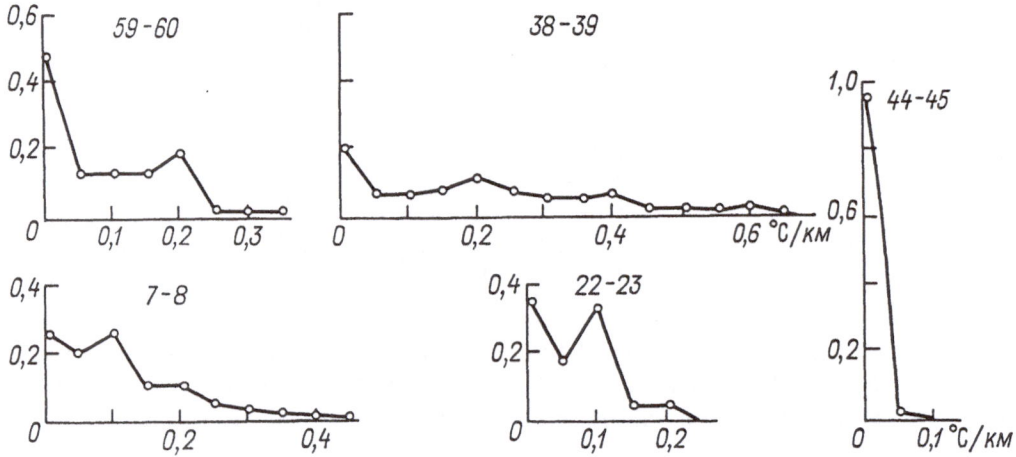

Fig. 2.6. Histograms of horizontal temperature gradients for individual recorded segments in the Indian Ocean. Histograms calculated with a step of $0.05°C/km$ (with the same numbering of segments).

In the Indian Ocean and adjacent areas (fig. 2.4), as in the Pacific Ocean, the segments with highly monotonic temperature changes such as segments $44 - 45$, $51 - 52$ or $35 - 36$ (see fig. 2.2) alternate with segments that abound in sharp temperature gradients, e.g. segments $36 - 37$ or $34 - 35$. It is interesting to note that the last three segments mentioned are the segments of one and the same track which extends almost in the zonal direction from the coast of Africa north of Madagascar (see fig. 2.2). Numerous traverses of sharp gradients are also observed on segments $48 - 49$, $70 - 71$ and $72 - 73$. An abundance of sharp gradients characterizes segments $5 - 6$ and $7 - 8$ in the Mediterranean Sea, $78 - 79$ and $80 - 81$ in the English Channel and $84 - 85$ in the Baltic Sea.

This simple visual analysis provides the basis for an important qualitative conclusion, namely that the frequency of sharp horizontal temperature gradients in the ocean apparently varies significantly from area to area, and that the probability of their appearing in the coastal areas of the ocean, or in marginal and inland seas, is much greater than in the open parts of the ocean.

The histograms shown in figs. 2.5 and 2.6 confirm to some extent the qualitative conclusion stated above. The towing areas with a weak monotonic change in temperature produce histograms with one strong peak (up to $0.7 - 0.9$) near the zero gradient (segments $45 - 46$ and $52 - 58$ in fig. 2.5, and $44 - 45$ in fig. 2.6). Such histograms are characteristic of the open parts of the ocean. Areas with frequently recurring high gradients are characterized either by histograms with two significant peaks (segment $2 - 3$ in fig. 2.5 and segment $59 - 60$ in fig. 2.6), or by histograms which are highly drawn out in the direction of high gradients and do not show any significant peaks even in the vicinity of zero gradients (segments $7 - 8$ and $38 - 39$ in fig. 2.6). Such histograms are the most typical for coastal areas or marine basins, though there are exceptions, such as segment $38 - 39$ (fig. 2.6), which occur in the open ocean.

As indicated by the lower series of histograms in fig. 2.5 b, which were plotted with narrower intervals (0.02 instead of the $0.05°C/km$ in fig. 2.5 a), gradients varying from 0.01 to $0.03°C/km$, rather than zero gradients, prevail in the majority of areas. With the characteristic scales of temperature inhomogeneities of about 10 km, such gradients correspond to amplitudes of temperature variability in the neighbourhood of $0.1 - 0.3°C/km$, which is quite plausible and characterizes the patchiness of surface layer temperature distribution usually encountered in the ocean.

The second peak on two-peak histograms (21 out of 84, i.e. 25%, are 2-peak histograms) is usually found between the absolute values of the gradient $0.15 - 0.25°C/km$. The frequency of such gradients in this cases reaches $0.1 - 0.2$. If we accept that the second peak corresponds to the frequently recurring sharp gradients on the background of a weak gradual temperature change depicted by the first peak, then we can conclude that the value $0.2°C/km$, selected by us as a criterion, is of both formal and real significance. However, the second peak can also appear in cases where an area has only one wide zone with a mean gradient of approximately $0.2°C/km$ and a width equal to $10 - 20\%$ of the total length of the area. Consequently, histograms cannot serve as a means of obtaining information on either the number of individual frontal zones in an area, or the average distance between them.

In order to obtain such information for a time series, we calculated with the help of an electronic computer the number of cases n for each area where $|G_f(T)| \geq 0.2°C/km$ in one or simultaneously in several successive points of a series. The mean distance $\bar{\ell}$ between zones corresponding to such cases was then calculated as $\ell/(n+1)$, where ℓ denotes the length of the area in km. Then, the routes on the maps (figs. 2.1 and 2.2 a, b) were marked with fractional notations in which the denominators give the values of $\ell(km)$, and the numerators denote the percentage, of the total length of each section, of segments with $|G(T)| < 0.2°C/km$, (with an accuracy of up to 1%). Such maps clearly show that the mean distance between zones of high horizontal temperature gradients varies from 2 to 1100 km. The highest values of $\bar{\ell}$ are characteristic of the open parts of the ocean. In such cases, the fraction of the segments where $|G(T)| < 0.2°C/km$ approximates 100%. The lowest values of $\bar{\ell}$ are mostly encountered in the coastal areas. In such cases, the fraction of weak gradients can fall to $24-26\%$. However, regions where the values of $\bar{\ell}$ varied from $2-5$ to $10-20$ km were also encountered in the open parts of the Pacific (fig. 2.1) and Indian (fig. 2.2 a) Oceans. It is possible that these were cases of inhomogeneity of the temperature field which are characteristic under the conditions of intensive solar heating in calm weather [69] (see also section 3.5).

At first, we [31] did not consider it possible to attribute thermal inhomogeneities of this type to fronts. This was because their physical nature was incomprehensible to us. We now believe that there are no significant differences in the mechanisms that support these inhomogeneities and many of the larger frontal interfaces, apart from the dynamic nature of the motions generating them. The kinematics of the process remain the same. Warm zones of stationary inhomogeneities appear as a result of convergences of the horizontal components of the orbital velocities of internal waves near the surface, while cold zones appear as a result of their divergence. Sharp horizontal temperature gradients (fronts) are located between these zones. It is not impossible that, after a day's heating, closed large-scale (1 km) convective cells may appear because of the resulting modulations of the thickness of the heated layer. However, the characteristic ratio of the transverse dimensions of these cells should not exceed 10^{-2} (10 m : 1000 m), which is very unlikely. Therefore, this assumption requires proof. In any case, the duration of the thermal inhomogeneities in calm weather and the small-scale (but very sharp) fronts associated with them does not exceed $\frac{1}{2}$ day. We should note that the even more frequent alternation of convergence lines and thermal gradients on the surface of the ocean may be associated with Langmuir cells.

The fairly high frequency of recurrence of thermal fronts in the open ocean ($\bar{l} = 10 \ldots 30 \ km$) and coastal areas ($\bar{l} = 2 \ldots 10 \ km$) may also be caused by coincidence if the research vessel is following the main direction of a highly meandering front. The point is that the towing of a temperature sensor is always carried out blindly when no additional information (e.g. satellite data) is available. A false picture of crossing numerous fronts is created when the crossings of meanders of one and the same front are recorded. According to Woods [261], the mean wave length of instability at oceanic fronts is equal to 10 km. According to R.D. Pingree [208], the characteristic dimensions of cyclonic eddies at the shelf fronts around Europe are equal to $20 - 40 \ km$. If we take the actual spatial patterns of fronts obtained by analysis of satellite IR images, we shall find that up to 6-7 crossings of the same front ($\bar{l} = 30 \ km$) may be recorded over 200 km of the route when travelling along the subtropical convergence front in the Sargasso Sea. Travelling along the front eastward from Flamborough Head in the North Sea [208] will give approximately 16 crossings in 100 km of track ($\bar{l} = 6 \ km$). It is quite clear that in the absence of additional information on the spatial structure of the temperature field, the data collected by a research vessel can easily mislead an inexperienced researcher about the actual frequency of recurrence of fronts in various regions of the ocean.

The situation may be relatively simple when only one well-defined front is meandering. On the other hand, the solution may prove to be quite ambiguous when dealing with the complex "multifrontal" structure of the frontal zone under study (see section 4.1). This type of structure characterizes the frontal zones of the Kuroshio Current [249] and the Gulf Stream [250], as well as the frontal zones in the areas of intensive coastal upwelling (see fig. 2.7).

Fig. 2.7 depicts the location of fronts (sharp, weak and diffuse) near the Pacific coast of North America between 36 and $48°N$, based on an analysis of an IR image with 1 km spatial resolution, obtained on 8 September 1976 from the NOAA-5 satellite [195]. We can clearly see that the front separating the cold water of the upwelling from the open ocean at a distance of about 100 km from shore is not a continuous one and, generally speaking, is not even parallel to the shoreline. Its sharpest parts, which are associated with transverse circulation on the shelf, run perpendicular to the shore and extend for $200-300 \ km$ in the open ocean. The whole picture, especially when viewing the original IR image, resembles the surface of a highly turbulent stream (e.g. the wake of a vessel under high magnification). The characteristic dimensions of individual "turbulent elements" of this picture vary from 20 to 100 km. Travelling along an imaginary route AB for approximately 1300 km (fig. 2.7), a vessel would record 33 crossings of fronts, which corresponds to $\bar{\ell} \approx 40$ km. Over the entire cloudless area of the image covering $430,000$ km^2, we can count approximately 40 poorly interrelated fronts measuring from $30 - 40$ to $400 - 500$ km in length. The results of our analysis of a satellite IR image are in good agreement with Holladay and O'Brien's spectral analysis of aerial IR measurements in the same area [141]. With a somewhat higher spatial resolution and using data collected along frequent tracks, they established that the peak of the spatial spectral density of the temperature field at the ocean surface occurred on wavelengths of about $16 - 40$ km in the vicinity of the Oregon upwelling.

It should be said that the same characteristic scales of the phenomenon are observed when examining the IR images, of other coastal areas of the ocean, found in various publications (e.g. [174]).

If we now turn to the regions of the ocean located $300 - 1000$ km from shore, our field of view will take in the frontal zones of the Gulf Stream, the Kuroshio, the Brazil Current, the Humboldt Current, the Agulhas Current, the Somali Current, the East Australia Current and other main boundary currents of the World Ocean. Satellite IR images corresponding to these regions can also be found in refs. [174, 195] and other sources. Analysis of similar images for the Gulf Stream area [50, 139, 173] together with the results of investigations on the finestructure of its frontal zone by means of conventional sampling have revealed a great diversity of structural elements in this zone, e.g. meanders, separated warm and cold rings, small cyclonic eddies, alternating streams of warm and cold water which penetrate to depths of $100-200$ m, etc. (see sections 3.1.3 and 4.1). In the given area, all of these structural elements create an extremely anisotropic spatial finestructure of the temperature field, with characteristic scales from $5 - 10$ to $200 - 300$ km and the alternation of frontal interfaces at the same distances. As indicated by the available IR images [11, 174], the picture in other areas of boundary currents is apparently similar.

In the region of the Atlantic subtropical convergence in the Sargasso Sea (about 1500 km from shore), the characteristic horizontal scales of temperature inhomogeneities and motion vary from $30 - 50$ to 200 km [21, 34, 172, 251]. These scales and, consequently, the distances between the fronts on the background of the overall high horizontal temperature gradient in the convergence zone, are determined by the synoptic-scale eddies which migrate through this area from east to west [21, 253]. Fig. 2.8 depicts the position of the centres of the eddies at the POLYMODE polygon in September 1977, together with the position of the tongues of warm and

53

cold water and the fronts. On the basis of our measurements of temperature and
salinity made while underway during the summer and autumn of 1977, the distances
between the fronts in this area may vary from 40 to 100 *km* at this time of the year,
which is in good agreement with the picture depicted in fig. 2.8, based on the data
of a hydrological survey and a towed thermistor.

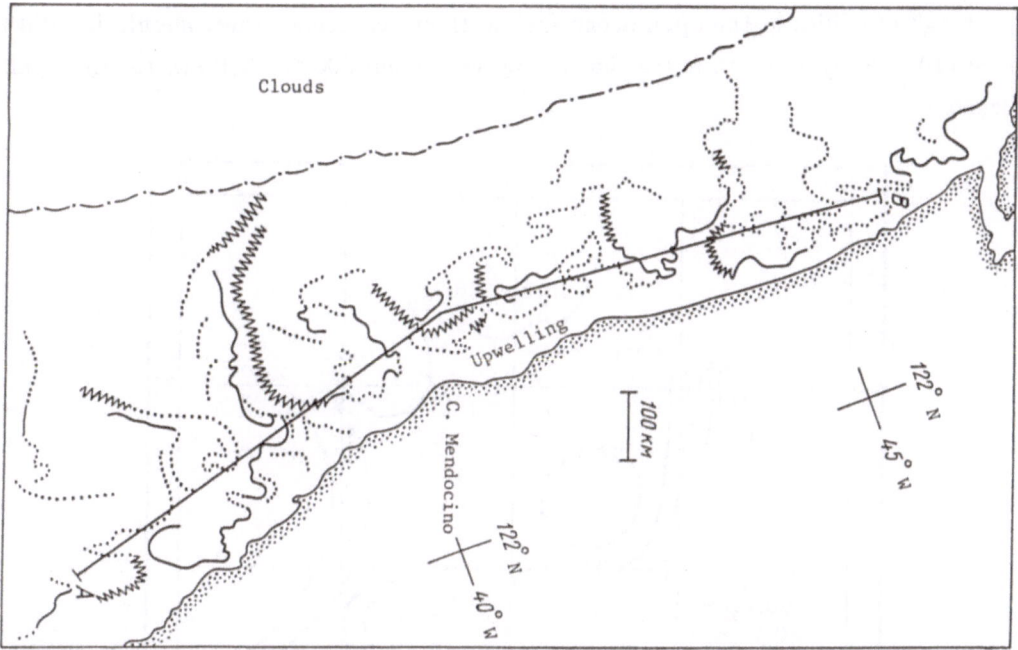

Fig. 2.7. Example of an analysis of the position of thermal fronts on the basis
of an IR image of the frontal zone of the Oregon upwelling, obtained from the
American satellite NOAA-5 [195]. Solid lines correspond to sharp fronts, zig-zag
lines to moderately sharp fronts, and dotted lines to weak or diffuse fronts.

So far, the results of analyses of IR images obtained from satellites have been quite consistent with the earlier results from statistical processing of data collected by research vessels, with the exception of only the lowest values of $\bar{l}$. There is nothing with which we can compare the highest values of $\bar{l}$ for the central regions of the oceans, as no adequate IR images have been published for these regions as yet. One can assume that the frequency of appearance of thermal fronts in the regions of the World Ocean which are far from the main frontal zones of climatic origin can be associated only with the frequency of eddy formation, and therefore cannot be less than $100 - 200 \ km$. Indeed, considering that one can hardly expect "close packing" of eddies in the open ocean and so their occurrence there should be quite rare $(10 - 20\%)$, one can agree that $\bar{l}$ may vary from 500 to 1000 km for the open ocean.

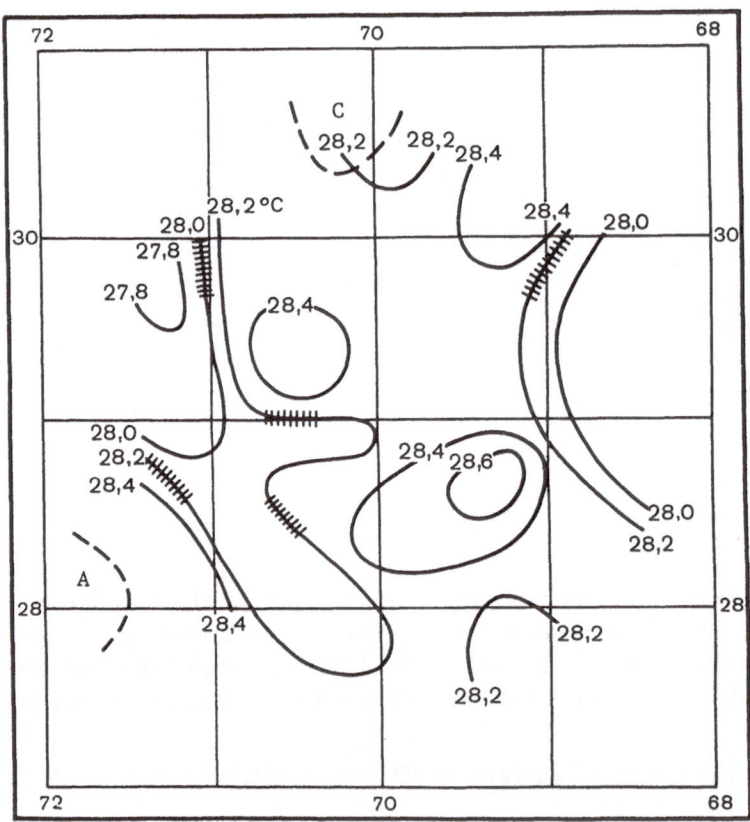

Fig. 2.8. Distribution of temperature at the 10 m level on the polygon of the POLY-MODE experiment on 18-29th of September 1977. Dashes indicate the position of the fronts at the boundaries of the warm and cold tongues, based on the data from visual observations. Broken lines depict the positions of the centres of eddies: C – cyclonic, A – anticylonic.

The author's observations from the research vessel "Akademik Kurchatov" during the crossing of the Pacific and Atlantic oceans on its 34th cruise have shown that in the western parts of the ocean where the dynamics of the general circulation is of a basically convergent nature, sharp frontal interfaces are encountered much more frequently ($\bar{l} \sim 100\ km$) than in the eastern parts where the dynamic background is predominantly divergent. One can even say that in the framework of the criteria discussed in section 1.2, the eastern parts of the oceans are in many areas characterized by rather diffuse frontal zones without any frontal interfaces at all. This type of situation may be of a seasonal nature, as the local frontogenetic mechanisms within frontal zones may be associated with seasonal processes (e.g. coastal upwelling).

Examples in this case are the frontal zone of the Peru upwelling and the equatorial frontal zone east of the Galapagos Islands, which is characterized by a sharp frontal interface only during the summer of the northern hemisphere.

As a result of this discussion, the question posed in the heading can be answered in the following way.

1. In coastal areas, especially in places of intensive upwelling, in inland seas and straits, and also in the frontal zones of powerful boundary currents (Gulf Stream, Kuroshio, etc.), the mean distance ($\bar{l}$) between thermal fronts may be only several tens of kilometres ($30 - 50\ km$), with variations from 5 to $100\ km$ in certain situations. Values of $3 - 5\ km$ obtained for $\bar{l}$ over a distance of hundreds of kilometres on the basis of data collected by research vessels in coastal areas may have resulted from the crossing of numerous meanders of one and the same front.

2. In regions of the open ocean where frontal zones of climatic origin are encountered (e.g. subtropical convergence), the characteristic values of $\bar{l}$ approach $100\ km$, though the elements of the spatial finestructure of these frontal zones may in some cases (in the presence of synoptic eddies) be characterized by smaller scales as well ($40 - 50\ km$).

3. In the open parts of the ocean which are far from the main frontal zones, the values of $\bar{l}$ increase to $500 - 1000\ km$.

4. In the western parts of the ocean, sharp thermal frontal interfaces are on the whole encountered more frequently than in the eastern parts. However, this difference may be altered as a result of seasonal phenomena and processes.

5. Due to internal waves, small-scale inhomogeneities of the thermal field with horizontal frontal temperature gradients that alternate with a $1 - 5\ km$ interval may appear in the upper $3 - 8\ m$ layer practically everywhere at temperate and low

latitudes under conditions of intensive solar heating in calm weather. The distances between convergences in the Langmuir cells which appear during moderate winds may be even smaller $(10 - 50 \; m)$.

In this section, we had to deal with a whole series of questions not yet touched upon in this book in order to make our discussion more complete. These questions include the frontogenetic role of eddies in the ocean, information on the spatial finestructure of frontal zones, as well as questions related to the thermohaline variability in the frontal zones of coastal upwelling.

All of these questions will be discussed in greater detail in subsequent sections and chapters of the book.

We would now like to augment what we already know about the frequency of thermal oceanic fronts with the scarce and fragmentary field data on salinity fronts, which can be obtained from the literature, archives and our own observational data.

As we have already mentioned, we still do not have enough reliable data on continuous underway salinity observations to form any generalized ideas about the frequency of distribution of salinity fronts at the surface of the World Ocean. Roden [214] indicates that according to his observations, salinity fronts do not by any means always coincide with thermal fronts. This statement holds true for the entire ocean as a whole, where various frontogenetic mechanisms are in effect under various specific frontogenetic conditions. My own study of a number of specific local oceanographic situations has shown that in places where there are parallel sources of spatial variability in temperature and salinity, the unified frontogenetic process most frequently generates **thermohaline frontal zones** in which the spatial temperature and salinity gradients practically coincide. Such is the situation off the Oregon coast of North America where upwelling is the source of cold and saline waters, and the run-off of the Columbia River is the source of fresher and warmer surface waters. Here, in the coastal frontal zone, the number of crossings of individual temperature fronts calculated by us (on the basis of data obtained during continuous recording of temperature and salinity [101]) coincided within $\pm 10\%$ with the total number of crossings of salinity fronts. The overwhelming majority of the fronts in this area consequently had a high negative T, S-correlation, and represented sharp density fronts (see section 3.2 for more detail). However, in a small number of cases (not more than 5% of the total number which approximated 100), the sharp salinity fronts were not accompanied by any noticeable spatial changes in temperature. This can be attributed to the pure effect of river run-off in the absence of upwelling. Similar overall conditions are observed near the zone of the

Brazil upwelling where the run-off of freshened waters from Sepetiba Bay creates sharp salinity fronts with horizontal salinity gradients of up to $1 - 1.6\ ppt/km$ [183]. However, they do not resemble the discharge fronts described in subsection 3.3.1, but are reminiscent of the specific frontogenetic mechanisms in the region of strong currents on the periphery of a zone of intensive upwelling.

On the whole, river run-off is the main source of the purely saline fronts at the surface of the ocean, not only in estuarine regions, but far out in the open ocean, if we are dealing with large rivers like the Amazon, Orinoco, Congo, Ganges, Brahmaputra, Irrawaddy, etc. Salinity fronts associated with the run-off of these rivers and isolated lenses of freshened waters bounded by sharp salinity fronts have been encountered by various researchers hundreds and even thousands of kilometres from shore. The salinity jumps (ΔS) across such fronts amounted to 1 ppt and more (see [81, 123, 222], as well as subsection 3.3.1).

The situation in the parts of the ocean not subjected to the effect of river run-off deserves special attention. We have some data from the area of subtropical convergence in the Atlantic. Cooler and less saline waters are located north of the convergence. The salinity and temperature increase rather sharply across the zone of convergence in the southern direction. Since the subtropical convergence itself therefore has a common positive T, S-correlation, any frontogenesis in the deformation field of synoptic eddies results in the appearance of thermohaline fronts with a positive T, S-correlation and partial compensation of the mutually opposing density contributions of temperature and salinity [34, 35]. In some cases, this compensation near the surface of the ocean is almost complete, and the density gradient at the front is practically imperceptible (fig. 2.9 a). However, in this area, we often encounter pure temperature fronts, as well as (in a smaller number) purely saline fronts. The results of our underway recording of temperature and salinity [68] near the surface ($\sim 3\ m$) in this area during June-September 1977 are as follows:

	n	%
Total number of crossings of frontal gradients of temperature and/or salinity in 1500 km of the track in the Sargasso Sea	35	100
Number of thermohaline fronts with a positive T, S- correlation	17	49
Number of thermohaline fronts with a negative T, S- correlation....	4	11
Number of purely thermal fronts ($\Delta S = 0$)	13	37
Number of purely saline fronts ($\Delta T = 0$)	1	3
Number of fronts with $|\beta \Delta S| > |\alpha \Delta T|$	8	23

(α and β are the specific contributions of temperature and salinity respectively to the changes in density ρ).

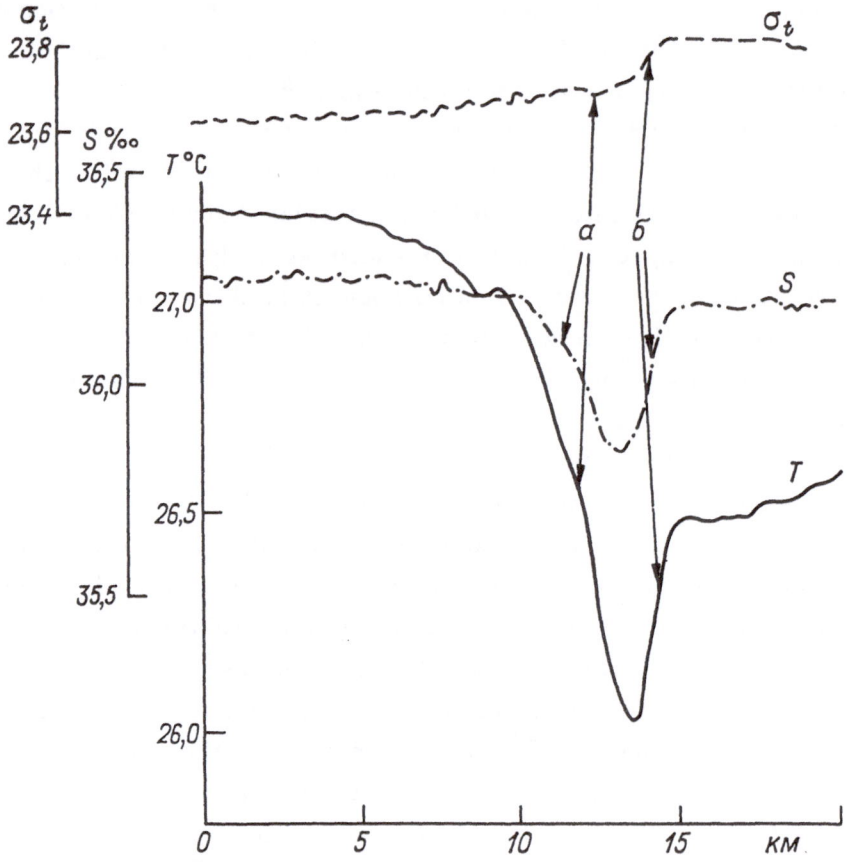

Fig. 2.9 Example of the recording of horizontal temperature (T) and salinity (S) gradients in the Sargasso Sea in September 1977. a – area without changes in density σ_t; b – area with a small change in density σ_t due to the dominance of the salinity contribution.

We cannot claim that these data are complete, or representative. However, they definitely indicate that in regions of the ocean not subjected to the effect of river run-off, the total number of thermal fronts (with and without the contribution of salinity) is significantly greater than the total number of purely saline fronts and fronts in which the salinity contribution $\Delta \rho_S = \beta \Delta S$ to the density change exceeds the corresponding temperature contribution $\Delta \rho_T = \alpha \Delta T$ in absolute value, so that $|\Delta \rho_S| > |\Delta \rho_T|$.

2.2 General background of spatial variability
in temperature and salinity near the surface of the ocean

We observe fronts near the surface of the ocean against a background of a natural space-time variability in temperature and salinity, which may be quite significant even beyond the frontal zones. For example, in fig. 2.10, we can see a sharp (about 1 *km* wide) thermohaline front which was recorded by the author in the Sargasso Sea in September 1977 with the help of an AIST temperature-salinity probe operating in a continuous-flow system [68].

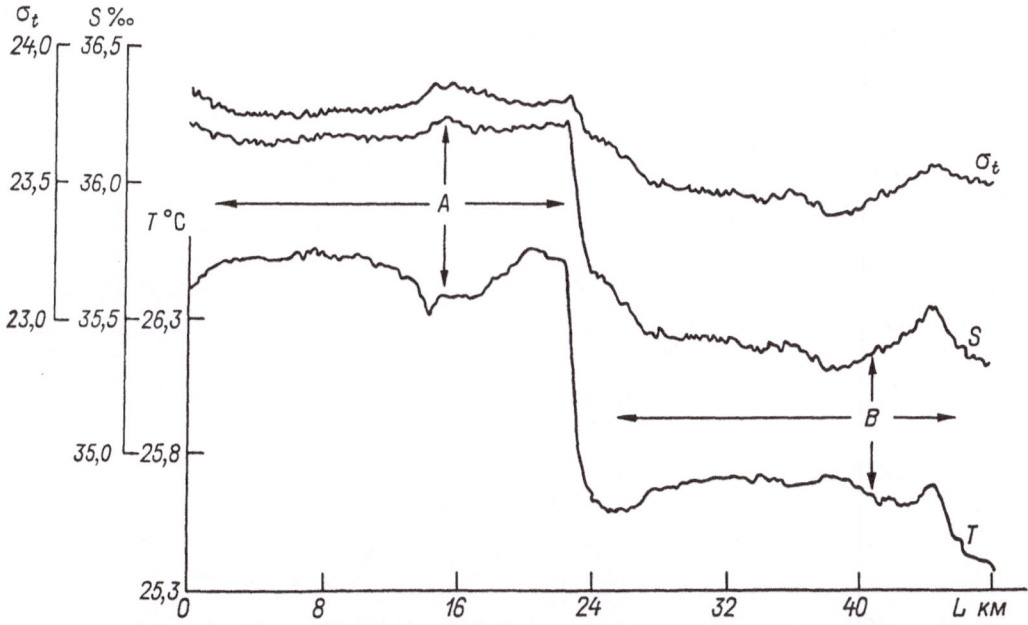

Fig. 2.10. An example of regions with a negative T, S- correlation (A and B) on both sides of a positive T, S- correlated front in the Sargasso Sea.

Significant changes in temperature and salinity are observed along both sides of the front for a distance of at least 20 miles. These changes differ from the sharp drop in temperature and salinity from north to south at the front not only by their smaller amplitude, but also by the fact that there is a negative correlation between them, i.e. away from the front, an increase in temperature corresponds to a drop in salinity and vice versa. Basically, this does not come as a surprise, as the front itself and the changes away from the front may be caused by entirely different factors, or, despite the close interrelation, may be of a completely different physical nature. For example, we often observe cases where intensive temperature or salinity fluctuations of a clearly oscillatory (quasi-periodic) nature take place as we come close to the front. Fig. 2.11 depicts an example of such fluctuations (regions A and B) as we approach the main front of the Gulf Stream along 70°W from the north (the author's observations during the 27th cruise of the research vessel "Akademik Kurchatov" in September 1978).

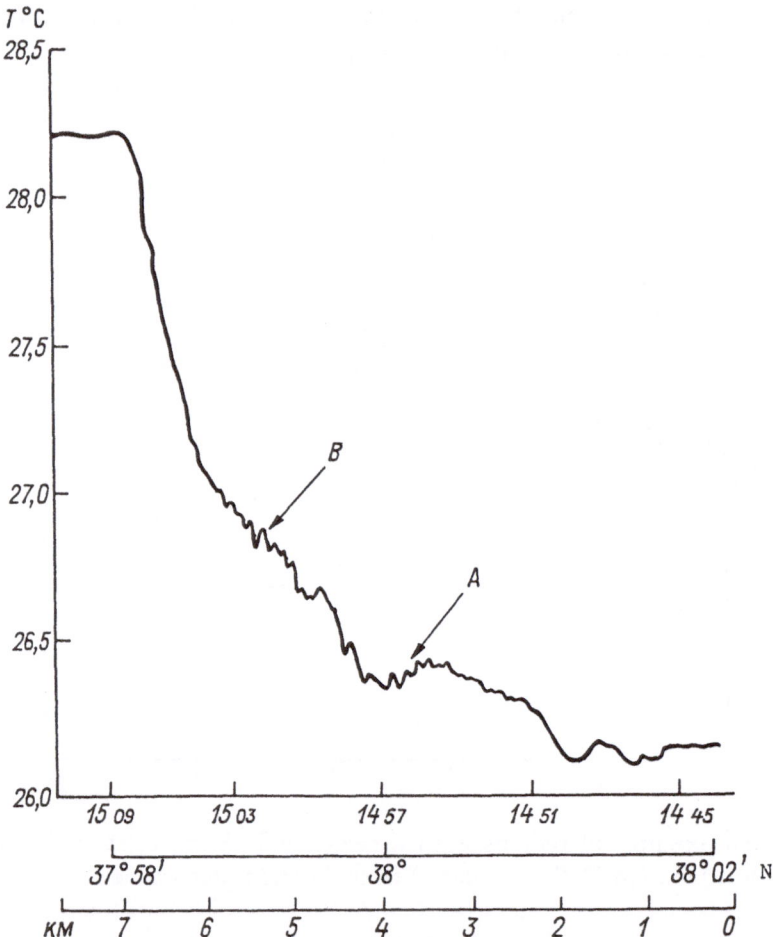

Fig. 2.11. Quasi-periodic fluctuations in temperature (regions *A* and *B*) as the main front of the Gulf Stream is approached from the north along 70°W on 11 September 1978.

The various natural causes of the patchiness of temperature distribution in the surface layer of the ocean and the typical characteristics of spatial variability in the temperature of the surface layer under various conditions are discussed in detail in "Mesoscale Variability of the Temperature Field in the Ocean", prepared and published under the supervision of the author and edited by him in 1977. Although many new and interesting results were accumulated in the course of later investigations [18, 34, 35, 69], the above- mentioned book may still be of interest to those who are dealing with the physics of this question. In the context of this section, it would make sense to repeat and somewhat expand the discussion of the composite spectrum of spatial variability in the ocean temperature field, plotted on the basis of all published data.

2.2.1 Composite spectrum of spatial variability of the temperature field in the ocean

Fig. 2.12 depicts the composite spectrum of SLT (surface layer temperature) variability for horizontal scales from 10 m to 1000 km, plotted on the basis of data available in the literature. It includes the following spectra.

I – Saunders [225]. Measurements of the radiation temperature in the IR spectrum from airborne remote sensing over the Mediterranean Sea. In the wave-length band $\lambda = 3 \dots 100$ km, the spectrum is approximated by the power law $k^{-2.2}$ (k – wave number) which does not satisfy any of the known models of turbulence. We should note that the area in which the measurements were carried out covered a part of the thermal front of the Ionian Basin [176], which, though it did not come to the surface at the time of the measurements, could probably have affected the surface temperature. Woods [260] believes that the exponent 2.2 reflects the combined effect of perfect geostropic turbulence (k^{-3}) and a system of sharp horizontal temperature gradients or "jumps" (k^{-2}). "Jumps" of this type may be fronts, but they may be of an artificial nature related to the method of measurement and computation.

II – Holladay and O'Brien [141]. Measurements of radiation temperature from airborne remote sensing in the vicinity of a coastal upwelling (Oregon coast of the USA). The isotropic part of the two-dimensional spectrum is approximated by the function k^{-3} both for the instantaneous temperature values, and for the mean values (3-week average) and deviations from them in the wave band $\lambda = 4 - 20$ km. As we have already mentioned, this type of relationship is characteristic of geostrophic turbulence, and it should be especially well-defined in areas of strong baroclinic instability (see section 2.8). This is apparently observed in the area of upwelling where the available potential energy of baroclinicity created by the upwelling of the waters is transformed into the kinetic energy of an eddy. It is interesting to note that the one-dimensional spectrum calculated on the basis of the same data displays, as the authors maintain, a power law similar to the one obtained by Saunders [225]. This difference points to the anisotropy of the field, which is clearly seen in fig. 2.7.

III – McLeish [181]. Measurements of radiation temperature. McLeish's spectra (of which only two are shown) display significant variability in the spectral level and the slope of the curves. Perhaps this is due to the effect of "noise" caused by measurement inaccuracy (see below). The temperature field in the vicinity of strong coastal currents shows the highest variability. The spectra diminish with an increase in the wave number k; however, beginning with $k \approx 2 \ldots 5$ $cycles/km$ ($\lambda \approx 200 \ldots 500$ m), the spectra reach a plateau. The enhanced spectral density is attributed to surface "slicks" (band-like inhomogeneities) which are presumably due to the cellular circulation of the Langmuir-cell type. However, it is possible that the increase in spectral density at high wave numbers is related to the "noise" caused by measurement inaccuracy. The spectral density of "white noise" is $E_0 = \sigma_0^2 \Delta \ell$, where σ_0^2 denotes the noise variance, and $\Delta \ell$ the sampling interval. The estimation of this additional value, which is hardly ever determined in practice, may prove significant, especially for low-precision measurements (e.g. remote sensing measurements where the accuracy during aircraft observations is not higher than $0.2 - 0.25°C$).

IV – Naumenko et al. [45]. Measurements by towed thermistors in the surface layer (at a depth of 20 cm) of the Atlantic Ocean. Curve 2 corresponds to an area of upwelling (Cape Blanc). The power law behaviour of the spectrum is between k^{-2} and k^{-3}.

V – Zenk and Katz [273]. Measurements by towed thermistors in the upper mixed layer (at the 26 m level). The spectra display different levels of temperature variability in the vicinity of a thermal front. The upper curve 3 comes from the frontal zone itself, while curves 1 and 2 come from the areas adjacent to the front on different sides.

VI – Bernstein and White [88]. Data from a series of standard hydrographic stations in the thermocline of the Pacific Ocean (down to 160 m). The spectrum is approximated by the function k^{-2}. The authors attributed the recorded temperature fluctuations to synoptic eddies, the origin of which is related to baroclinic Rossby waves.

VII – Spectrum based on measurements taken by means of a towed thermistor during the 13th cruise of the research vessel "Akademik Kurchatov" (at the 5 m level) in the Atlantic Ocean as calculated under the supervision of the author by co-workers of his laboratory.

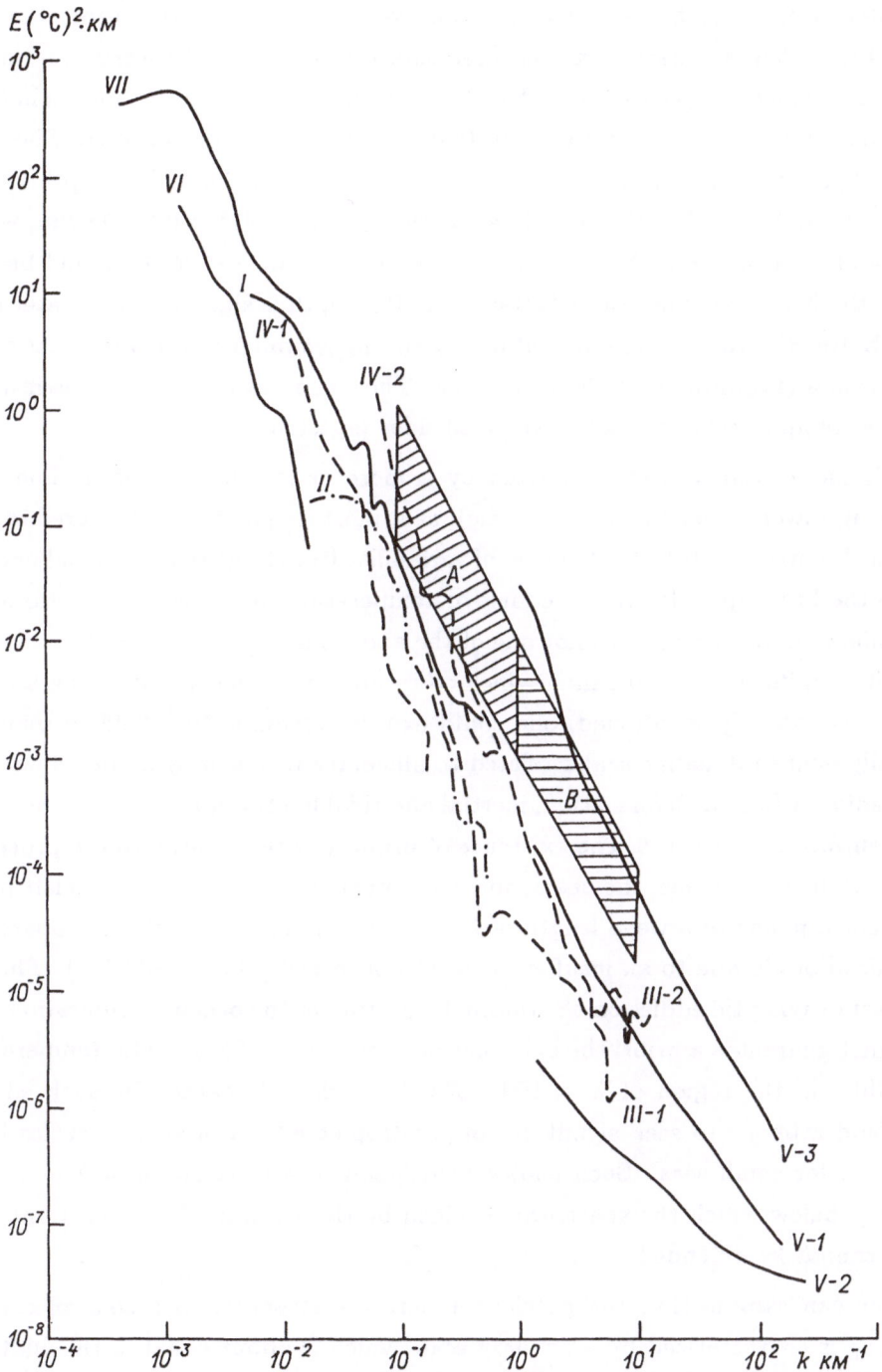

Fig. 2.12. Composite spectrum of horizontal temperature variability in the upper quasi-homogeneous layer of the ocean . The shaded area corresponds to variability which can be expected as a result of oceanic fronts. See text for further explanation.

Interesting spectra are given by Keunecke and Magaard [160] (not shown in fig. 2.12). They are based on measurements taken by towed thermistors in the seasonal thermocline (at the 30 m level) in the Baltic Sea. The section included a small quasi-stationary geostropic eddy (with a diameter of $d \approx 20\ km$). The local radius of Rossby deformation was $R \approx 13 \ldots 18\ km$. The slope of the spectrum is approximated by the function k^{-3} in a wavelength band of $\lambda = 0.5 - 12\ km$, which is proof of geostrophic turbulence developing on scales of $\lambda < R$. It should be said that in the North Sea and the Atlantic where the depth is significantly greater than in the Baltic Sea, the authors did not detect anything similar to the quasi-stationary disturbances encountered in the Baltic Sea. The temperature fluctuations in these basins were apparently caused by internal tides (spectrum k^{-2}).

All the spectra are characterized by a more or less monotonic decline with increasing wavenumber (decreasing scale), as is most frequently encountered in geophysics. It corresponds to the transfer of variability from large scales to smaller ones, and to the breakup of turbulent eddies or temperature inhomogeneities. However, it is somewhat surprising that not one of the above-mentioned spectra has any statistically significant local maxima which correspond to "energy-containing scales". As we have already mentioned, one could expect a temperature field to manifest naturally defined dynamic scales related to planetary waves or synoptic eddies, the local radius of Rossby deformation, inertial and tidal fluctuations, and to convection in the surface mixed layer. The existence of primary scales related to the processes of heat exchange between the ocean and atmosphere is also possible. The left peak, which corresponds to a wave length of $\lambda \sim 800\ km$ on spectrum VII, is apparently of artificial origin due to an insufficient length of record (about 1800 km). Though the effect of synoptic eddies on the thermal structure of the ocean is unquestionable, we cannot guarantee a priori the existence of a maximum of horizontal temperature variability, in the region of $\lambda = 100 - 200\ km$, which is caused by such eddies. One could attempt to seek signatures of geostrophic eddies of smaller scales [160] in spectra for small seas. Such eddies correspond to a wavelength of $\lambda \sim 40\ km$ ($\lambda = 2d$), below which the spectrum obtained by the authors of ref. [160] unfortunately cannot be extended.

One can assume that the patchy temperature structure in a zone of coastal upwelling is characterized by a primary scale which is about equal to the width of this zone ($20 - 40\ km$). Patches of this size are seen on the radiation temperature maps obtained by Holladay and O'Brien [141]. However, the spectral method did not enable us to prove the dominance of this scale because of the insufficient size of the surface area studied.

The spectra (on scales ranging from several kilometres to 100 km) also showed no fluctuations related to the heat exchange between the ocean and atmosphere, which should at least affect the structure of the radiation temperature field. A number of other investigations have also confirmed the absence of sources related to heat exchange on scales up to 50 km. Therefore, we still do not know whether the processes of heat exchange with the atmosphere are characterized by dominant spatial scales within the given range of the spectrum. It is possible that these preferred scales occur at scales of more than 100 km and less than 5 km.

The monotonic nature of the spectra may be an indication of the "inertial" transfer of variability across the spectrum. Is it possible to establish the real nature of this transfer, i.e. is it continuous (due to weak interactions) or discrete (due to strong interactions) [260]? Both processes show a continuous spectrum when averaged. An abnormally high variability of the hydrophysical fields in frontal zones can serve as indirect proof of the existence of discrete transfer. This known fact has also been confirmed with respect to the horizontal temperature field (see spectra IV-2 and V-3 in fig. 2.12). If we attempt to extrapolate the spectral density of these spectra to large scales, we find that it is considerably higher than in spectrum VI for example, and cannot comply with the generally accepted ideas regarding the level of possible mesoscale (synoptic) and large-scale variability. We can therefore assume that the horizontal variability of a temperature field due to the range of synoptic eddies encompassed by spectrum VI is transferred by a jump to the range of smaller scales related to the frontal zones without "losses" in the intermediate scales. This can take place as a result of the enstrophy cascade [262] (fig. 2.13).

Let us discuss in greater detail whether the frontal zones and the patchiness of spatial temperature distribution can be seen (manifest themselves) in the spectra of spatial SLT variability. Let us look at the parts of the composite spectrum marked by the letters A and B respectively. Segment A represents a region of characteristic horizontal scales of parts of the frontal temperature gradients (in their cross-section), i.e. $1 - 10$ km. Segment B includes a variability associated with elements of the finestructure (horizontal) near the frontal interfaces as such (100 $m - 1$ km). No noticeable peaks can be seen anywhere on the various spectra that fall into these segments. All of these spectra are similar in slope to k^{-2}. The spectrum of Zenk and Katz [273], marked V-3, is located on segment B which in the level of its spectral energy is a whole order of magnitude higher than the mean level of the other spectra. This should not come as a surprise, as the data taken for analysis in this case applied to a section executed in the frontal zone, and were characterized by a high spatial resolution which determined the maximum resolved Nyquist wave number $k_N = (2\Delta x)^{-1} = 0.24$ *cycles/m*. Apparently, it is only this type of resolution that enables us to present a good spectral image of the high level of spatial variability associated with the horizontal finestructure of the frontal zones. However, even this type of high spatial resolution apparently does not result in the appearance on the spectra of significant peaks which would correspond to some kind of defined scales associated with the fronts, e.g. modulation effects in the temperature field due to internal waves which are encountered in abundance near fronts (see section 3.5).

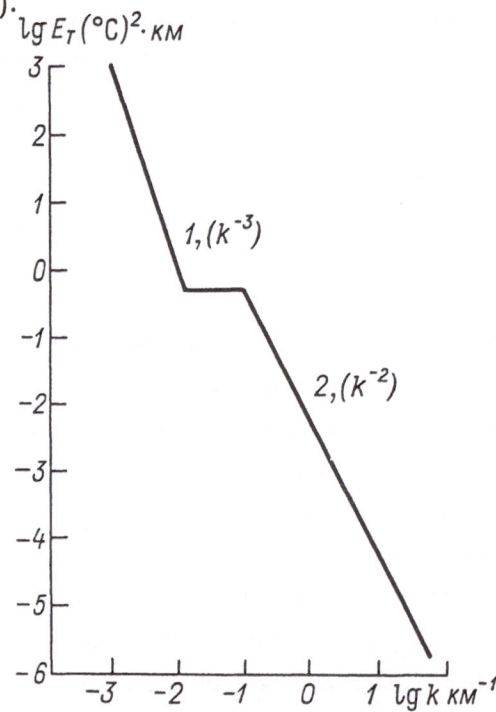

Fig. 2.13. Conjectural composite spectrum of horizontal temperature variability with an abrupt transition from synoptic (eddy) scales (1) to frontal ones (2).

One could assume that the continuity of the spectrum of spatial temperature variability near the surface of the ocean is due to the presence of inhomogeneities on practically all scales within the range of 100 m to 1000 km. As a result, those who have not considered the frequency of oceanic fronts might assume that fronts are likely to show up better as a result of spectral analysis of data collected by the towing of temperature sensors at deeper levels (say, $50-150\ m$) where, as we know, small-scale patchiness of the temperature field associated with near-surface effects

[18, 31, 69] does not exist. On the other hand, the spectra for the 100 and 150 m levels discussed by Moseley and Del Balzo [190] are totally monotonic, decreasing approximately like k^{-2} within the given range. Like Zenk and Katz's spectrum V-3 [273], these spectra are plotted on the basis of relatively short series (not more than 30 km) with a spatial resolution of $0.25 - 0.5$ m. As expected, the level of their spectral density $E_T(k)$ is an order of mangitude lower than in the spectrum of V-3 for the same values of k. However, longer series also do not produce spectral peaks on any of the established scales which could correspond to fronts or temperature "patches" of any predominant size. Basically, this is what was to be expected, as the relative frequency of oceanic fronts is such (see section 2.1) that we can count on an average 1-2 crossings of a front per tow when recording temperature over a distance of $30 - 50$ km in coastal areas. In the open ocean, a front is crossed 1-2 times over a distance of $500 - 1000$ km. It is quite clear that the statistical effect of such rare events can in no way dominate the results of spectral analysis. On the other hand, the "patchiness" of the temperature field, which manifests itself over a wide range of scales and greatly affects the general level of the spectral density $E_T(k)$, does not clearly determine the type of relationship between the slope of the spectrum and k. The inadequacy of the method of spectral analysis in such cases has been noted many times in the literature [112, 245].

To summarize, we can say that long series of temperature measurements over long distances ($100 - 1000$ km) with a resolution of $\Delta x \approx 100 \ldots 200$ m do not enable us to obtain a noticeable peak corresponding to oceanic fronts in the spectra, because of the relatively low frequency of the latter (only several crossings per series). On the other hand, short series ($30 - 50$ km) obtained directly in the frontal zones by means of measurements with a resolution of about 1 m are distinguished by a high level of spatial temperature variability, which gives us abnormally high values of $E_T(k)$ in the wave number interval of $1 - 100$ $cycles/km$. We do not know of any long series of measurements which would enable us to resolve both small-scale and large-scale spatial temperature variability within the $100 - 10^{-3} cycles/km$ range equally well. It cannot be ruled out that a spectrum plotted on the basis of such a series would resemble the one depicted in fig. 2.13.

In the relevant range of wave numbers from 10^1 to $10^{-1} cycles/km$ (segments A and B in fig. 2.13) which Moseley and Del Balzo call "buoyant-convective", the spectrum of horizontal temperature variability in the ocean should, according to their investigations [190], take the form of

$$E_T(k) = A_1 k^{-\frac{5}{3}} + A_2 k^{-3}, \tag{2.1}$$

where k is the wave number, and the dimensional coefficients A_1 and A_2 are functions of depth due to their dependence on the rate of dissipation of kinetic energy ϵ, the rate of dissipation of temperature fluctuations ϵ_T and the Väisälä-Brunt frequency N. The values A_1 and A_2 are determined empirically by the results of spectral analysis of data. A knowledge of the value of N enables us to determine the values of ϵ and ϵ_T, as well as their variation with depth, if the data are collected with a towed sensor at different levels. Therefore, we can say that we already have a model [190] which can be used to describe the background spatial temperature field in the interval of interest to us and includes fronts, but it should be checked against a large number of field data. However, we can already say that the above equation (2.1) can explain the resultant values of the slopes of the spectra $E_T(k)$ in the given interval, which fluctuate between $k^{-1.7}$ and k^{-3} according to the results of various authors.

2.2.2 Spatial variability of salinity

With regard to the spatial variability of salinity near the surface of the open ocean, there have as yet been no statistical studies in the literature, except for the spectrum of spatial variability in salinity at the 3 m level obtained by us over a fairly narrow range of scales (from 1 to 10 km) and published in refs. [34, 35]. This spectrum also has no peaks, and diminishes with a relationship approximating k^{-3}. It is quite clear that a single spectrum obtained from a relatively short series of readings (80 km) can in no way serve as a reliable unified model of the spatial variability of the salinity field in the near-surface layer of the open ocean.

It would be interesting to analyze our own and published data on the spatial variability of the salinity field in the near-surface layer of the ocean, together with the information available on the temperature in this layer.

We should point out that the inhomogeneities of temperature and salinity distribution on scales from 10 to 100 km are of a significantly different nature in different parts of the ocean. In the coastal zone where there is no perceptible upwelling, the changes in temperature and salinity are usually poorly correlated, and are characterized by significantly greater variance than in areas of the open ocean. Measurements in the open ocean have shown that the temperature and salinity of the near-surface layer are highly correlated. As a rule, the spatial fluctuations in temperature and salinity show either a positive, or a negative correlation on scales from several kilometres to tens and hundreds of kilometres [109, 152].

Some years ago, Evans [109] demonstrated how a positive T, S- correlation observed over a distance of more than 1000 *km* alternated with a negative correlation characteristic of an area just as great. An equally characteristic example (fig. 2.14), borrowed from Ikeda et al. [152], demonstrates an analogous occurrence over a distance of about 100 *km*. These illustrations, taken together with our own example mentioned earlier (fig. 2.10), show that this type of situation is quite typical of different parts of the World Ocean.

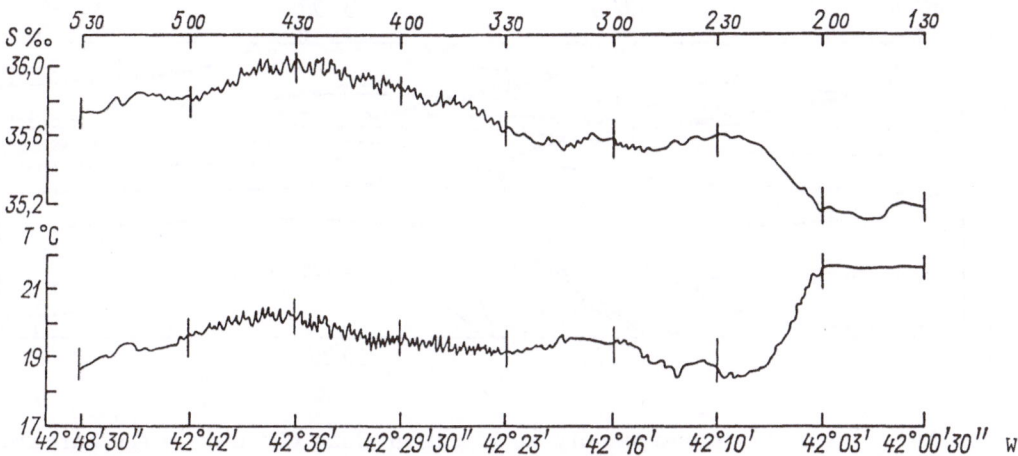

Fig. 2.14. An example of how a negative T, S-correlation changed to a positive correlation at the 3 *m* level along a section at $23°04'S$ in the vicinity of Cabo Frio (Brazil upwelling), based on the data of Ikeda et al. [152].

The appearance of regions with a negative T, S-correlation on the background of a positive correlation should be attributed either to the spatial nonuniformity of vertical mixing [54, 60] with a varying ratio of negative and positive vertical temperature and salinity gradients in the near-surface layer of the ocean, or to the upwelling of waters from great depths on the outskirts of fronts, which compensates the usual downwelling of waters along the frontal interface. The vertical distribution of temperature and salinity in the upper layer may be such that the spatially inhomogeneous vertical motions or mixing will inevitably result in horizontal thermohaline inhomogeneities with negative correlation.

Fig. 2.15 depicts a short section across a zone of sharp horizontal temperature and salinity gradients, plotted on the basis of readings taken by the research vessel "Viktor Bugayev" along $69°40'W$ on the polygon of the POLYMODE buoy stations in the Sargasso Sea in July 1977. The position of the temperature and salinity isolines at this section is such that any local horizontal advection will result in positively-correlated thermohaline anomalies, whereas a nonuniform vertical advection or spatially inhomogeneous mixing will result in the appearance of negatively-correlated thermohaline inhomogeneities in the horizontal plane (including at the surface of the ocean).

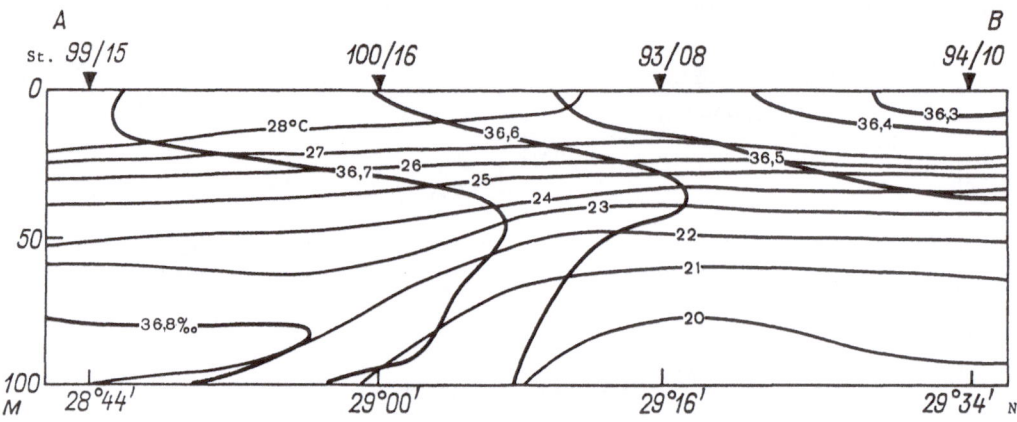

Fig. 2.15. Temperature and salinity for a section along 69°40′W in the Sargasso Sea.

As we know, the precipitation and especially the heavy rains that often fall at low latitudes greatly affect the nature of the horizontal and vertical distribution of salinity in the near-surface layer of the ocean [18]. The variations in salinity due to precipitation can fluctuate from several hundredths per thousand to $1 - 1.5$ *ppt* in extreme conditions, depending on the intensity of the rain, the depth of mixing (which depends on wind speed) and on the initial near-surface stratification. Such changes in salinity are very significant from the point of view of the typical scales and variability of this property. However, it is impossible to record them reliably from satellites with the help of microwave methods sensitive to salinity changes [62,77], since the resolution threshold of these methods is not higher than $1 - 2$ *ppt*. Because of this, we do not have any remote sensing information similar to fig. 2.7, which would enable us to study the characteristic spatial scales of salinity variability and the mean distance between salinity fronts (see section 2.1). Researchers are left only with direct methods of measurement. A comparatively small number of special measurements of salinity variability near the surface of the ocean has been carried out so far; most of them have to do with the effect of precipitation, and can be credited to the author and his co-workers A.I. Ginzburg, A.G. Zatsepin and V.Ye. Sklyarov [18]. The information thus obtained on the spatial "patchiness" of salinity in the near-surface layer of the ocean can be summed up in the following way:

1) with weak winds, the depth of freshening during heavy rainfalls seldom exceeds $1-2$ m, which means that the salinity in this layer diminishes by $0.5-1$ ppt with 30 mm of precipitation;

2) moderate winds increase the thickness of the freshened layer to $5-10$ m, which means that the salinity in this layer decreases by only $0.1-0.2$ ppt with each 30 mm of precipitation;

3) rainfall is mostly commonly accompanied by a simultaneous slight temperature drop of $0.1-0.2°C$ near the surface of the ocean; on the other hand, there may be no simultaneous thermal effect, as the changes in the heat-exchange conditions at the ocean/atmosphere boundary during rainfall are quite complex and ambiguous;

4) the characteristic horizontal dimensions of the surface patches freshened by precipitation correspond to the typical dimensions of cumulonimbi $(1-5$ $km)$. However, during heavy and long rainfall, these patches may combine, forming "saucers" of $0.5-1$ ppt freshened water with a diameter of several hundreds of kilometres near the surface of the ocean (Fig. 2.16);

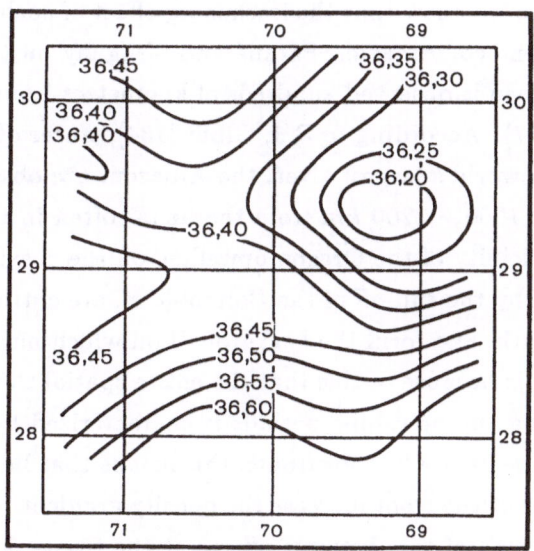

Fig. 2.16. Distribution of salinity (ppt) near the surface in the Sargasso Sea on 18-26th of July 1977.

5) the thermodynamic effect of such extensive freshened areas is determined mainly by the suppression of turbulent exchange of the freshened layer with the underlying strata, which may lead to overheating of the ocean surface during the day, and overcooling during the night. During high atmospheric humidity and nocturnal cloudiness, the daytime heat can accumulate in the freshened layer over several days, which at some latitudes can help generate and spread tropical cyclones and hurricanes;

6) even very heavy rainfalls do not usually result in the formation of sharp frontal interfaces at the margins of the freshened patches. This is most probably due to the small thickness of the layer freshened by the rain. However, the seasonal prevalence of precipitation over evaporation, and evaporation over precipitation in certain parts of the ocean can result in the formation of salinity fronts at the boundaries of these areas when convergent currents occur near the surface [213, 214].

The freshening of coastal areas of the ocean by the run-off of large rivers can reach very high levels, the horizontal salinity changes over small distances amounting to $10 - 15$ *ppt* and more. The formation of fronts in this case is more a rule than an exception to it (see section 3.3). Despite the theoretical possibility of mapping salinity changes in such areas from satellites with the help of ultrahigh-frequency equipment [77], we still have no adequate spatial salinity-field data obtained by remote sensing methods in areas where large rivers flow into the ocean.

At the same time we know that the freshening effect of such large South American rivers as the Orinoco, Amazon, Paraná and Uruguay on the structure of the upper layer of the ocean is observed hundreds of kilometres from the shore [59, 123, 271] (see also fig. 2.17). According to R.J. Gibbs [134], traces of oceanic freshening by the run-off of the world's largest river, the Amazon, are observed near the surface at a distance of $1000 - 1200$ *km* from the delta, often in the form of isolated lenses [222]. In the vicinity of the Oregon upwelling off the coast of North America, the waters freshened by the run-off of the Columbia R. are entrained by the coastal current far to the south, and form the background on which numerous fronts of the upwelling emerge. It is because of this that the entire spatial thermohaline variability in this area during the upwelling periods is characterized by a strong negative T, S-correlation (see section 3.2). For fronts, this means that both the temperature and salinity gradients at the front increase the density gradient. This in turn should lead to an intensification of the dynamic effects at the fronts, and consequently to the sharpening of the latter. Obviously, the final result of frontogenesis, regardless of how the local deformation field originated, depends greatly on the nature of the background spatial variability of temperature and salinity; furthermore, the variability of salinity can be especially important. In the author's opinion, thermal fronts with an additional salinity contribution to the density gradient and salinity fronts with or without an additional thermal contribution to the density gradient are much sharper in all of their external and dynamic manifestations, than purely thermal fronts or fronts with a partial or full density compensation of temperature and salinity contributions (during a positive T, S-correlation). This question will be discussed in section 3.2.

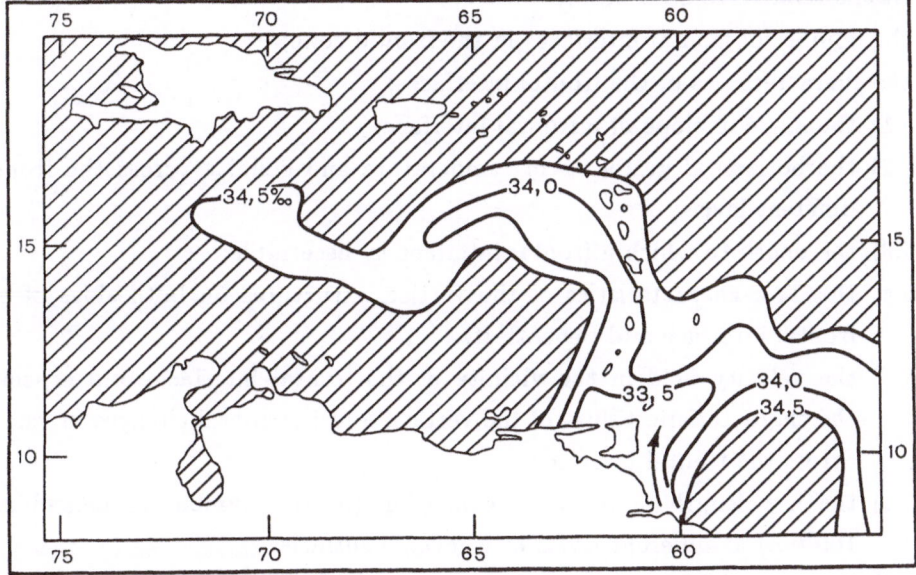

Fig. 2.17. Large-scale freshening effect of the Orinoco R., based on the data of G. Wüst [271].

2.3. Main physical parameters of frontal zones and interfaces

On the strength of the fact that our data on frontal zones and fronts in the ocean cannot yet be regarded as sufficiently exhaustive and because the descriptive stage in the study of oceanic fronts is still far from complete while the theoretical description requires a solid base of facts for its development, one of the main tasks of modern research into oceanic fronts lies in the collection and organization of reliable physical information about the phenomenon. This information should be quantitative in places where this is possible, and should interpret the most typical aspects of the phenomenon, as well as all the deviations from that which is considered "normal".

At the present time, data on the following physical parameters are regarded as the most important for a satisfactory physical description of frontal zones and frontal interfaces:

1) the width of the frontal zone at the surface of the ocean, or at any characteristic surface;

2) the along-front direction, the range and depth limits of the front;

3) the general T, S-characteristic of the frontal zone. The mean horizontal gradients of temperature and salinity across the zone. The spacing of isopycnals, isotherms and isohalines in a cross- section of the frontal zones, and their positions relative to each other;

4) the number of frontal interfaces in the frontal zone;

5) the characteristics of the frontal interfaces;

   5.1) the angle of slope of a frontal interface;

   5.2) the horizontal gradients of temperature and salinity across the front at different levels;

   5.3) the depth to which a frontal interface is discernible;

   5.4) the T, S-characteristics of the vertical structures on both sides of each frontal interface and beneath it;

   5.5) the velocity field in the vicinity of the frontal interface, and especially the characteristic values of the vertical and horizontal changes in velocity across the frontal interface;

   5.6) the wavelength of frontal meandering (or the number of meanders in 100 km) at different levels or isopycnal surfaces;

   5.7) the radius of curvature of the most curved parts of the front;

   5.8) the velocity and direction of general migration of a frontal interface;

   5.9) the velocity and direction of migration of meanders;

   5.10) the observed tendency towards sharpening or relaxation of the frontal interface;

   5.11) the actual time of frontogenesis (or frontolysis) if the process was observed;

6) the statistical and physical characteristics of the vertical thermohaline fine-structure near frontal interfaces and on the periphery of the frontal zone.

All the characteristics listed above can in principle be observed by means of direct measurements in the ocean. Some of the characteristics can be measured more easily with the help of remote sensing methods (e.g. from satellites) than from research vessels. On the basis of the above characteristics, one can obtain indirect values of other equally important but unmeasurable physical parameters such as the vertical component of velocity at a frontal interface or the coefficients of turbulent transfer of heat and salt across the surface of a frontal interface (see section 4.4), etc. On the whole, the above list contains the most important elements of a 3-dimensional physical description of oceanic frontal phenomena. However, the real possibility of successfully executing all the measurements required for this purpose in the ocean is still problematic. This is why there are practically no descriptions of this nature in the literature as yet. We can mention only a number of papers [50, 159, 163, 188, 254] which already contain some elements of the 3-dimensional approach to describing and analyzing frontal phenomena (see also section 4.3). It seems that a planned combination of remote measurements from Earth-orbiting artificial satellites with a survey carried out simultaneously by several vessels and airplanes (with the use of expendable bathythermographs and aerobathythermographs) can help organize the necessary measurement programs in the near future.

To the above list of what it is desirable to know about frontal zones and interfaces in order to carry out an adequate 3- dimensional description of them, we must add a number of comments based on the results of our own observations. In accordance with the criteria discussed above, (see section 1.2), the concept of "width" makes sense for a frontal zone, but not for a front. However, the actual nature of the temperature change across a frontal zone may be such that the maximum horizontal temperature (or salinity) gradient is not reached at some point, but is observed within a narrow band of finite width (e.g. see fig. 2.18). In this case, it is best to use the term "actual width of the front" and give this width in the description as one of the physical characteristics. For the case depicted in fig. 2.18 for example, the actual width of the thermohaline front A is equal to $1 - 1.2 \ km$ with the total width of the frontal zone B equal to 22.5 km.

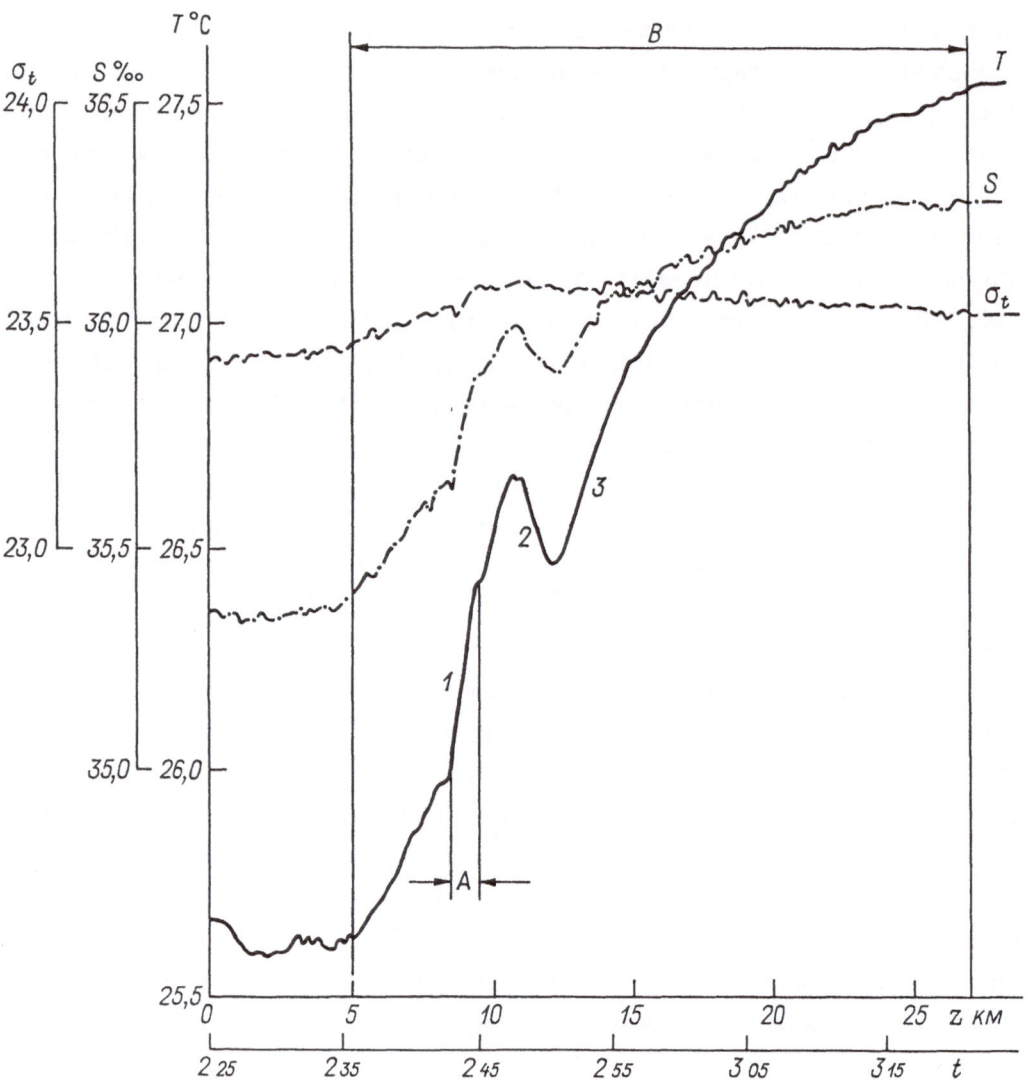

Fig. 2.18. An example of the crossing of a frontal zone in which several frontal interfaces (1, 2, 3) are observed. The actual width of one of the fronts is marked by the letter A. The total width of the zone (B) exceeds 20 km. These data were collected by the author during the 25th cruise of the research vessel "Akademik Kurchatov" in the Sargasso Sea on 14 Sept 1977. The frontal zone presumably bounds the cold wake of the tropical cyclone "Clara" from the south. The cyclone passed through this area a day before the vessel.

As we shall see below (see section 2.4), despite the significant physical similarity of frontal phenomena in the ocean and atmosphere, their similarity in many important details has not yet been confirmed by observations, while the importance of some of the phenomenological differences may at first glance appear to be exaggerated. Observations of oceanic fronts (when properly organized) can help clear up the debatable and vague questions in this area. Among other things, the determination of the direction in which the frontal interface or a part of it is moving gives us the right to regard the corresponding front as a "warm" or a "cold" one. The phenomenon of overturning at a "cold" front, which has been noted by Knauss [163], was also observed by us during the 34th cruise of the research vessel "Akademik Kurchatov" near the northwestern extremity of Isabela Is. (Galapagos Islands) where a very sharp thermal front with a temperature change of $5°C$ separates the upwelling waters of the Cromwell Current with a temperature of about $19°C$ from the typical surface waters of the archipelago with a temperature of more than $24°C$. During our observations, the frontal interface moved quite rapidly from the shore toward the open ocean (probably under the effect of the tidal current). According to the data of salinity-temperature-depth profiles at three stations, the isotherms at a section across the frontal interface (fig. 2.19) had a configuration which indicated that cold waters were overriding warm waters. The observations carried out during this cruise in other regions of the World Ocean give us reason to believe that thermal fronts which follow the pattern of a "warm" front are significantly less defined than "cold" fronts.

The most typical manifestation of a thermoclinic near-surface frontal interface on a T, S- or θ, S-diagram is the Y-shaped (or fork-shaped) distribution of points which correspond to two closely spaced hydrographic stations (fig. 2.20). The left branch usually corresponds to the fresher and colder waters, and the right one to the more saline and warmer waters. The depth at the point where the branches join corresponds to the deepest level reached by the frontal interface. The "fork" can have a different slope on a T, S-diagram, depending on the correlation of the thermal and saline contributions to the horizontal density gradient across the front. We should note that the T, S-curves on both sides of a purely baroclinic frontal interface are identical.

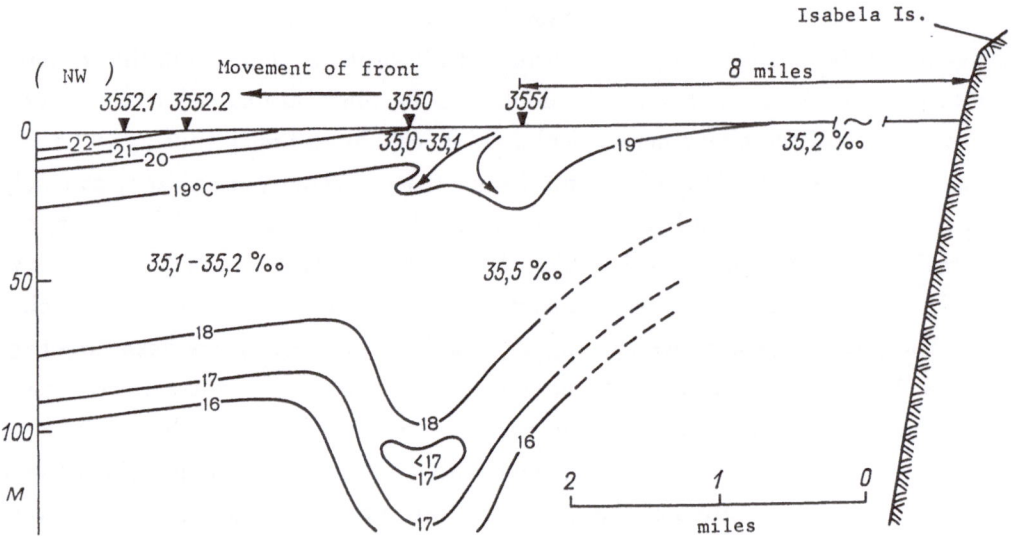

Fig. 2.19. An example of overturning when cold waters override warm waters, in the area where the Cromwell Current comes to the surface near Isabela Is. (Galapagos Islands).

Highly noteworthy are the cases where three different thermohaline structures were detected in the same frontal zone near one or several nearby frontal interfaces. It is possible that occlusion-type fronts similar to atmospheric ones are also encountered in the ocean, but nothing definite is known about such fronts as yet. On the other hand, frontal structures with three types of waters, like the one described by Katz [159], can be mistaken for occlusion structures, whereas the third "water mass" is simply the result of the sinking of the products of cross-frontal mixing of the two main types of waters separated by the front.

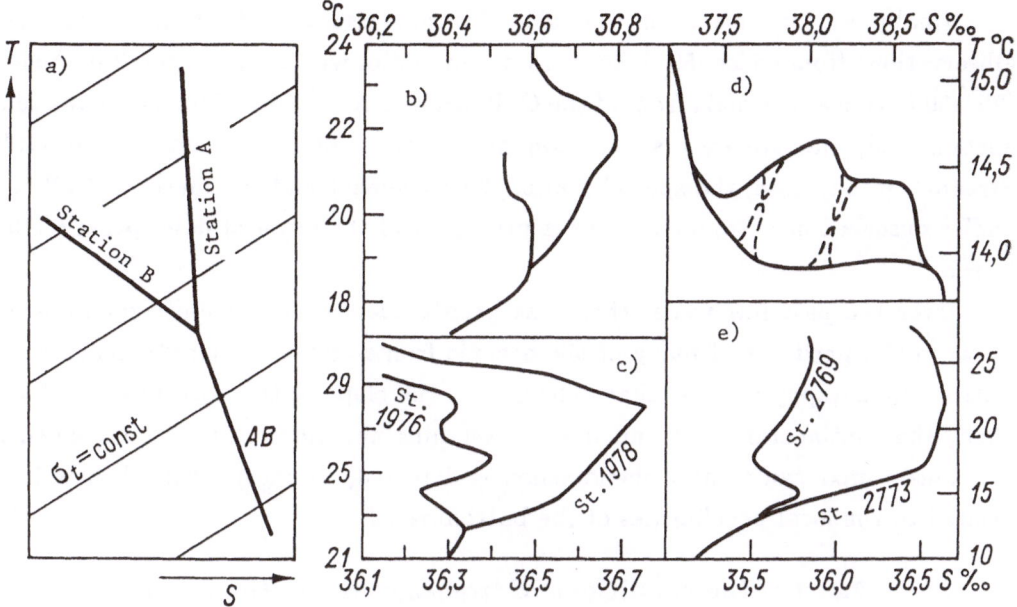

Fig. 2.20. Examples of the typical Y-form of a T, S-curve for pairs (or groups) of stations executed on different sides of a frontal interface. a – schematic; station A – in warm and more saline water, station B – in cold and less saline water. The maximum difference in temperature and salinity across the front is observed at the surface of the ocean; b – one of the frontal interfaces of the subtropical convergence in the Sargasso Sea, based on the data of ref. [159]. The curves have been plotted by averaging several pairs of stations; c – frontal interface at the outlet from the Gulf of Aden to the Red Sea, based on the data of the 22nd cruise of the research vessel "Akademik Kurchatov" from 7-8th July 1976; d – abyssal frontal interface in the Strait of Sicily which separates the waters of the western and eastern basins of the Mediterranean Sea, based on the author's observations during the 18th cruise of the "Akademik Kurchatov" in April 1974. The curves have been plotted by averaging the data of two groups of stations located on both sides of the frontal interfaces (see section 4.2, fig. 4.6 and fig. 4.7). The temperature and salinity contrasts are not observed at all in the near-surface layer, due to heat exchange with the atmosphere. The dashed line depicts the cross-frontal quasi-isopycnic intrusions which are frequently observed; e – frontal interface of a cold (cyclonic) ring of the Gulf Stream, based on the data collected by the author [66] during the 27th cruise of the "Akademik Kurchatov" in the Sargasso Sea in September 1978.

We should also take into account the type (sign) of the T, S- correlation in the small details of the horizontal inhomogeneities on both sides of the front, as mentioned in section 2.2. It is possible that this indirect method could help to determine the position of the upwelling zone which is associated with the frontal interface and is found at some distance from it.

Finally, we draw attention to eddies of various sizes close to frontal interfaces. Observations from space [50] help us to detect eddies with a rather small diameter $(25-30\ km)$ near the main front of the Gulf Stream. Are eddies of smaller diameters encountered, and are they always confined to such intensive fronts as the Gulf Stream? In any case, the spatial scales, the movement and the lifetime of all the eddies observed near fronts are an essential part of the physical description of the latter.

Over the past few years, there has been evidence for a definite relationship between the positions of many of the oceanic fronts and characteristic features of bottom topography such as sharp changes in the slope of the bottom around the shelf, the continental slope and underwater hills and ridges. Therefore, when a previously unknown frontal phenomenon is detected, we must establish how it is related to the local peculiarities of the bottom relief.

2.4. On the conditions of frontogenesis in the ocean and in the atmosphere

Fronts are a characteristic feature of the general circulation in the atmosphere and in the ocean. However, the significant differences between these two media (in physical characteristics, the equations of state, the nature of stratification, scale factors, etc.) do not permit us to expect a perfect analogy between atmospheric and oceanic fronts. We are more surprised that, despite these differences of the two media, the fronts in them have very many similar features both in their dynamics and in the process of their formation.

Comparison of the main features of oceanic and atmospheric dynamics has always been fruitful [193, 194, 204, 206, 219, 220] for many reasons of which we mention the two main ones. The first, purely scientific one, is that by using the same fundamental physical laws and equations of motion for the analysis of phenomena in two media with different characteristics, we can better understand the role of this or that physical factor in the dynamics of each of the media. Rossby [219] suggests that a comparison of circulation patterns in the atmosphere and ocean provides us with the most useful alternative to experiments with a controlled change of the fundamental parameters. The other reason stems from the earlier and more rapid development of dynamic meteorology as a physicomathematical science as compared with oceanology which, for a long time, was regarded as one of the descriptive-geographic sciences. Moreover, soon after World War II, the development of radar technology enabled meteorologists to carry out three-dimensional analysis of certain atmospheric processes [219], whereas oceanographers had to rely mainly on a two-dimensional approach (conventional vertical sections). The first large-scale synoptic surveys in the ocean, e.g. "Operation Cabot" [120] (see also section 1.1), were at that time a novelty and could not be carried out very frequently because of their high cost. Therefore, it was basically oceanographers that profited from these comparisons, due to the more extensive methodological, experimental and theoretical experience of meteorologists.

In order to better understand the limitations which nature inevitably places on all comparisons of this type, let us sum up the differences between atmosphere and ocean (apart from the differences in the main physical properties which determine the characteristic scales) which in our view cannot but affect the nature of fronts and the processes of their formation and dissipation. The differences are as follows.

1. The atmosphere is heated from below, which often gives it a hydrostatic instability and makes its near-ground layer less stable on the average than the near-surface pycnocline of the ocean. The ocean at temperate and low latitudes is heated from above, which results in a sharp jump in temperature and density under a comparatively thin layer. This jump is hydrostatically quite stable.

2. The circulation of air in the atmosphere is generated by thermal effects, i.e. internal forces, whereas the main part of the energy of oceanic circulation is associated with the external generating force of the wind. Because of this, the different origins of the zonal jet streams in the two media are apparently very important.

3. The Ekman layer in the atmosphere is found at the base of the fronts, and the convergence of the Ekman transport in this layer contributes greatly to frontogenesis in the atmosphere [206]. In the ocean, fronts can appear and persist without the direct participation of the wind, and the wind-driven Ekman layer in such cases is neither significant, nor necessary for the creation and maintenance of the front. On the other hand, frontogenesis associated with the convergence of the Ekman drift in the ocean can create very shallow fronts. However, the internal Ekman layers that appear in the near- frontal currents in the ocean can play a very important role in maintaining the sharpness of oceanic frontal interfaces (see section 2.9).

4. In the ocean, the density of the water is determined not only by the tempera-ture, but also by the salinity, for which there is no perfect analog in the atmosphere. Whereas purely saline fronts exist in the ocean, a drop in the moisture content alone of the atmosphere is not enough for the formation of a front. Furthermore, ther-

mohaline fronts with a practically zero horizontal density and pressure gradient are possible in the ocean.

5. The ocean has impermeable lateral boundaries which are not found in the atmosphere. The oceanic thermocline is always enclosed by these boundaries from the sides, whereas the tropopause in the atmosphere is located much higher than even the highest mountains.

6. Because of the combination of factors listed in points 2 and 5, the main currents in the ocean and atmosphere may have significantly different and even opposite directions which must generate differences in the movement of synoptic-scale eddies in the given media.

7. Atmospheric cyclones together with their characteristic frontal systems can be regarded as unstable waves which appear as a result of baroclinic instability of zonally aligned jet streams [164, 198]. Large-scale currents in a zonal direction do exist in the ocean (the Gulf Stream, Kuroshio, Circumpolar Antarctic Current and the Equatorial currents) and their instability also generates eddies, meanders and fronts [55, 192, 223]. However, for the reasons stated in point 6, the movement of these eddies, their own frontogenetic effect and their relationship with fronts differ greatly from what we are accustomed to seeing in the atmosphere. For example, among the variety of eddies detected in the ocean, nobody has ever observed an eddy which is similar in every respect to an atmospheric cyclone with a characteristic warm sector bounded by a warm and cold front. Also, atmospheric occlusive fronts apparently have no complete analog in the ocean.

On the other hand, we should also mention the factors which lead to similarity in the dynamics of fronts in these media.

The rotation of the Earth affects atmospheric and oceanic fronts in the same way. Therefore, in cases where the scales of the fronts are large enough and the fronts are well-defined in the density field, we observe a geostrophic balance of motion and the mass field, determined by the velocity and density values on both sides of the front. At the same time, the fronts acquire a slope proportional to the velocity gradient across the front, i.e.

$$tg\,\alpha = \frac{f\bar{\rho}(u_1 - u_2)}{g(\rho_2 - \rho_1)}\,, \qquad (2.2)$$

where g is the acceleration of gravity, f – the Coriolis parameter, u_1 and ρ_1 – the current velocity along the front and the density in the lighter water, and u_2 and ρ_2 – the current velocity along the front and the density of the denser water.

From equation (2.2), which is known as the Margules formula, we see that a cyclonic vorticity is always observed at fronts both in the atmosphere and in the ocean [206]. In the ocean only small-scale fronts are likely to be exceptions. However, it should be said that oceanic fronts are actually far from being in a perfectly balanced state. The intensive ageostrophic motions developing at them give researchers reason to believe that they are "semi-geostrophic" phenomena [262].

We should also note that the field of motion in both the atmosphere and the ocean is such that so-called deformation fields, which in meteorology are regarded as the basis of frontogenesis, occur. The concept of a deformation field was introduced into meteorology by T. Bergeron [87], and later successfully developed by S. Petterssen [204, 205]. A deformation field (see section 2.5) is a configuration of streamlines, which in combination with a field of any other scalar variable (e.g. temperature) results in the compression or spreading out of the lines of concentration of this scalar in the direction known in meteorology as the stress axis (fig. 2.21). A pattern of flow with which horizontal deformation fields are inevitably associated is one in which the velocity field contains large-scale horizontal eddies with a vertical axis. Eliassen [108] refers to this flow field as Rossby's planetary regime. In Eliassen's opinion, the horizontal deformation fields in the atmosphere are only the "initiators" of the frontogenetic process. Atmospheric fronts sharpen and are then maintained in a sharp state due to the appearance of transverse motions ("vertical deformation fields") which are caused by the initial frontogenesis. The same is apparently observed in the ocean. C.W. Newton [194] believes that vertical motions effectively intensify the baroclinicity of primary frontal disturbances, and contribute to the sharpening of frontal gradients during frontogenesis in both the atmosphere and the ocean.

In Rossby's opinion [218, 219], vertical frontogenetic motions and particularly the downwelling of cold masses of air in the atmosphere and water in the ocean during cyclogenesis in the jet stream zone represent one of the fundamental analogies between the atmosphere and the ocean. The acceptance of this analogy leads to acceptance of the fact that oceanic currents like the Gulf Stream are at least partly maintained by internal forces [219] generated by the release of potential energy during the downwelling of cold water masses in the process of meandering and eddy formation.

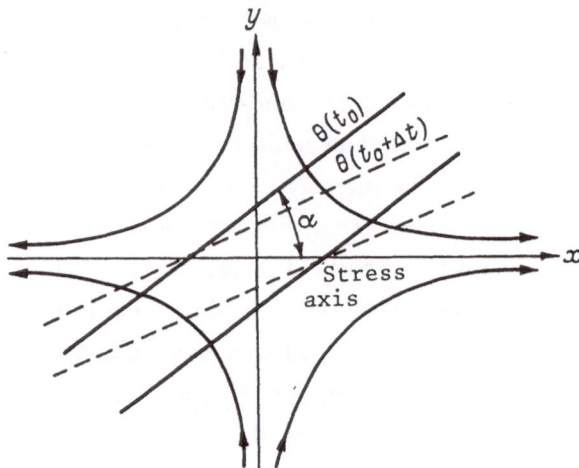

Fig. 2.21. Diagram of a hyperbolic deformation field and changes in the position of the isolines of potential temperature θ in this field after a time interval Δt.

It should be said that the mechanism of the "vertical deformation field", proposed by Eliassen in the development of Rossby's ideas [218, 219], is only one of many possible physical interpretations of the sharpening of fronts that appear in the primary horizontal deformation field. There is no doubt that the meanders and eddies appearing as a result of the instability of zonal jet streams in the ocean and atmosphere are characterized by having significantly more intensive horizontal deformation fields (with a deformation rate $D \approx 10^{-5}s^{-1}$ for the ocean, see section 2.5) than large-scale planetary gyres or circulations ($D \approx 10^{-6} \ldots 10^{-7}s^{-1}$ for the ocean). Therefore, on the basis of our own criteria and terminology, fronts as such (and frontal interfaces) could be regarded as due to eddy formation, whereas synoptic-scale eddy formation itself could be attributed to the instability of planetary frontal zones. Such an interpretation is just as applicable to the ocean as it is to the atmosphere. In the ocean, we observe well-defined large-scale gyres which include zonal jet streams and western boundary currents of the Gulf Stream or Kuroshio type. It can be said that these planetary gyres are formed by the combined effect of the wind, the rotation of the Earth and the lateral boundaries of the oceanic basins. Though the driving forces of atmospheric jet streams are of an entirely different nature (as we have already mentioned), the end result of their instability is generally similar in both cases. This contradiction was somewhat disturbing to C.G. Rossby [218, 219], whose scientific view would have required a more complete dynamic analogy between atmospheric and oceanic jet streams. This contradiction has to some extent remained unsolved to this day. The opinion that the formation of sharp frontal interfaces within oceanic frontal zones is mostly associated with meandering and eddy formation in these zones has not yet been accepted universally. Basically, such a concept is quite compatible with the natural sequence of processes that lead to continuously decreasing scales, and is in accord with the concept of a flow of enstrophy from larger scales to smaller ones in the spectrum of two- dimensional (geostrophic) turbulence subject to potential vorticity conservation [41]. It is in this spirit that Woods formulated his hypothesis [262] on the double cascade of enstrophy and energy across the spectrum of oceanic turbulence "from megametre to millimetre", i.e. from large-scale oceanic gyres across synoptic-scale geostrophic eddies and across "semi-geostrophic" fronts (transferring enstrophy) to characteristic scales of dissipation, with the participation of internal waves (transferring energy) and finestructure. Woods introduced this hypothesis for the ocean with the characteristics of its circulation, structure and scales taken into account. It is definitely a step forward in our understanding of the inter-relation between the various classes of motion in the ocean. However, the genesis of this hypothesis arose from a careful study of the above-mentioned analogies and differences between the ocean and atmosphere, while the concept of two-dimensional turbulence, which serves as the basis of Woods' hypothesis, was naturally borrowed from atmospheric physics.

The two frontogenetic mechanisms discussed above (horizontal and vertical deformation fields) do not exhaust all the possible conditions that could lead to frontogenesis. Hoskins and Bretherton [146] found at least eight frontogenetic mechanisms in the atmosphere, these being

1) a horizontal deformation field;

2) a horizontal shearing motion;

3) a vertical deformation field;

4) differential vertical motion;

5) release of the latent heat of evaporation;

6) friction against the Earth's surface;

7) turbulence and mixing;

8) radiative effects.

When applying this list of mechanisms, one should eliminate point 5, as there is no analog of atmospheric humidity in the ocean. The analog of friction at the Earth's surface (point 6) is bottom friction, which plays an important role in shallow areas where fronts are often formed by tidal mixing (see section 3.4). The seventh mechanism (turbulence and mixing) can be related both to near-bottom friction, and to wind stress at the sea surface. As our observations have shown, oceanic fronts appear under conditions of extremely strong interaction between the ocean and atmosphere (storm mixing, the effect of typhoons and hurricanes on the upper layer of the ocean). Apart from this, turbulence and mixing during frontogenesis in the ocean also play a secondary role, intensifying or weakening fronts as a result of, say, turbulent entrainment across a frontal surface. As we have already shown, vertical motions in different directions (point 4) and vertical deformation fields (point 3) play an important role during the frontogenesis associated with eddy generation in jet streams. These mechanisms can be, and really are, effective in the regions of upwelling of deep waters, especially near shore.

On the whole, one can say that almost all of the oceanic mechanisms mentioned by Hoskins and Bretherton [146] can be interpreted as a type of deformation field, or as mechanisms which are conducive to the frontogenetic effect of a deformation field. To these we can add the specifically oceanic mechanism of convergence of the Ekman drift associated with a spatially variable wind field above the ocean and nonuniformities of heat and mass exchange with the atmosphere [214].

When dealing with the processes of frontogenesis for synoptic- type fronts which are not associated with external sources of heat (or salt in the surface layer of the ocean), we can regard the motion as adiabatic. This assumption is widely applied in dynamic meteorology,and can be successfully used to study oceanic fronts. If we disregard viscosity, we can say that adiabatic movements in the ocean and atmosphere occur along isentropic surfaces. Since the potential temperature of the air is preserved in the atmosphere during adiabatic motion, the atmospheric isentropes coincide with the lines or surfaces of equal potential temperature θ.

In the ocean, not only the potential temperature and salinity, but also the potential density ρ_θ is preserved during adiabatic motion. Consequently, oceanic isentropes should practically coincide with the isopycnals. Since oceanic isopycnals are most generally intersected by surfaces of equal pressure, equal potential temperature and equal salinity, the adiabatic motion of water along isopycnal surfaces in the presence of deformation fields should and does cause significant sharpening of the temperature and salinity gradients on the isopycnal surfaces.

The conventional and most orthodox way (when making comparisons with the atmosphere) of determining the position of a frontal interface in the ocean is to single out the surfaces (loci) of all maxima of the pressure gradient (of maximum baroclinicity*) at the isopycnal surfaces. It is in this case that the near-frontal jet stream associated with maximum baroclinicity will coincide with the surface of the frontal interface. As a rule, one can expect the maximum of baroclinicity to coincide with the maximum of thermoclinicity on isopycnal surfaces, and then establish the frontal position using the latter. This is precisely what the author did [59, 61, 72] when analyzing the data of the "Polygon-70" expedition in the tropical Atlantic in 1970. Isopycnic analysis as applied to the study of oceanic fronts is particularly characteristic of the investigations of Woods and coauthors [178, 265]. However, far from all oceanographers are applying it in practice. Basically, it is conceivable that in the ocean (in contrast to the atmosphere), there may be no baroclinicity, at least theoretically, despite the presence of thermoclinicity and haloclinicity. MacVean and Woods [178] believe that in such cases the baroclinicity is actually never equal to zero, but is simply small and undetectable against the background noise of internal waves. Whereas three-dimensional analysis of measurements from an aircraft permitted the study of atmospheric fronts in three dimensions back in 1964 [102], oceanographers are only beginning to apply special methods of measurement [106, 265] which make it possible to obtain a three-dimensional picture of temperature and salinity on selected isopycnal surfaces. The difficulty here lies in obtaining a spatial resolution adequate enough for frontal analysis. This is in the neighborhood of several hundred metres horizontal separation between measurement points, with rapid coverage of sufficiently large areas. From a vessel moving at a speed of $8 - 10$ knots, modern towed equipment of the "Batfish" type makes it possible to carry out quasi-synoptic measurements in a $50 - 100$ m layer over an area of $500 - 900$ km^2 in

---

* The expressions "baroclinicity", "thermoclinicity" and "haloclinicity" are used here in accordance with Woods' terminology [262, 263] with the meaning of isopycnal surfaces intersected by surfaces of equal pressure, temperature and salinity, respectively.

about 20 hrs [265]. However, the "microsurvey" method with an AIST STD-probe enabled the author as early as 1970 to carry out frequent measurements (every $2-3$ miles) over an area of 340 km^2 and to a depth of 500 m in approximately 24 hrs. As a result, it was possible to map a thermohaline front with local isopycnal temperature gradients of up to $0.2 - 0.3°C/km$ in the sharpest parts on an isopycnal surface with $\sigma_t \approx 25.00$ [59, 61].

When studying atmospheric and oceanic frontogenesis, one can apply the principle of potential vorticity conservation [41], examining the process in a system of coordinates related to the potential temperature or potential density respectively. This approach is even more convenient since frontal zones in the atmosphere and ocean are zones in which relative vorticity is concentrated [206]. The use of the principle of conservation of potential vorticity helped Danielsen [102] to detect the displacement of the tropopause during atmospheric frontogenesis (fig. 2.22). This is an important mechanism for air exchange between the stratosphere and troposphere. The analog of this process in the ocean is apparently the deflection of the thermocline at frontal interfaces. This leads to the formation of an intrusive finestructure and an effective heat and mass exchange across the frontal interface [178, 263] (see sections 4.2 and 4.4). In their most complete model of atmospheric frontogenesis, Hoskins and Bretherton [146] showed that an increase in relative vorticity in frontal zones could lead to a breakdown of the geostrophic balance (as a result of the dominance of the relative vorticity over the planetary vorticity in the equations of motion), and to the development of ageostrophic motions, including vertical ones, along frontal interfaces. In this case, the frontal Richardson number diminishes to critical values, which results in the appearance of "clear-air turbulence" in the atmosphere and "billow turbulence" in the ocean due to Kelvin-Helmholtz instability.

To conclude this comparative analysis, I would like to present two tables from Newton's paper [194], which clearly compare the typical characteristics of the Gulf Stream with the corresponding characteristics of an atmospheric jet stream (table 2.1), as well as the characteristics of unstable waves (meanders) formed in both currents (table 2.2).

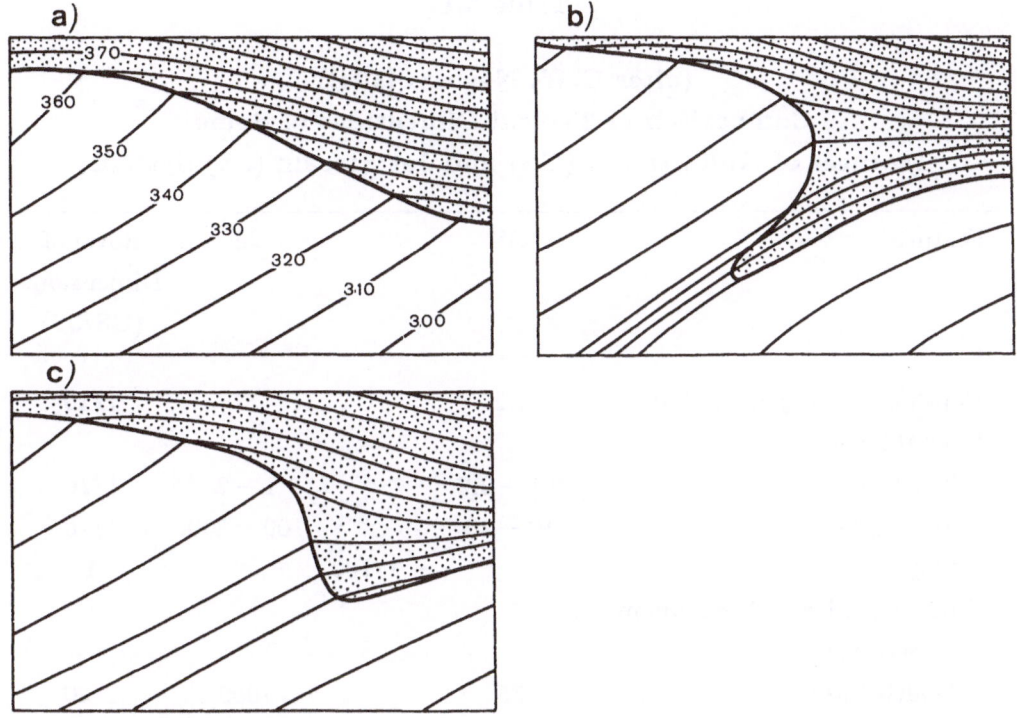

Fig. 2.22. Different phases (a, b, c) of tropopause displacement above an atmospheric frontal interface as the latter becomes sharper. Diagram borrowed from ref. [102].

Table 2.1

(after C.W. Newton [194])
Comparison of Typical Dimensions of Some
Features of Gulf Stream (GS) and Jet Stream (JS) Systems

Feature	GS	JS	Ratio of Dimensions (GS/JS)
Depth of troposphere,* km	1.2	12	1/10
Frontal layer			
Thickness, km	$0.1 - 0.2$	$1 - 2$	1/10
Width, km	$10 - 20$	$100 - 200$	1/10
Slope	$\sim 1/100$		1
Current (at level of maximum velocity)			
Width,† km	35	1000	1/30
Speed maximum, ms^{-1}	2	60	1/30
Anticyclonic shear	$0.7f$ ‡		1
Cyclonic shear	$1 - 2f$		1
Vertical shear in frontal layer, $10^{-2}s^{-1}$	$1 - 2$		1

* For the Gulf Stream east of Blake Plateau.

† Distance between points on either side where current speed falls off to half the speed in the core for a sharply defined current.

‡ Here f denotes Coriolis parameter.

Table 2.2

(after C.W. Newton [194])
Comparison of Some Properties of Waves on
the Gulf Stream (GS) and Jet Stream (JS)
($f = 0.9 \cdot 10^{-4} s^{-1}$ at 38° Latitude)

Feature	GS	JS	Ratio of Dimensions (GS/JS)
Assumed wave dimensions			
Wavelength, km	200	4000	1/20
Current speed (core), ms^{-1}	2	60	1/30
Phase speed, ms^{-1}	0.4	8	1/20
Amplitude/wavelength	1/10		
Derived properties			
Radius at bends, km	50	1000	1/20
$KV/f (= V_0/V)$	0.4	0.7	$\sim 1/2$
Divergence,* $10^{-1} s^{-1}$	1.8	4	$\sim 1/2$
Vertical motion,† $cm \ s^{-1}$	0.5	8	$\sim 1/20$
$[\beta L^2/4\pi^2]/V, \%$	1	12	

* Mean divergence over a half wavelength at current core in upper troposphere.
† In middle troposphere beneath current core.

A comparison of the figures for the Gulf Stream and atmospheric jet stream contained in tables 2.1 and 2.2 shows what great care should be taken when drawing quantitative analogies between the ocean and atmosphere. Indeed, the synoptic spatial scale in the atmosphere is 20 times greater than the synoptic scale in the ocean (table 2.2), and numerically corresponds more to the planetary scales of oceanic circulation, whereas the synoptic scales of disturbances in the ocean approximate the characteristic dimensions of atmospheric fronts. These correlations should always be kept in mind.

2.5 The concepts of deformation field and frontogenesis

We have already said that a deformation field is the main cause of atmospheric frontogenesis. This also holds true for the ocean. The rate of isoline compression of the concentration c of any passive scalar (e.g. ozone in the atmosphere, or dissolved oxygen in the ocean) in a deformation field located in a horizontal plane can be described by the equation

$$\frac{d}{dt}\left(\frac{\partial c}{\partial n}\right) = -\frac{\partial v_n}{\partial n}\frac{\partial c}{\partial n}, \qquad (2.3)$$

where n is in the direction of the normal to the isolines of c, v_n is the horizontal component of the velocity along the normal n, and t denotes time. From the diagram in fig. 2.21 and equation (2.3), we see that with an ideal axi-symmetric deformation field, compression of the isolines of c along the axis of compression will take place for any angle α, between the axis of stretching and the isolines of c, less than 45°. At the same time, the isolines of c will turn until they become parallel to the axis of stretching. At angles $\alpha > 45°$ the isolines of c will spread out.

In the case where the isoline compression of c occurs as a result of horizontal motion with a shear (fig. 2.23), the equation for the rate of isoline compression of c is similar to (2.3), i.e.

$$-\frac{d}{dt}\left(\frac{\partial c}{\partial y}\right) = \frac{\partial u}{\partial y}\frac{\partial c}{\partial x}, \qquad (2.4)$$

and the process is nothing more than a type of deformation. In this case we were dealing with horizontal motion along a randomly oriented x-axis with a speed $u(y)$. The deformation is observed in the zone where $\partial u/\partial y \neq 0$.

It is quite clear that the situations described by equations (2.3) and (2.4) are realistic not only for atmosphere, but for the ocean as well.

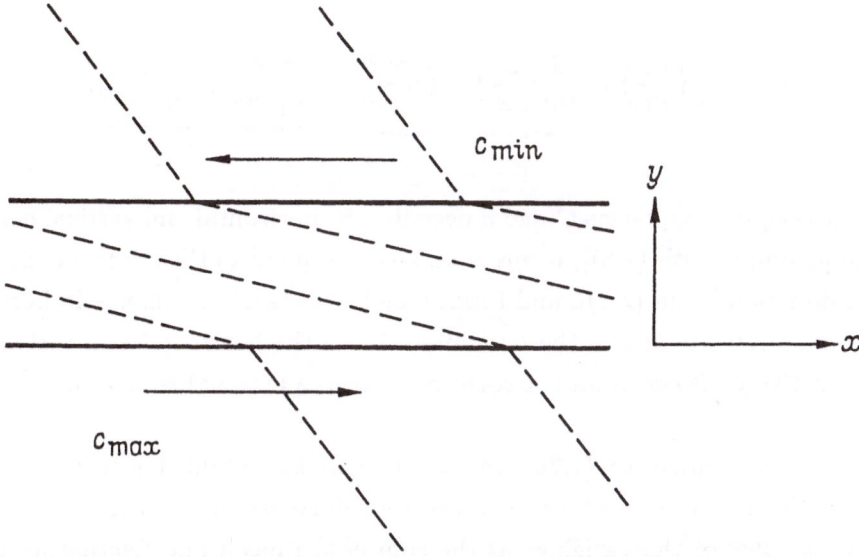

Fig. 2.23. Diagram demonstrating the increase in the horizontal gradient of the concentration of the scalar c across a frontal interface under the effect of an along-frontal current with a velocity shear.

In meteorology, the left hand sides of equations (2.3) and (2.4) are called the frontogenetic function F. In a general three-dimensional case this can be expressed as

$$F_3 = \frac{d}{dt}(\nabla_3 c).\tag{2.5}$$

After treating d/dt and ∇ term by term and rearranging the terms, we get

$$F_3 = \nabla_3 \frac{dc}{dt} - \left(\frac{\partial c}{\partial x}\nabla_3 u + \frac{\partial c}{\partial y}\nabla_3 v + \frac{\partial c}{\partial z}\nabla_3 w\right),\tag{2.6}$$

where w is the vertical component of velocity.

If the x-axis is chosen parallel to the front, then the following two components should be of interest to us:

$$F_z = \frac{d}{dt}\left(\frac{\partial c}{\partial z}\right) = \underbrace{\frac{\partial}{\partial z}\left(\frac{dc}{dt}\right)}_{a} - \left(\underbrace{\frac{\partial c}{\partial x}\frac{\partial u}{\partial z}}_{b} + \underbrace{\frac{\partial c}{\partial y}\frac{\partial v}{\partial z}}_{c} + \underbrace{\frac{\partial c}{\partial z}\frac{\partial w}{\partial z}}_{d}\right); \qquad (2.7a)$$

$$F_y = \frac{d}{dt}\left(\frac{\partial c}{\partial y}\right) = \underbrace{\frac{\partial}{\partial y}\left(\frac{dc}{dt}\right)}_{A} - \left(\underbrace{\frac{\partial c}{\partial x}\frac{\partial u}{\partial y}}_{B} + \underbrace{\frac{\partial c}{\partial y}\frac{\partial v}{\partial y}}_{C} + \underbrace{\frac{\partial c}{\partial z}\frac{\partial w}{\partial y}}_{D}\right). \qquad (2.7b)$$

In these equations, terms C and d describe the horizontal and vertical deformation field [compare with (2.3)], terms B and b correspond to the effects of shear in a different direction [as in (2.4)], and terms c and D express the change in horizontal and vertical gradients due to the nonuniformity of the horizontal and vertical motions respectively. Terms a and A correspond to the external sources of the scalar c.

Expressions similar to (2.7a) and (2.7b) can be obtained directly from the equation of heat (T) and salt (S) conservation; furthermore, having taken the instantaneous value of the variables as the sum of the mean and fluctuating values, one can obtain additional terms, which describe the effects of turbulence, on the right hand side of the equations. Roden [214] did so when he studied the contributions of various processes to the balance required for the persistence of quasi-stationary large-scale fronts of climatic origin.

Expressions like (2.7a) and (2.7b), or modifications similar to the ones used by Roden [214], can serve as the basis for modelling oceanic frontogenesis to the extent that the salinity S or temperature T in the ocean can be regarded as passive scalars. The latter stipulation is associated with the fact that both temperature and salinity can to some extent be regarded as scalar properties at the initial stages of oceanic frontogenesis [178] (see section 2.6). Their local distribution in this case has no effect on the external deformation field. Only at a certain stage of the sharpening of the local salinity and temperature gradients do we begin to observe the local dynamics of the frontal interface, which is what determines its further behavior and existence. These local dynamics are apparently associated to a large extent with the increasing effect of turbulent viscosity.

Equation (2.3) can be rewritten as

$$\frac{d}{dt}\left(\frac{\partial c}{\partial n}\right) = -D\frac{\partial c}{\partial n}, \qquad (2.8)$$

where $D = \frac{\partial v_n}{\partial n} = -\frac{d/dt\,(\partial c/\partial n)}{\partial c/\partial n}$ is called the rate of deformation, and has a dimension of $|t^{-1}|$. MacVean and Woods [178] have shown that the distance L between any two points located on the same axis along n (along the axis of compression) changes according to the following relationship with $D = const$:

$$L(t) = L_0\, exp(-Dt). \tag{2.9}$$

This can be called the relationship of ideal frontogenesis. An interesting result can be derived from this relationship, i.e. a linear scale which Woods* refers to as "the width of an equilibrium front". This scale B is deduced on the assumption that, at some point in time, the rate of change in the width of the frontal zone dL/dt due to frontogenetic convergence of temperature or salinity isolines becomes equal in its absolute value to the speed of relaxation (diffusion) of the frontal zone due to turbulent mixing, characterized by the coefficient of horizontal exchange K_ℓ. In this case, the process of frontogenesis should cease, and the width of the frontal zone should acquire its equilibrium value B. Any subsequent deformations should begin with this equilibrium state when L_0 and $t = 0$. Assuming that

$$K_\ell = \frac{d}{dt}(L^2) \tag{2.10}$$

at each point in time, $dL/dt = K_\ell/(2L)$, and when $L = L_0 = B$

$$\frac{dL}{dt} = \frac{K_\ell}{2B}. \tag{2.11}$$

On the other hand, from (2.9) we derive that

$$\frac{dL}{dt} = -DB \tag{2.12}$$

when $L_0 = B$ and $t = 0$.

---

* Report at the 2nd International Symposium on Ocean Turbulence, Liège (Belgium), 7-18th of May 1979.

Having equated the absolute values of the right hand sides of (2.11) and (2.12), we get

$$B = \left(\frac{K_\ell}{2D}\right)^{\frac{1}{2}}. \tag{2.13}$$

By substituting the values $K_\ell \simeq 26 m^2/s$ [142] and $D = 10^{-5}s^{-1}$ [178] in (2.13), we get $B = 1.14 \cdot 10^3 m \simeq 1$ km. The initial values of K_ℓ and D selected above characterize both intensive local frontogenesis, and comparatively intensive cross-frontal turbulent exchange. Consequently, we can conclude that in the case of intensive local frontogenetic and frontolytic processes, the characteristic width of the zone of maximum horizontal gradients of any scalar should be of the order of 1 km. For weak frontogenesis on planetary scales ($D = 10^{-7}s^{-1}$) and the values of $K_\ell = 10^4 m^2/s$ characteristic of general oceanic circulations, we get $B \approx 220$ km, i.e. the characteristic width of the climatic frontal zones in the ocean. At the present time, it would probably be difficult to say anything more specific with regard to the physical significance of the scale B. We only know that such important parameters as the rate of deformation, the actual width of a front (see section 2.3) and the intensity of horizontal turbulent exchange are most probably associated in some way with the local dynamics of the frontal interface. In other words, the values K_ℓ and D in equation (2.13) cannot be assigned arbitrarily, at least at the stage where the local temperature or salinity gradients begin to affect the local field of motion more than the weaker external deformation field which is the origin of frontogenesis. For example, one can picture a locally convergent motion, due to an increase in density during mixing, which results in local values of D that are many times greater than the intensity of the background deformation field. However, it is possible that having determined the value of K_ℓ from observations on the intrusive fine structure in the vicinity of a specific frontal interface, as was the case in refs. [142] and [154], one can reliably determine the intensity of the local frontogenetic process (find the local values of D) by the scale B which was measured under natural conditions. These values will probably prove to be correct, as it is very difficult to imagine a reliable method of direct measurement of local values of D in the ocean. The use of satellite data on the temperature field of the ocean can be recommended for larger scales. However, this method will become truly accessible only after considerable improvement in the precision and reliability of satellite measurements of the temperature field of the ocean.

2.6 On numerical modelling of oceanic frontogenesis

Despite the obvious importance of the problem, theoretical studies of frontogenesis in the open ocean are not very extensive in the literature. Researchers began to fill in this gap actively only in recent years.

The papers by P.S. Lineikin [39] and V.S. Maderich [40] on the evolution of large-scale ($\sim 1000~km$) density anomalies in the ocean have shown that as a result of nonlinear processes associated with long divergent non-dispersive Rossby waves, fronts eventually emerge in the density field from initially "smooth" density anomalies, and these fronts are accompanied by an increasing concentration of currents. It takes several years for a frontal interface to form. Calculation of the downward vertical velocity at the surface of the ocean led to more rapid frontogenesis, but did not alter the result significantly. The task amounted to analyzing the quasi-linear hyperbolic equation for density at the ocean surface, the solution of which represented a Riemann wave. A front is the result of the "steepening" of an initial disturbance. The speed of frontal motion, like the velocity of Rossby waves, was equal to $2.4 \cdot 10^3 km/yr$ [39].

A wavy view of the nature of fronts in the ocean is of undoubted interest. Such an approach, characteristic of works [39] and [40], can be regarded as one method for the study of frontogenesis and the circulation regime of oceanic fronts. However, other approaches too are needed for the investigation of fronts as a geophysical phenomenon, and also for the description of the current and density fields in frontal zones.

For example, we can study the redistribution of scalars in the ocean, or only in its upper mixed layer, under the influence of stationary (or transient) disturbances of the velocity field. In this case, it is convenient to examine the superposed velocity field from four connected eddies, two cyclonic and two anticyclonic. Woods [261] was apparently the first to present a hypothesis on the formation of oceanic fronts during the passage of eddies. Woods suggested that eddies be regarded as a quasi-horizontal deformation field. At that time, it was known quite well that a deformation field was the basis of atmospheric frontogenesis (see sections 2.4 and 2.5). It is currently assumed that the intensity of a deformation field is particularly high at the boundary of mesoscale oceanic eddies of a cyclonic and anticyclonic nature, and that the formation of sharp discontinuities in the temperature and salinity field is most likely to occur in these boundary areas. The earliest evidence for this assumption is found in ref. [72] which describes a thermohaline front at the boundary of an anticyclonic and a cyclonic eddy, detected during the Soviet hydrophysical experiment "Polygon-70" in the tropical Atlantic. Within the framework of these concepts on deformation fields, we can also assume that fronts are generated on the periphery of an isolated stationary eddy in two diametrically opposed regions [38]. On the whole, we can say that many of the oceanic frontogenetic mechanisms can be interpreted as types of a three-dimensional deformation field [70].

The first attempts to describe the frontogenetic effects of oceanic eddies with the help of comparatively simple numerical models were made in refs. [239] and [47]. This question will be discussed in greater detail in section 3.1. We shall now discuss the possibility of applying to the ocean the numerical methods of frontogenetic research developed at the beginning of the past decade for the atmosphere (see [36] and [38]).

When modelling atmospheric fronts, meteorologists usually replace the complex deformation fields with a simple horizontal deformation field of motion. This makes it simpler to solve the problems. The simplest case of a deformation field is the hyperbolic field of streamlines $\Psi = Dxy$, commonly used in meteorology, where x and y are the horizontal coordinates as usual, and D is the characteristic strength of the deformation field. For the ocean, D varies from 10^{-6} to $10^{-5}s^{-1}$ [178]. A hyperbolic-type deformation field can be regarded as an acceptable schematization of the field of motion for four adjoining eddies of a cyclonic and an anticyclonic nature.

The most interesting and characteristic features of the open ocean are the frontal zones in the surface layer. The analog of these oceanic fronts are the tropospheric fronts which were quite recently modelled by Hoskins and Bretherton [145, 146] who discussed frontogenesis in a semi-geostrophic approximation (semi-geostrophic theory). The significance of the semi-geostrophic approximation lies in the fact that only the motion along the front is regarded as geostrophic, while the motion directed along the normal to the front is ageostrophic. At the same time, turbulent mixing is ignored. For instance, Hoskins' calculations confirmed the previously known fact that the tropopause undergoes deflection during the formation of a tropospheric front [102] (see section 2.4 and fig. 2.4 and fig. 2.22). The displacement of the tropopause at atmospheric fronts is regarded as one of the important mechanisms for air exchange between the stratosphere and troposphere. In the ocean, the role of the tropopause is played by the pycnocline, i.e. the layer of maximum vertical density gradient. Therefore, oceanic fronts should also be regarded as effective mechanisms for heat and salt exchange between the upper mixed layer and the underlying layers across a hydrostatically stable pycnocline.

The model of oceanic frontogenesis developed recently by MacVean and Woods is the analog of Hoskins and Bretherton's atmospheric model [146] as applied to the ocean. It is also based on the semi-geostrophic approximation in the absence of turbulent friction. The first results of calculations carried out with the help of this model were discussed in 1977, whereas a detailed description of the redistribution of scalars in a deformation field during frontogenesis in the upper layer of the ocean was published only in 1980 [178]. Simultaneously with the appearance of MacVean and Woods' first results, N.P. Kuz'mina and B.Ya. Kutsenko [38] published their paper which discussed a physically similar model of oceanic frontogenesis with the assignment of a somewhat more realistic initial temperature (density) field. Therefore, in order to avoid unnecessary repetition, further discussion of the question is based on N.P. Kuz'mina's final interpretation of the material [36].

Within the framework of the quasi-geostrophic equations, we shall discuss frontogenesis as a result of the action of a hyperbolic deformation field on a zonally extended baroclinic oceanic region. Unlike the results obtained in ref. [47] on the basis of an integral model, the object of our analysis will primarily be the vertical structure of the currents and temperature (density) field in the area of study, as well as the evolution of the density gradient during the formation of a frontal interface. Why is this important? First of all, in order to explain better the occurrence of fronts at the surface of the ocean, it is necessary to visualize clearly the vertical structure of the frontal interface, as well as the changes in current velocity and temperature gradients with depth. Secondly, we have reason to believe [62, 259, 261] that synoptic-type frontal interfaces are the most probable and effective mechanism of heat and salt transfer across the oceanic pycnocline. From this point of view, it is necessary to visualize the behavior of the layer of maximum density gradient during the formation of a frontal interface.

For adiabatic motion with the hydrostatic and Boussinesq approximations, the quasi-geostrophic equations of frontogenesis can be given as

$$\frac{\partial u}{\partial t} + V\frac{\partial u}{\partial y} - \frac{\partial V}{\partial y}u - fv = 0; \tag{2.14}$$

$$\frac{\partial P}{\partial y} + fu = 0; \tag{2.15}$$

$$\frac{\partial \rho}{\partial t} + V\frac{\partial \rho}{\partial y} + \frac{\partial \overline{\rho(z)}}{\partial z}w = 0; \tag{2.16}$$

$$\frac{\partial v}{\partial y} + \frac{\partial w}{\partial z} = 0; \qquad (2.17)$$

$$\frac{\partial P}{\partial z} = -\frac{g\rho}{\rho_0}, \qquad (2.18)$$

where u, v and w are the zonal, meridional and vertical components of the deviations of velocity from barotropic flow, with the stream function $\Psi = Dxy$ (a hyperbolic deformation field), $V = -Dy$ is the convergent velocity of the deformation field, $P = p/\rho_0$ is the deviation of pressure (p) from the rest value divided by the mean oceanic density, ρ is the deviation of density from the mean rest oceanic density ρ_0, and $\overline{\rho(z)}$ is the mean oceanic density as a function of depth.

To derive the system of equations (2.14)–(2.18), we assumed that the isolines of the density field were initially located along the axis of stretching (zonal direction), and the disturbances of the density, velocity and pressure field did not depend on x. This and the fairly simple structure of the deformation field made it possible to proceed from a three-dimensional problem to a two-dimensional one in the yz section. The system (2.14)–(2.18) was solved with boundary conditions $w = 0$ at $z = 0, H$, where H stands for the depth of the ocean. At the same time, it was assumed that the contribution of salinity to density was insignificant, and so the linearized equation of state was determined by the relationship

$$\rho = -\alpha T \qquad (2.19)$$

where α is the temperature coefficient of density, and T is the deviation of temperature from the mean rest state T_0.

The initial system of equations was reduced to a more convenient form. We derived the following equation for the conservation of potential vorticity

$$\left(\frac{\partial}{\partial t} - Dy\frac{\partial}{\partial y}\right)\left(\frac{\partial^2 T}{\partial y^2} + f^2\frac{\partial^2}{\partial z^2}\frac{T}{N^2}\right) = 0, \qquad (2.20)$$

where $N^2 = (\alpha g/\rho_0)(d\,\overline{T(z)}/dz)$ is the square of the Väisälä-Brunt frequency.

The boundary conditions for equation (2.20) were derived from the density conservation equation (2.16) with $w = 0$ at the levels $z = 0$ and $z = H$, i.e.

$$\left(\frac{\partial}{\partial t} - Dy\frac{\partial}{\partial y}\right)T = 0. \qquad (2.21)$$

For the velocity of the geostrophic flow u along the frontal interface, we used the thermal wind equation derived from (2.15) and (2.18), i.e.

$$\frac{\partial u}{\partial z} = -\frac{\alpha g}{\rho_0 f} \frac{\partial T}{\partial y}. \tag{2.22}$$

The velocities v and w were found by deriving the stream function ϕ in the yz plane, i.e.

$$v = -\frac{\partial \phi}{\partial z}, \quad w = \frac{\partial \phi}{\partial y}. \tag{2.23}$$

From the set of equations (2.14)–(2.18), we derived an elliptical equation for function ϕ,

$$\frac{N^2}{f^2} \frac{\partial^2 \phi}{\partial y^2} + \frac{\partial^2 \phi}{\partial z^2} = \frac{2D\alpha}{f^2 \rho_0} g \frac{\partial T}{\partial y}, \tag{2.24}$$

which was solved for boundary conditions $\phi = 0$.

The equation (2.20) and the boundary condition (2.21) were then simplified by eliminating the operator $(\partial/\partial t - Dy\partial/\partial y)$. In place of (2.20), we obtained an elliptic-type equation

$$\frac{\partial^2 T}{\partial y^2} + f^2 \frac{\partial^2}{\partial z^2} \frac{T}{N^2} = \Gamma(ye^{Dt}, z), \tag{2.25}$$

and in place of (2.21), the boundary condition

$$T = \Gamma'\left(ye^{Dt}\right), \quad z = \begin{cases} 0 \\ H \end{cases}. \tag{2.26}$$

The functions $\Gamma(ye^{Dt}, z)$ and $\Gamma'(ye^{Dt})$, which characterize the initial baroclinicity of the temperature (density) field at $t = 0$, were derived from the initial conditions for temperature deviations from the mean state. The initial distribution of temperature $T_1 = \overline{T(z)} + T(y)$ was assigned in two ways which corresponded to constant and variable stratification of the water, i.e.

$$a) \quad T_1 = a(z - 0.5H) - b\frac{2}{\pi}\text{arctg}(\text{sh}\gamma y); \tag{2.27}$$

and

$$b) \quad T_1 = c\,\text{arctg}[20(z/H - 0.8)] - b\frac{2}{\pi}\text{arctg}(\text{sh}\gamma y), \tag{2.28}$$

where $a = dT_\phi(z)/dz = 0.01°C/m$ and $b = 0.4°C$ are constants selected on the basis of data typical of the Atlantic Ocean, and $\frac{\gamma}{\pi} = fH^{-1}\left(\frac{\alpha}{\rho_0}g\frac{dT_\phi(z)}{dz}\right)^{\frac{1}{2}}$ is a value which is the inverse of the Rossby deformation radius. The constant $c = 3.527°C$ for the temperature field with variable stratification was selected in such a way that the mean vertical temperature gradient for case (b) would be equal to the constant vertical temperature gradient in case (a). By substituting the initial temperature deviation from the mean state $T = T_1 - \overline{T(z)}$ in equation (2.25) and the boundary conditions (2.26), we derived the equations for functions Γ and Γ', i.e.

$$\Gamma = b\frac{2}{\pi}\gamma^2\frac{\text{sh}(\gamma e^{Dt}y)}{\text{ch}^2(\gamma e^{Dt}y)} - b\frac{2}{\pi}\text{arctg}[\text{sh}(\gamma e^{Dt}y)]f^2\frac{d^2}{dz^2}\left(\frac{1}{N^2}\right);$$

$$\Gamma' = -b\frac{2}{\pi}\text{arctg}[\text{sh}(\gamma e^{Dt}y)].$$

Equations (2.24) and (2.25) were solved numerically. The values of T at the lateral boundaries when solving equation (2.25) are given by $T = \Gamma'(\pm\frac{1}{2}Le^{Dt})$ at $y = \pm L/2$, where L is the horizontal scale of the computation region. Equation (2.22) for the geostrophic velocity was integrated on the assumption that the level of no motion was $z = 0.5\ H$. The study region was limited horizontally to a distance of 250 km, while the depth of the ocean H was taken as 1 km. The solutions along the y and z axes are equal to $\Delta y = 5\ km$ and $\Delta z = 20\ m$ respectively. The characteristic intensity of the deformation field D was accepted as $0.2 \cdot 10^{-5}s^{-1}$, and the values $f = 10^{-4}s^{-1}$ and $\alpha = 0.2\ kg/m^3$ per $1°C$ were selected for the Coriolis parameter and the temperature coefficient of density.

Figs. 2.24 a and b depict the isolines of the field of temperature perturbations away from the mean state in the ocean in the yz plane at the initial moment and at $t = 240\ hrs$ for constant stratification. The formation of frontal interfaces is observed in the central part of the region in the upper and near-bottom $200 - m$ layers. In this case, the surface of the maximum horizontal gradient is vertical, whereas observations of oceanic fronts have shown that the surfaces of frontal interfaces in quasi-geostrophic equilibrium are always inclined, and the tangent of the angle between the frontal interfaces and the horizontal varies from 0.001 to 3 [70]. Strong geostrophic currents form in the area of sharp horizontal gradients (fig. 2.25 a, b). For instance, the geostrophic current u in the core increases from $18 \cdot 10^{-2}$ to $30 \cdot 10^{-2}m/s$ in 240 hrs. On one side of the frontal interface (in the region of warm water), we observe rising, and on the other side (in the region of cold water) sinking. The vertical velocities on the edges of the front grow insignificantly, and do not exceed $10^{-5} m/s$.

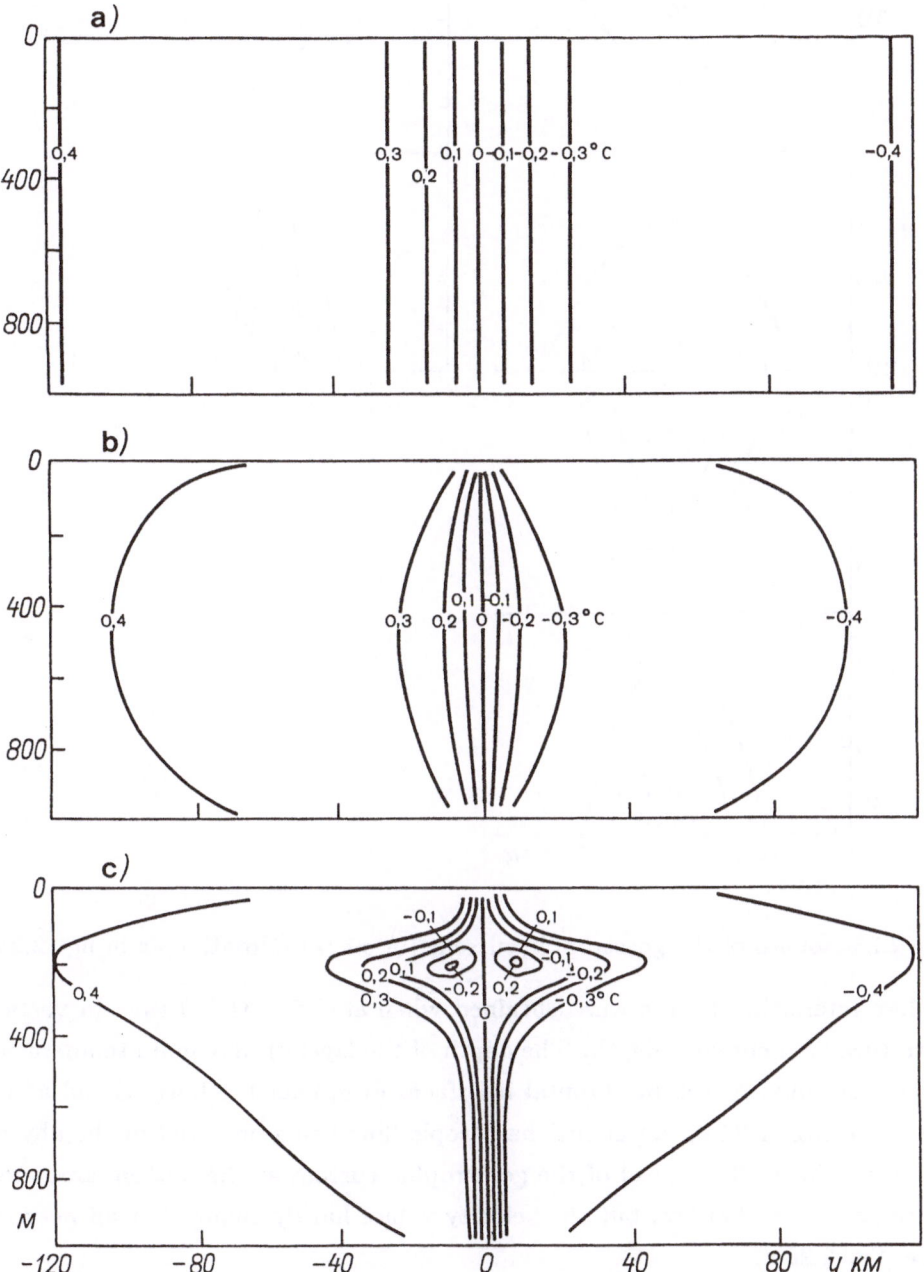

Fig. 2.24 Isolines of the field of temperature perturbations away from the mean state in the ocean. a – initial distribution ($t = 0$); b – for $t = 240$ hrs (constant stratification); c – for $t = 240$ hrs (variable stratification).

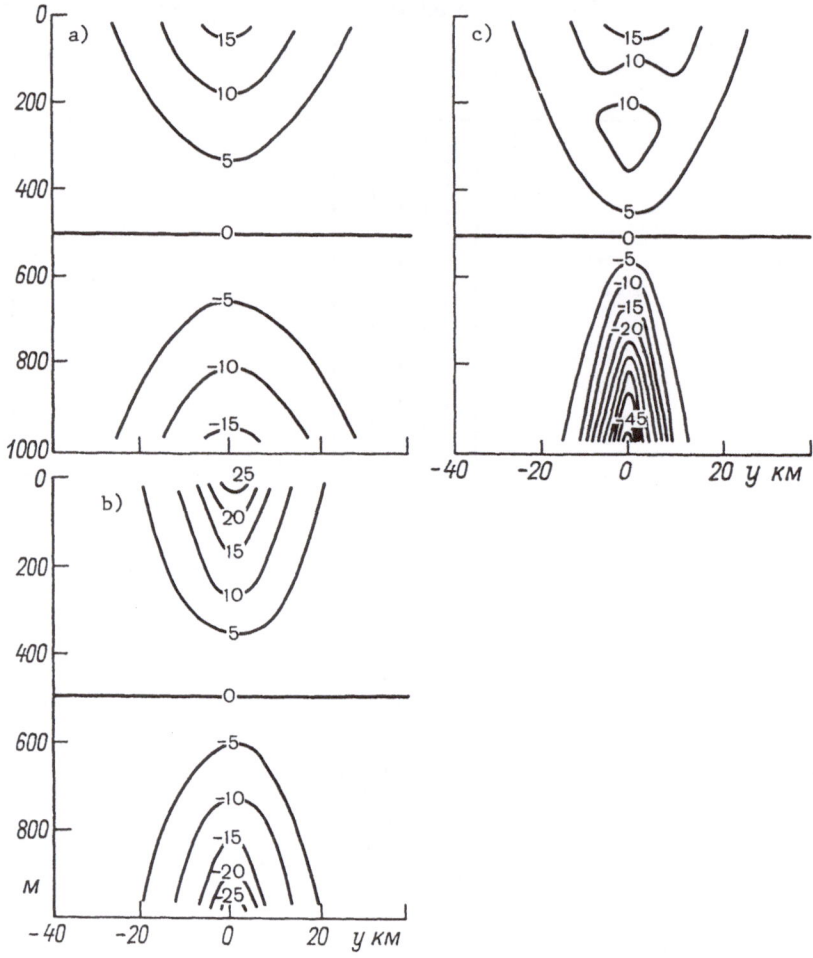

Figure 2.25 Isolines of the geostrophic current (in *cm/s*) (Notation as in fig. 2.24).

More interesting results were obtained when studying the change in vertical temperature gradient with depth. The depth of the layer of maximum temperature gradient was equal to 200 *m*. Frontal interfaces form near the bottom and at the very surface (fig. 2.24 c), an almost barotropic flow being observed in the abyssal layer below 500 *m*. The speed of the geostrophic current at the bottom amounted to 50 *cm/s* for $t = 240$ *hrs*, but the velocity values hardly changed at all near the surface (fig. 2.25 c).

The upper $100 - 300$ *m* layer is of greatest interest to us. In the layer with the maximum Väisälä-Brunt frequency, we observe a decrease in the horizontal gradient and a change of sign. This effect is clearly seen in the field of temperature deviations from the mean unperturbed state (fig. 2.24 c). The change in the sign of the horizontal gradient at the 200 m level corresponds to a change in the slope of the layer with the maximum Väisälä-Brunt frequency. Fig. 2.26 depicts the distribution of the entire temperature field T_{ent} at the initial moment $(t = 0)$ and at $t = 240 \ hrs$. This diagram clearly shows that, below the surface frontal interface, there exists a characteristic deflection of isotherms, which is an indication of differential vertical motion. Perhaps such motions can lead to an intensive heat exchange between the upper quasi-homogeneous layer and the underlying water masses. In this case, the existence of local upwelling can help generate sharp finestructure and horizontal and vertical temperature and salinity inhomogeneities on the outskirts of frontal interfaces.

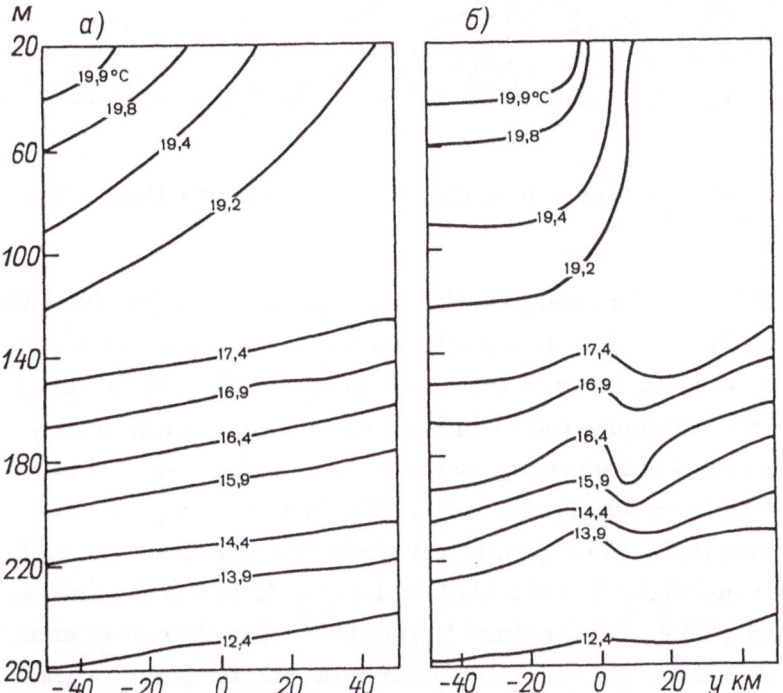

Fig. 2.26 Temperature distribution in the upper $250 - m$ layer of the ocean: a – for $t = 0$, b – for $t = 240 \ hrs$ (variable stratification).

Fig. 2.27 depicts a vertical section of the temperature field across a cold eddy, based on the data of F.M. Vukovich [255]. It is possible that the wave-like deformation of the layer with the maximum horizontal temperature gradient (marked with arrows) in the vicinity of the right hand front can be explained in terms of the model output.

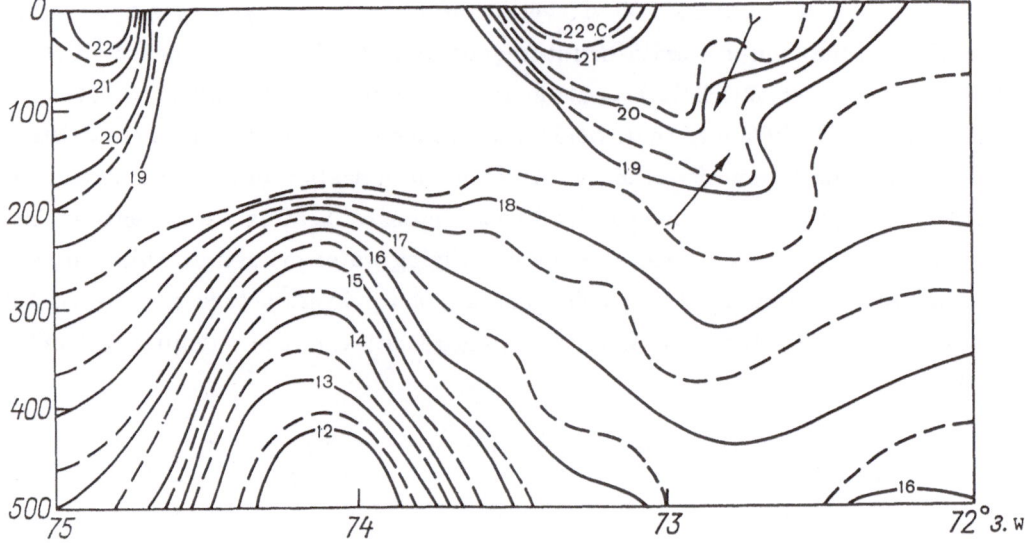

Fig. 2.27. Vertical section of temperature field in an eddy (based on the data of F.M. Vukovich [255].

It would be useful to compare N.P. Kuz'mina's results [36, 39] with those of MacVean and Woods [178]. Essentially, a two-dimensional problem was solved in both models, as it was assumed that no changes took place in the direction of the front. The first example studied in both cases was an initial weakly baroclinic zone with a pycnocline emerging at the surface of the ocean. The deformation rate across the front was $2 \cdot 10^{-6}s^{-1}$ in the first case [36, 39] and $10^{-5}s^{-1}$ in the second case [178]. Consequently*, it would take approximately 83 hrs for the characteristic horizontal frontal scale L to decrease 20–fold in the case examined by MacVean and Woods, and five times longer (416 hrs) in the case examined by N.P. Kuz'mina. In 240 hrs (10 $days$), the mean horizontal temperature gradient at the surface increased only 5-fold in the case studied by Kuz'mina (see fig. 2.26 a). It would take only 2 days for the same sharpening of the front in the case discussed in ref. [178]. The calculations carried out in [178] have shown that the velocity vectors of the convergent motion in the plane across the front are parallel to the isopycnals in the water masses (fig. 2.28), and become horizontal only near the free surface. The characteristic speeds of these motions in a zone 25 km wide and up to 125 m deep are given as $(1 - 10) \cdot 10^{-2}m/s$ in ref. [178]. Consequently, they

---

* see equation (2.8) in section 2.5.

should be 5 times smaller in the case examined by Kuz'mina. As we have already mentioned, all of these differences depend on the choice of the various values of D, and therefore are not fundamental. The fact that the convergent motions in the uppermost layer are parallel to the free surface can be explained by the highly intensive frontogenesis in this layer. The calculations of Woods and MacVean show that when $D = 10^{-5}s^{-1}$, a discontinuity with an infinitely large horizontal density gradient can appear near the surface, from a comparatively diffused baroclinic zone, in $3 - 4$ days. In one of the examples given by Woods and MacVean [178, 263], where temperature manifests itself as a passive scalar, the intensive near-surface frontogenesis results in the bowing of the thermocline and a characteristic deflection of isotherms, which leads to the formation near the surface of intrusions of water from one side of the front into the water masses on the other side (see section 4.2). In Kuz'mina's model, this type of isotherm deflection (fig. 2.26 b) is associated with differential vertical motion, the effect of which is particularly noticeable where the initial vertical temperature gradient is locally intensified due to depth-variable stratification. In this case, the second of Kuz'mina's examples [with depth-variable stratification corresponding to (2.28)] is to some extent equivalent to the example with an upper homogeneous layer described in [178].

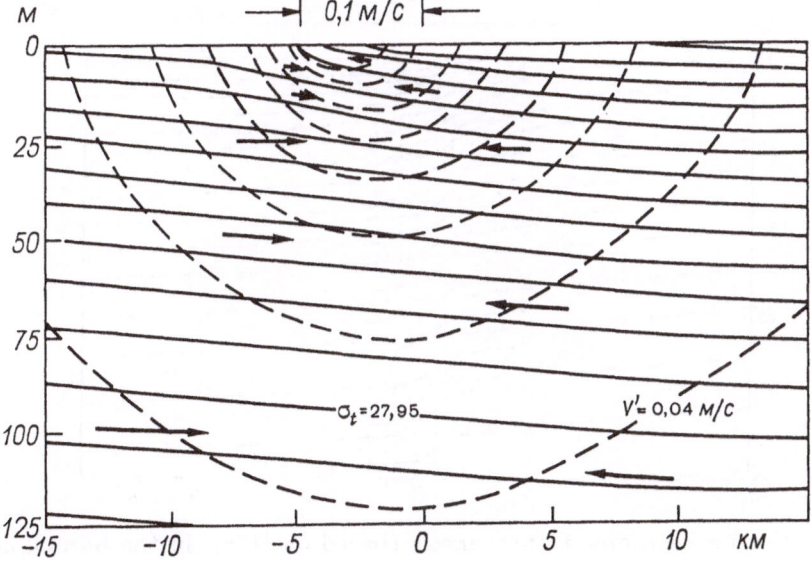

Fig. 2.28. Result (from [178]) of the frontogenetic deformation of an initially weak baroclinic zone after 57 and 76 hrs with $D = 10^{-5}s^{-1}$. Arrows – cross-frontal velocity vectors. The velocity scale is indicated above. Dashed lines – isolines of the geostrophic along-front current directed into the diagram, plotted at intervals of 0.04 m/s. Continuous lines – isopycnals plotted at intervals of 0.2 σ_t units.

It should be said that, unlike the authors of [36, 38], MacVean and Woods used a transformed transverse coordinate which makes it possible to account for the nonlinear inertial effects in the vertical plane and to obtain a systematic displacement of the front near the free surface in the direction of denser water, and a sloping of the entire frontal interface, which is clearly seen in figs. 2.28 and 2.29 b. In Kuz'mina's quasi-geostrophic model, this effect is not observed, and because of this, the frontal interface is perpendicular to the free surface; this can be regarded as a significant shortcoming of the model.

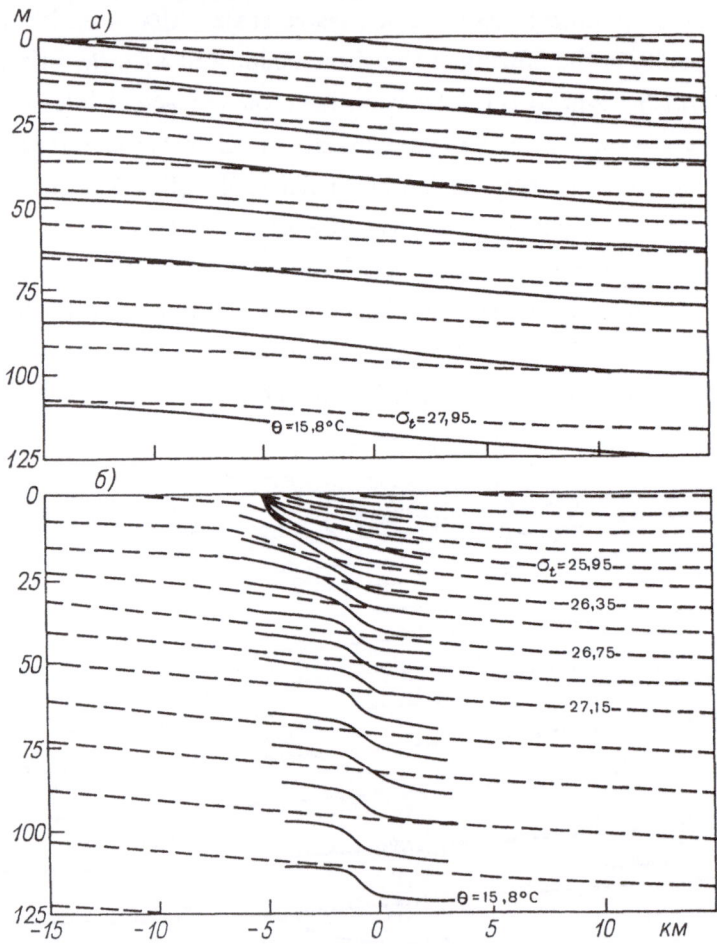

Fig. 2.29. Semi-geostrophic frontogenesis (based on [178]) in the baroclinic region where density is determined not only by temperature, but also by salinity: isotherms (continuous lines, interval $0.8°C$) intersect isopycnals (dashed line, interval 0.2 units σ_t). a – initial state; b – after 83.21 hrs of the action of a deformation field with $D = 10^{-5}s^{-1}$.

Also interesting are MacVean and Woods' examples of frontogenesis, in which the initial distribution of salinity at a section normal to the front is given in an implicit form by prescribing the initial slope of the isotherms relative to the isopycnals (figs. 8–14 in ref. [178]). It is with this type of formulation of the problem that temperature manifests itself as a passive scalar to a greater extent than in cases where the density field is wholly determined by the temperature field. On the basis of these examples, it appears that the maximum intensification of the near-surface and abyssal horizontal temperature and density gradients as a result of frontogenesis should be expected in cases where the horizontal temperature and salinity gradients make a contribution of the same sign to the horizontal density gradient (both gradients intensify baroclinicity). In this case, the slopes of the isotherms and isopycnals have the same sign initially, but the isotherms are more sharply inclined to the horizontal than the isopycnals (see fig. 2.29 a and b). It is this situation that characterizes the fronts in the coastal upwelling zone off Oregon (see section 3.2).

Earlier, we discussed the process of synoptic-scale oceanic frontogenesis within the framework of the quasi-geostrophic approximation. Actually, a quasi-geostrophic model describes only the initial stages of the formation of frontal interfaces, and then only qualitatively. Therefore, it would be advisable to analyze oceanic frontogenesis using the nonlinear "primitive" equations of hydrodynamics. As before, we shall examine solutions which do not depend on the coordinate directed along the front. A hyperbolic-type deformation field with intensity D will again be used as the simplest approximation to the field of motion of adjoining cyclonic and anticyclonic eddies (see fig. 2.21).

The equations of adiabatic motion with the hydrostatic and Boussinesq approximations are written as follows:

$$\frac{\partial u}{dt} + \frac{\partial(uu)}{\partial x} + \frac{\partial(vu)}{\partial y} + \frac{\partial(wu)}{\partial z} - fv = -\frac{\partial P}{\partial x}; \qquad (2.29)$$

$$\frac{\partial v}{\partial t} + \frac{\partial(uv)}{\partial x} + \frac{\partial(vv)}{\partial y} + \frac{\partial(wv)}{\partial z} + fu = -\frac{\partial P}{\partial y}; \qquad (2.30)$$

$$\frac{\partial P}{\partial z} = -g\frac{\rho}{\rho_0}; \qquad (2.31)$$

$$\frac{\partial u}{\partial x} + \frac{\partial v}{\partial y} + \frac{\partial w}{\partial z} = 0; \qquad (2.32)$$

$$\frac{\partial \rho}{\partial t} + \frac{\partial(u\rho)}{\partial x} + \frac{\partial(v\rho)}{\partial y} + \frac{\partial(w\rho)}{\partial z} = 0, \qquad (2.33)$$

The symbols used are similar to those in equations (2.14)– (2.18).

The horizontal velocity

$$\mathbf{V} = Dx\mathbf{i} - Dy\mathbf{j}, \tag{2.34}$$

where $\mathbf{i}$ and $\mathbf{j}$ are the unit vectors in the directions x and y respectively, corresponds to the initial barotropic field of motion chosen, with hyperbolic-type streamfunction Ψ.

By substituting (2.34) into the system of equations (2.29)–(2.33), we can find the barotropic pressure field P, which corresponds to the given velocity:

$$P = -D^2(x^2 + y^2)/2 - fDxy. \tag{2.35}$$

We shall need the solutions of (2.29)–(2.33) for the deviations, from the barotropic deformation field (2.34) and (2.35), which are independent of the along-front coordinate x. The unknown value is marked by a tilde ($\tilde{}$).

a) $u = Dx + \tilde{u}(y, z, t)$;

b) $v = -Dy + \tilde{v}(y, z, t)$;

c) $w = \tilde{w}(y, z, t)$; $\qquad\qquad\qquad\qquad\qquad$ (2.36)

d) $\rho = \tilde{\rho}(y, z, t)$;

e) $P = [-D^2(x^2 + y^2)/2] - fDxy + \tilde{P}(y, z, t)$.

By substituting (2.36) into the system (2.29)–(2.33), we obtain the system (2.37)–(2.41) which describes the perturbation flow during frontogenesis in the vertical plane yz which is normal to the front, i.e.

$$\frac{\partial \tilde{u}}{\partial t} + \frac{\partial(\tilde{v}\tilde{u})}{\partial y} + \frac{\partial(\tilde{w}\tilde{u})}{\partial z} - \frac{\partial V}{\partial y}\tilde{u} + V\frac{\partial \tilde{u}}{\partial y} - f\tilde{v} = 0; \tag{2.37}$$

$$\frac{\partial \tilde{v}}{\partial t} + \frac{\partial(\tilde{v}\tilde{v})}{\partial y} + \frac{\partial(\tilde{w}\tilde{v})}{\partial z} + \frac{\partial}{\partial y}(\tilde{v}V) + \frac{\partial \tilde{P}}{\partial y} + f\tilde{u} = 0; \tag{2.38}$$

$$\frac{\partial \tilde{\rho}}{\partial t} + \frac{\partial(\tilde{v}\tilde{\rho})}{\partial y} + \frac{\partial(\tilde{w}\tilde{\rho})}{\partial z} + V\frac{\partial \tilde{\rho}}{\partial y} = 0; \tag{2.39}$$

$$\frac{\partial \tilde{v}}{\partial y} + \frac{\partial \tilde{w}}{\partial z} = 0; \tag{2.40}$$

$$\frac{\partial \tilde{\rho}}{\partial z} = -g\frac{\tilde{\rho}}{\rho_0}, \tag{2.41}$$

where $V = -Dy$ is the convergent component of the barotropic deformation field.

The following boundary conditions were set at the boundaries of a rectangular region with scales H (depth of the ocean) and L (horizontal dimension of the region):

$$\tilde{w} = 0 \text{ at } z = 0, H; \tag{2.42}$$

$$\tilde{u}\left(\pm\frac{L}{2}, z, t\right) = \tilde{u}\left(\pm\frac{L}{2}, z, 0\right); \tag{2.43}$$

$$\tilde{\rho}\left(\pm\frac{L}{2}, z, t\right) = \tilde{\rho}\left(\pm\frac{L}{2}, z, 0\right); \tag{2.44}$$

$$\tilde{v}\left(\pm\frac{L}{2}, z, t\right) = \tilde{v}\left(\pm\frac{L}{2}, z, 0\right). \tag{2.45}$$

The system (2.37)–(2.41) is conveniently integrated by splitting the system into equations which describe the barotropic and baroclinic parts of the motion. The use of "rigid lid" approximation (2.42) makes this simple. Let us introduce vertical averaging by the formula $\langle(\ )\rangle = \frac{1}{H}\int_0^H (\ )dz$ and then integrate the hydrostatic equation (2.41) with respect to z. After subtraction of the averaged equation from it, (2.41) acquires the form

$$\tilde{P} - \langle\tilde{P}\rangle = -\frac{g}{\rho_0}\left[\int_0^z \tilde{\rho}dz - \left\langle\int_0^z \tilde{\rho}dz\right\rangle\right]. \tag{2.46}$$

After carrying out the same procedure with equation (2.38), we get an equation for the velocity component v in a form convenient for calculation, i.e.

$$\frac{\partial\tilde{v}}{\partial t} + \frac{\partial}{\partial y}(\tilde{v}\tilde{v} - \langle\tilde{v}\tilde{v}\rangle) + \frac{\partial}{\partial z}(\tilde{w}\tilde{v}) + \frac{\partial}{\partial y}(\tilde{v}V) +$$

$$\frac{\partial}{\partial y}(\tilde{P} - \langle\tilde{P}\rangle) + f(\tilde{u} - \langle\tilde{u}\rangle) = 0. \tag{2.47}$$

Equations (2.37), (2.39), (2.40), (2.46) and (2.47) represent a closed system of equations for modelling frontogenesis. N.P. Kuz'mina used finite-difference methods to solve this system [37]. Before discussing the initial conditions for solving the above system, we would like to note that, for the sake of convenience, we again assumed that the change in density is determined only by the change in temperature, and is a linear function of it as in (2.19).

The initial distribution of temperature $T_o = T(z) + T$ was again prescribed as in (2.27), with the same values of a, b and γ. The initial value of the along-frontal geostrophic component of velocity $u(y, z, 0)$ was derived for (2.27) from the thermal wind equation (2.22) with the assumption that the surface of zero velocity was located at the level $z = H/2$. The initial values of the velocity components $v(y, z, 0)$ and $w(y, z, 0)$ were derived from the quasi-geostrophic equations obtained from (2.37)–(2.38). The calculation procedure is similar to the one described in detail in refs. [36] and [38].

The region of study was limited to a zone extending for 200 km in the horizontal direction across the front. The depth of the ocean H was assumed to be 1 km. The resolution in the y and z directions was taken as 4 km and 20 m respectively. The intensity of the deformation field D was this time taken to be $10^{-5}s^{-1}$. The values $f = 10^{-4}s^{-1}$ and $\alpha = 0.2 \ kg/m^3 per \ 1°C$ were selected for the Coriolis parameter f and the coefficient of thermal expansion α. Numerical integration of the equations was carried out with a time step of 30 min.

As in the model of quasi-geostrophic frontogenesis discussed earlier [36, 38], the formation of frontal interfaces is observed, in the central part of the region, in the surface and near-bottom 200-metre layers. Fig. 2.30 a and b depicts the distribution of temperature in the surface layer at the initial moment (a) and at $t = 70 \ hrs(b)$. The horizontal gradient at the front diminishes rapidly with depth, and differs little from the initial one at levels exceeding 200 m. The period of frontogenesis, which is defined as the period of 10-fold sharpening of the initial horizontal temperature gradient near the surface $y = 0$ for a deformation field intensity of $D = 10^{-5}s^{-1}$, is only 3 days.

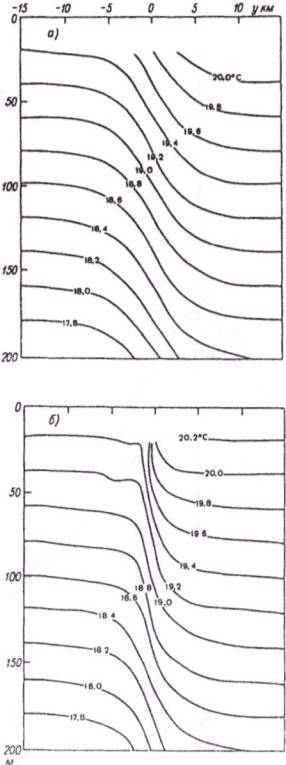

Fig. 2.30. Temperature distribution in the surface layer of the ocean a – at the initial moment $(t = 0)$, b – at $T = 70\ hrs$.

Fig. 2.31 a and b shows the change in temperature, and in the zonal and meridional velocity components at the level $z = 20\ m$, 70 hrs from the initial moment (a). On the lower graph (b), we see the displacement of the front in the direction of the cold water, this displacement reaching its maximum at the surface of the region and diminishing with depth. This effect draws attention to the deviation of the frontal interface from the vertical. It should again be said that in the case of quasi-geostrophic approximation, the surface of maximum horizontal temperature (density) gradient turned out to be vertical (see fig. 2.24). As calculations have shown, vertical velocities after 70 hrs of frontogenesis reach $10^{-4}m/s$ near the frontal interface (fig. 2.32). It is characteristic that upwelling occurred on the warm side of the frontal interface, and downwelling on the cold side, in all the frontogenetic models of both Kuz'mina, and Woods and MacVean. It should be said that in natural conditions, convergent circulation in the vertical plane perpendicular to the front is mostly characterized by downwelling on both sides of the frontal interface, at least in its immediate vicinity.

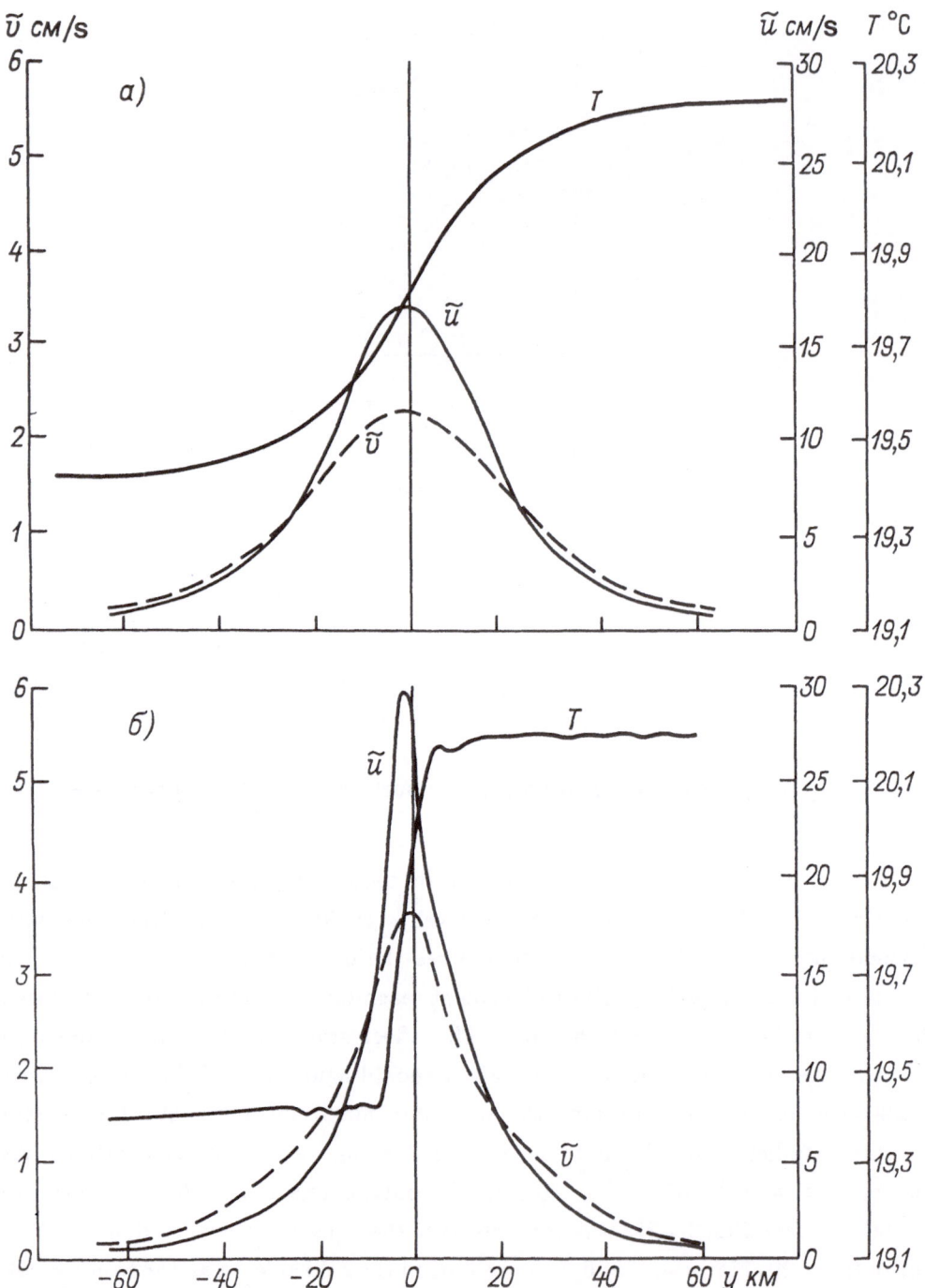

Fig. 2.31. Distribution of temperature $\tilde{T}$, and the along- front, $\tilde{u}$, and transverse, $\tilde{v}$, components of velocity at the 20 m level. a – at the initial moment ($t = 0$), b – at $t = 70$ hrs.

This and other differences of the model results from those observed in the real ocean brings us to the conclusion that neglect of turbulent mixing is an acceptable approximation only in the early stages of frontogenesis; at later stages, it can be justified only at a great distance from the frontal interface near the surface [178]. In any case, it has been shown (in section 2.5) that turbulent mixing is apparently a basic mechanism by which fronts attain a balanced quasi-stationary state in which horizontal density gradients cannot reach infinitely large values, but can on the other hand spread to greater depths than is allowed by semi- geostrophic or quasi-geostrophic models.

On the basis of this, we can say that the ability to model frontogenetic processes is not sufficient for providing understanding of the fundamental physics of oceanic fronts. It is also necessary to understand the dynamics of quasi- stationary oceanic fronts.

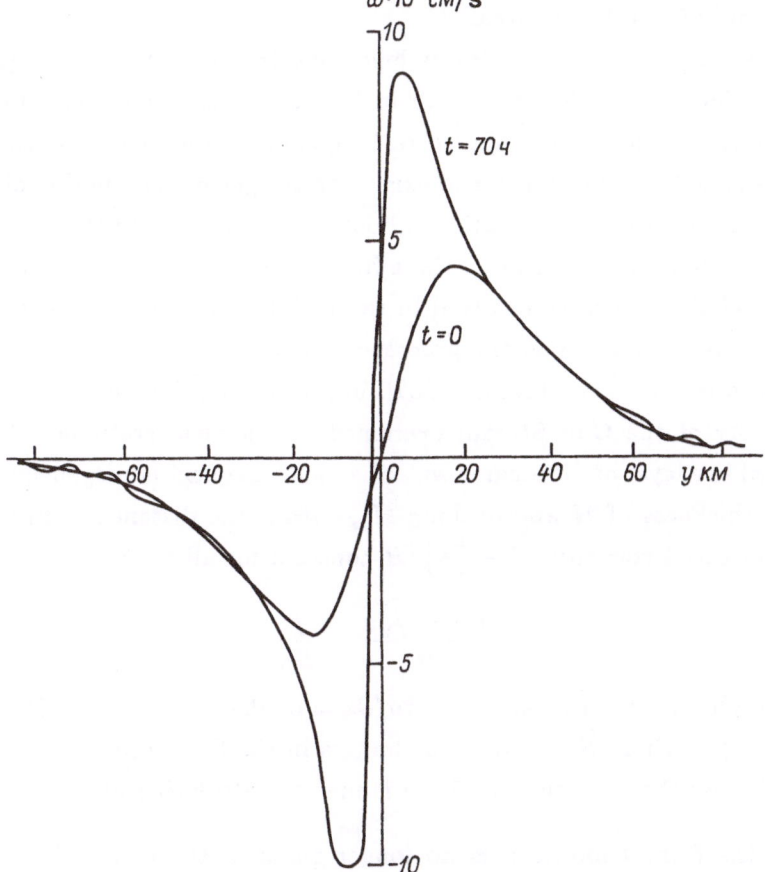

Fig. 2.32. Distribution of vertical velocity at the 20 m level at $t = 0$ and $t = 70$ hrs.

2.7. Problems of general frontal dynamics

In his analysis of frontal dynamics, Christopher Mooers [185] noted that the hydrodynamic models of (1) the generation of fronts (frontogenesis), (2) their stationary state and (3) their destruction (frontolysis) can be formulated differently, i.e. the first should be nonstationary, but may be devoid of any dissipative factors; the second makes it possible to disregard changes in time, but requires an estimation of dissipation and its main agent, turbulent viscosity; the third requires the study of both nonstationarity and dissipative forces.

Since, by analogy with the atmosphere (see section 2.4), the maintenance of sharp oceanic frontal interfaces is determined by the circulation in the plane normal to the front, a study of the balance of forces that sustains this transverse circulation should perhaps be regarded as the main task in the analysis of the dynamics of quasi-stationary frontal interfaces.

The majority of large-scale fronts belonging to the first category (see section 1.3) and having a climatic nature are in fact quasi-stationary ones in the sense that the frontogenetic process has no real significance for them. Quasi-stationary frontal zones and frontal interfaces co-exist with the general circulation of the World Ocean, and are subject to fluctuations about the mean state to the same extent as the general circulation. In studying these frontal interfaces, it is far more important to understand the dynamics involved in the maintenance of their quasi-stationary state, than it is to understand the process of frontogenesis.

This is precisely why Stommel [237], in formulating his now classic nonlinear inertial theory of the Gulf Stream examined first a quasi-stationary flow with a velocity $v(x)$ in a layer of "18-degree water" which was quasi-homogeneous in density and had a thickness of H approaching H_0, with x the distance from the origin*, and with potential vorticity $\left(f + \frac{\partial v}{\partial x}\right)/H$ constant for all x, i.e.

$$\frac{f + \partial v/\partial x}{H} = \frac{f}{H_0}. \tag{2.48}$$

The condition (2.48) means that $\partial v/\partial x = 0$ when $x \to \infty$ and $H \to H_0$. The origin ($x = 0$) is where $H = 0$ and no changes in the flow occur along y. Stommel checked whether the condition of (2.48) is met by calculating $v(x)$, i.e.

---

* Since the frontal interface is no longer zonal in the cases discussed in this section, the x coordinate here and from now on will no longer be along-frontal, but will be directed along the normal to the front, in accordance with the choice of the authors of [158] and [237].

$$v(x) = \int\limits_{x}^{\infty} f\left(\frac{H}{H_0} - 1\right) dx \qquad (2.49)$$

with $H(x)$ and H_0 taken from observations in the ocean, and by comparing the resulting profile with results from the usual dynamic calculations of Gulf Stream velocities (fig. 2.33). A comparison showed good agreement of the velocity profiles $v(x)$, and therefore confirmed the assumption of constant potential vorticity in the Gulf Stream jet, i.e. in the layer of 18-degree water moving in the core of the Gulf Stream along the periphery of the Sargasso Sea.

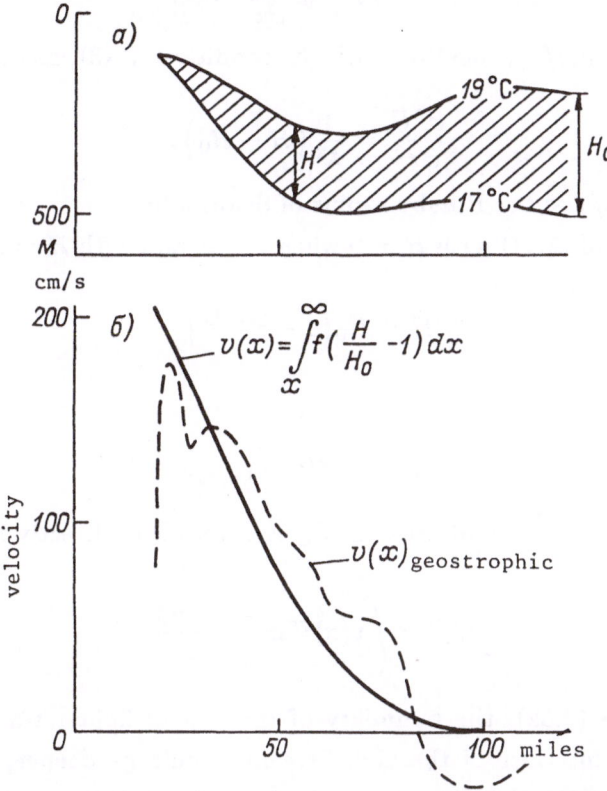

Fig. 2.33. Analysis of vorticity in a cross-section of the Gulf Stream in the vicinity of $39°40'N$ and $79°W$, after Stommel [237]. a – change in thickness of the layer of "18-degree" water between the 17 and $19°C$ isotherms across the main front of the Gulf Stream. The layer thins out ($H = 0$) near the frontal interface; b – change in velocity across the Gulf Stream: continuous line – calculated by integrating the potential vorticity assuming it constant along the x axis; dashed line – calculated by the usual dynamic method from hydrographic data.

This agreement prompted Stommel to simplify the problem as far as possible, and to set about determining whether the main characteristics of the Gulf Stream could be obtained from combining the condition of potential vorticity conservation with the condition of geostrophic motion in a homogeneous lens of lighter water with density ρ_1 and thickness H, overlying a deep layer of stagnant heavier water with density ρ_2.

The geostrophic relationship

$$fv = g'\frac{\partial H}{\partial x},\tag{2.50}$$

where $g' = g(\rho_2 - \rho_1)/\rho_2$, together with the condition (2.48), gives us the equation

$$\frac{\partial^2 H}{\partial x^2} = \frac{1}{\lambda^2}\left(H - H_0\right),\tag{2.51}$$

where $\lambda = \sqrt{g'H_0}/f$ is the Rossby radius of deformation.

The solution of (2.51) with $H = 0$ when $x = 0$, and with $H = H_0$ when $x = \infty$, gives us

$$H = H_0\left(1 - e^{-x/\lambda}\right)\tag{2.52}$$

and therefore

$$v = \sqrt{g'H_0}e^{-x/\lambda} = v_{max}e^{-x/\lambda}.\tag{2.53}$$

The total flow of the Gulf Stream Q_G is determined directly from (2.50) with

$$Q_G = \int\limits_0^\infty v(x)H\,dx = \frac{g'H_0^2}{2f}.\tag{2.54}$$

According to (2.52), the boundary of the lens of lighter water, which represents the frontal interface of the Gulf Stream, should go deeper, approaching H_0 exponentially far from the point ($x = 0$) where the Gulf Stream front emerges at the surface. From this point to the right looking downstream, i.e. in the same direction of positive x, the velocity of the geostrophic current $v(x)$ should diminish exponentially in accordance with (2.53). At the same time, the potential vorticity in the lens of light water should be constant and equal to f/H_0 for any value of x.

For the typical hydrographic Gulf Stream values of $H_0 = 800\ m, f = 10^{-4}$ and $(\rho_2 - \rho_1)/\rho_2 = 2\cdot 10^{-3}$, this simple theory gives the width $\lambda = 40\ km$ for the region in which the velocity of the Gulf Stream diminishes by a factor e, predicts the total flow of the Gulf Stream to be $Q_G = 64 \times 10^6 m^3/s$ (which is quite close to the actual value), and gives the maximum velocity of the Gulf Stream as $v_{max} = 4\ m/s$. It appears that the Rossby radius of deformation in this case is an important scale for the frontal jet stream.

From Stommel's simple model of the Gulf Stream, we can see that the geostrophic balance of the along-frontal jet stream and the constancy of potential vorticity in the lens of lighter water adjacent to the frontal interface and entrained into this jet stream play an important role in the dynamics of quasi-stationary large-scale fronts. The source of the lighter water and the process which initially brought it into contact with heavier water are not of critical importance in this case; the frontogenetic process remaining beyond the scope of this research.

This situation confused later researchers. It is not by chance that T.W. Kao [156, 157, 158] and Garvine [127, 129, 130] began their study of frontal dynamics not with large-scale fronts, but rather with small-scale ones where the source of the ambient water masses fitted in with the space and time scales of the problem. The starting point for both researchers was a freshwater river discharge into the ocean (see section 3.3). Unlike Garvine who first of all studied the purely viscous dynamics of a discharge front [127], Kao was interested from the start in the effect of the Earth's rotation on frontal dynamics, and came to the following conclusions [156, 157, 158]:

1) the time t_1 typically required for the establishment of a discharge front between light fluid spreading over the surface of heavier fluid under the effect of a pressure gradient is always less than the inertial time scale $1/f$, where f is the Coriolis parameter, i.e. $t_1 \ll 1/f$;

2) the effect of the Earth's rotation can be disregarded for all t values less than $1/f$. In this stage, a balance is established between the buoyancy forces which determine the horizontal pressure gradient, and the viscous tangential shear stress at the boundary between the two fluids. The relative motion of the two layers is uniform, with a constant velocity U_ϕ which can be called the forward velocity of the front;

3) for $t > 1/f$, the role of the Earth's rotation increases, and at $t_2 \simeq 2\pi/f$ becomes dominant in the balance of forces [158]. An along-front geostrophic current develops, while the forward velocity of the front greatly diminishes. This helps to decrease the frictional forces. A geostrophic balance is maintained throughout the frontal interface, except in the near-surface layer where the vorticity generated by baroclinicity can be so high that only viscous forces can balance it. Therefore, an Ekman layer should exist near the surface, regardless of whether there is a wind or

not.

These fairly general conclusions, which without a doubt oversimplify the real world (compare, for example, with Garvine's formulation of the problem in [127], see also subsection 3.3.1), serve as Kao's starting point for further generalization of earlier theoretical results for a large-scale region [156]. In this case, Kao uses similarity and dimensional methods for finding self-similar solutions.

Kao begins this generalization by retaining the previous initial condition that at the surface of the ocean, light water with a flow rate Q_e per unit length flows into quiescent ambient fluid with a density ρ_0. This inflow is characterized by a density deficit of $(\Delta\rho)_e$ with respect to the ambient fluid. Consequently, the reduced gravitational acceleration $g' = g(\Delta\rho)_e/\rho_0$ acts at the horizontal boundary between the two media. From g' and Q_e, we can derive a velocity scale $U_\phi = (g'Q_e)^{\frac{1}{3}}$ and a spatial scale $h_0 = (Q_e/g')^{\frac{1}{3}}$ for the buoyancy field. Together these give us a small time scale $t_1 = h_0/U_\phi$. The inertial time and length scales can be given as $t = 1/f$ and $L_0 = U_\phi/f$ respectively. In turn, the spatial scale of diffusion is written as $h_\nu = (\nu/f)^{\frac{1}{2}}$, where ν is the coefficient of diffusion, the nature of which Kao does not determine at all at this stage, though a turbulent diffusion of momentum is implied by the problem. It is obvious that ν has the kinematic viscosity dimension $[L^2T^{-1}]$. At the same time, Kao does not specify whether he means vertical or horizontal diffusion of momentum. The problem turned out to have three spatial scales with lengths h_0, L_0 and h_ν, which together yield two important numbers, 1) the densimetric Rossby number

$$\widetilde{Ro} = L_0/h_0 \tag{2.55}$$

and 2) the Ekman number

$$E = (h_\nu/h_0)^2. \tag{2.56}$$

The combination of these two values $\widetilde{Ro}/E$ gives us the Reynolds number

$$Re = \frac{\widetilde{Ro}}{E} = \frac{U_\phi h_0}{\nu}. \tag{2.57}$$

The coordinates are selected in such a way that the positive values of x are directed opposite to the inflow of light water to the front, while the z axis is directed vertically upward. The motion is considered to be independent of the coordinate y which from now on will be called the "along-front" coordinate, while coordinate x will be called "cross-front".

The initial system of equations is more conveniently described in dimensionless coordinates, i.e.

$$\xi = \frac{x}{L_0}, \ \eta = \frac{z}{h_0} \ \text{and} \ \tau = tf,$$

(2.58)

and with dimensionless variables

$$\tilde{u} = u/U_\phi, \ \tilde{v} = v/U_\phi \ \text{and} \ \tilde{w} = \frac{\widetilde{Row}}{U_\phi}.$$

(2.59)

Then the continuity equation will be

$$\frac{\partial \tilde{u}}{\partial \xi} + \frac{\partial \tilde{w}}{\partial \eta} = 0;$$

(2.60)

while the along-frontal component of vorticity

$$\varsigma = \frac{\partial u}{\partial z} - \frac{\partial w}{\partial x}$$

can be presented in a dimensionless form through the dimensionless stream function $\tilde{\Psi} = \Psi/Q_e$, where

$$\tilde{u} = \frac{\partial \tilde{\Psi}}{\partial \eta} \ \text{and} \ \tilde{w} = -\frac{\partial \tilde{\Psi}}{\partial \xi},$$

(2.61)

as

$$\tilde{\varsigma} = \left(\widetilde{Ro}\right)^{-2} \frac{\partial^2 \tilde{\Psi}}{\partial \xi^2} + \frac{\partial^2 \tilde{\Psi}}{\partial \eta^2}.$$

(2.62)

The density anomaly $\gamma = (\rho - \rho_0)/\rho_0$ has a maximum equal to $\gamma_e = -(\Delta \rho)_e/\rho_0$, and can be presented in a dimensionless form as

$$\tilde{\gamma} = \gamma/\gamma_e.$$

(2.63)

Therefore, the equation of mass conservation is

$$\frac{\partial \tilde{\gamma}}{\partial \tau} + \frac{\partial}{\partial \xi}\left(\tilde{u}\tilde{\gamma}\right) + \frac{\partial}{\partial \eta}\left(\tilde{w}\tilde{\gamma}\right) = E\left[(\widetilde{Ro})^{-2}\frac{\partial^2 \tilde{\gamma}}{\partial \xi^2} + \frac{\partial^2 \tilde{\gamma}}{\partial \eta^2}\right],$$

(2.64)

the equation for the conservation of the along-frontal component of vorticity is

$$\frac{\partial \tilde{\varsigma}}{\partial \tau} + \frac{\partial}{\partial \xi}\left(\tilde{u}\tilde{\varsigma}\right) + \frac{\partial}{\partial \eta}\left(\tilde{w}\tilde{\varsigma}\right) - \frac{\partial \tilde{v}}{\partial \eta} = \frac{\partial \tilde{\gamma}}{\partial \xi} + E\left[(\widetilde{Ro})^{-2}\frac{\partial^2 \tilde{\varsigma}}{\partial \xi^2} + \frac{\partial^2 \tilde{\varsigma}}{\partial \eta^2}\right],$$

(2.65)

and the equation for the along-frontal motion v is

$$\frac{\partial \tilde{v}}{\partial \tau} + \frac{\partial}{\partial \xi}\left(\tilde{u}\tilde{v}\right) + \frac{\partial}{\partial \eta}\left(\tilde{w}\tilde{v}\right) + \tilde{u} = E\left[(\tilde{R}o)^{-2}\frac{\partial^2 \tilde{v}}{\partial \xi^2} + \frac{\partial^2 \tilde{v}}{\partial \eta^2}\right]. \qquad (2.66)$$

The system (2.60), (2.64), (2.65) and (2.66) is a closed system of four equations for four unknowns $\tilde{u}$, $\tilde{v}$, $\tilde{w}$ and $\tilde{\gamma}$.

It should be said that the densimetric Rossby number $\widetilde{Ro}$, which is used in the system described above, differs from the Kibel-Rossby number discussed in [127] and [158] (see Section 1.3) in that the buoyancy spatial scale h_0 was used in it instead of the rather arbitrary transverse frontal scale L_ϕ. For this the authors of [127] and [158] took the width scale of the frontal zone, which is related to the thickness of the layer of light water and the slope of the frontal interface, i.e. $L_\phi \approx 100$ m for discharge fronts and $L_\phi \approx 100$ km for the Gulf Stream. The same approach is used in subsection 3.3.1 during the discussion of discharge fronts.

In the system of equations examined above, $\widetilde{Ro}$ is more indirectly related to the slope s of the frontal interface, so that with small values of E

$$s \approx 1/\widetilde{Ro}$$

Indeed, Margules' formula (2.2) can be written in the new notation as

$$s = tg\,\alpha = \frac{f\rho_0 \Delta v}{g(\Delta\rho)_e} \qquad (2.67)$$

where Δv is the difference between the geostrophic velocities on both sides of the front. It acquires the value $\Delta v \simeq U_\phi$ after the deflecting force of the Earth's rotation has replaced viscous stress in the balance of forces. Substituting this value in (2.67) and taking (2.55) into account, we get

$$s = tg\,\alpha \simeq \frac{fU_\phi}{g'} \simeq \frac{1}{\widetilde{Ro}}. \qquad (2.68)$$

Frontal interfaces in quasi-geostrophic balance are characterized by a mean slope of $s = 10^{-2}$. Consequently, $\widetilde{Ro}$ should be of the order of 10^2, as is quite apparent from (2.68). In turn, with such values of $\widetilde{Ro}$, the system (2.60), (2.64), (2.65) and (2.66) is greatly simplified, since the first term in square brackets on the right side of all these equations can be disregarded, and only one term will remain in each of them. On the basis of this, Kao maintains that the solutions of the initial system of equations (2.60), (2.64), (2.65) and (2.66) will be similar over a wide range of $\widetilde{Ro}$ values which are characteristic of the ocean, and depend only on E. In turn, this means that the dynamics and structure of the frontal interface remain the same for all fairly large values of $\widetilde{Ro}$ and small values of E. For example, the frontal interface in coordinates η and ξ will have universal form that does not depend on $\widetilde{Ro}$ for any value of τ after attaining quasi-geostrophic balance.

The validity of the similarity solution obviously depends largely on the ratio of the coefficients of vertical and horizontal viscosity, ν_V and ν_H. Kao maintains that similarity will be retained over a fairly wide range of values $\nu_H > \nu_V$ with an upper limit of $\nu_H/\nu_V > (\widetilde{Ro})^2$, which apparently results from the right part of (2.66). Confirmation of the similarity of the solutions of this system for this range of $\nu_H > \nu_V$ and their similarity to the results of oceanic observations does not indicate that horizontal turbulent diffusion for typical scales of the phenomenon can be disregarded. Rather, it applies only because the solutions of the system with small values of E differ little from each other, which in the end is equivalent to the well-known fact that viscous (turbulent) friction has nothing to do with the resulting geostrophic regime of motion. In the given problem, this applies everywhere, with the exception of the Ekman layer near the surface of the ocean.

We can also show that, in the light water, the potential vorticity in a quasi-stationary geostrophic regime (i.e. with low values of E and large values of τ) is constant. Indeed, in the light-water lens at some distance from the front in the direction of positive values of x, there should be a region where the geostrophic relation,

$$-\frac{\partial \tilde{v}}{\partial \eta} = \frac{\partial \tilde{\gamma}}{\partial \xi},$$

which comes from (2.65), becomes

$$-\frac{\partial \tilde{v}}{\partial \eta} = 0$$

because the density in the lens ceases to change horizontally.

In this region, Kao examines the vertical component of the absolute vorticity

$$\omega = \frac{\partial v}{\partial x} + f,$$

which, if represented in a dimensionless form with the help of the factor $U_\phi/L_0 = f$, becomes

$$\tilde{\omega} = \frac{\partial \tilde{v}}{\partial \xi} + 1.$$

After differentiating (2.66) with respect to ξ and using (2.60), we obtain the following equation when $\partial \tilde{v}/\partial \eta = 0$, $\partial \tilde{w}/\partial \eta = 0$ and $E \to 0$:

$$\frac{\partial \tilde{\omega}}{\partial \tau} + \tilde{u}\frac{\partial \tilde{\omega}}{\partial \xi} + \omega\frac{\partial \tilde{u}}{\partial \xi} = 0. \tag{2.69}$$

Assuming that the dimensionless thickness of the lens of lighter water is equal to H, we can rewrite (2.60) as

$$\frac{\partial \tilde{u}}{\partial \xi} + \frac{1}{H}\cdot\frac{dH}{d\tau} = 0, \tag{2.70}$$

where $d/d\tau = \partial/\partial\tau + \tilde{u}\partial/\partial\xi$.

The combination of (2.70) and (2.69) gives us $d(w/H)/d\tau = 0$, which in turn is identical to the law of conservation of potential vorticity

$$d(\omega/H)/dt = 0. \tag{2.71}$$

This result enabled Kao to say that Stommel's simple classic concept, of the steady nonlinear inertial theory of the Gulf Stream, is included naturally in the above interpretation of the problem.

Numerical integration of the initial system of equations was carried out by Kao by the procedure described in [158] in order to confirm the above scale analysis, which led to the conclusion that the solutions in the given region were similar. Kao carried out his calculations for several combinations of $\widetilde{Ro}$ values (from 10 to 210) and E values (from 0.025 to 2.10). An analysis of the results is of particular interest from the points of view of 1) a study of the process by which a quasi- steady balance is attained, 2) the determination of the density and dynamic structure of the frontal interface that results in this case, and 3) a numerical comparison with the corresponding characteristics of the Gulf Stream.

Transition to a quasi-steady regime. As expected, this process does not depend on $\widetilde{Ro}$, whereas it does depend qualitatively on the value of E. The development of the process is depicted in fig. 2.34 where the dimensionless speed of frontal advance $\tilde{u}_f$ along the x- axis is shown as a function of dimensionless time τ. The speed of the front decreases sharply during the first inertial period, during which the main process of adaptation to geostrophic balance takes place. The horizontal motion of the front later undergoes fluctuations with inertial frequency, the amplitude of which dies out more quickly at $E = 0.25$ than at $E = 0.025$. At the same time, the frontal interface deepens, as the condition of mass conservation requires, and we observe the lessening of vertical fluctuations which are in antiphase to the horizontal fluctuations. The along-frontal speed of the geostrophic current simultaneously reaches its maximum value, approximating U_ϕ, which is finally established over several inertial periods.

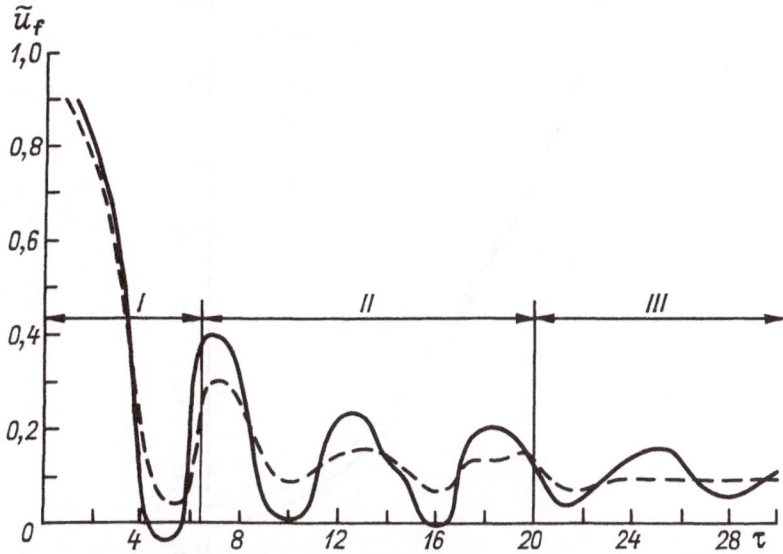

Fig. 2.34. Dependence of the dimensionless speed of frontal advance $\tilde{u}_f$ on the intensity of mixing, based on Kao's data [156]. Solid line – with $E = 0.025$; dashed line – with $E = 0.25$ (more intensive mixing). I – initial (nonstationary) regime; II – transitional regime; III – quasi-stationary regime.

Density and dynamic structure of a frontal interface. At $\tau = 7$, the form of the frontal interface marked by the isopycnal $0.1\gamma_e$ (dashed line in fig. 2.35) already differs very little from that observed in the quasi-steady regime, e.g. with $\tau = 30$ (solid line in fig. 2.35). As we can see in fig. 2.35, the form of the frontal interface is completely similar for all $\widetilde{Ro}$ values in the $10 - 210$ range. This similarity is best seen with respect to the coordinates $\xi_\alpha = \xi/\alpha$ and $\eta_\alpha = \eta/\alpha$, where α is the constant of proportionality which indicates the degree of change in time of the initial scale of the front L_0 and the buoyancy scale (lens thickness) h_0, so that $L(\tau) = \alpha L_0$ and $h(\tau) \sim \alpha h_0$. Kao's calculations and fig. 2.34 show that the asymptotic value of $h(\tau)$ when $\tau \to \infty$ is equal to $h \simeq 1.5\alpha h_0$, from which we can determine the asymptotic value of α if we use the value $\overline{h}$ taken, for example, from the Gulf Stream observations, i.e.

$$\overline{\alpha} \simeq \frac{h}{1.5 h_0} = \frac{h_0}{1.5(Q_e^2/g')^{\frac{1}{3}}}. \tag{2.72}$$

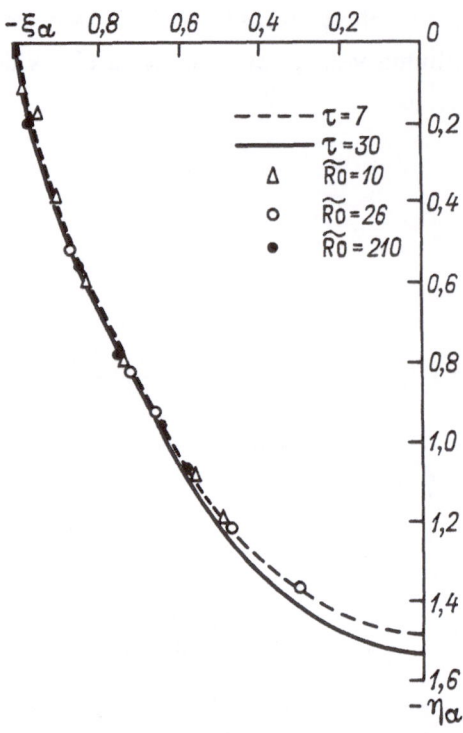

Fig. 2.35. Form of a frontal interface for various values of $\widetilde{Ro}$ and τ, based on Kao's model results [156]. The graph demonstrates a practically complete similarity of the profiles of the frontal interface with $\widetilde{Ro} = 10 - 120$ and the complete emergence to a quasi-steady regime after $\tau = 7 - 30$.

With $\overline{h} = 800\ m$ (Stommel's data) and the specified values $Q_e = 50\ m^2/s$ and $g' = 1.5 \cdot 10^{-2} m/s^2$, we get $\overline{\alpha} = 9.7$. This means that the process of geostrophic adjustment of the front in this case is a process which is qualitatively opposite to frontogenesis. In the latter case the frontal zone is sharpened, whereas in the course of adjustment to a geostrophic regime it expands. This difference is associated only with the fundamental difference in formulation of the problems in sections 2.6 and 2.7. In the first case, we observe a gradual sharpening of a weak baroclinic, geostrophically balanced frontal zone due to forced (externally applied) deformation; in the second case, the imposed ageostrophic frontal disturbance of small width L_0, defined at the initial moment by the flow rate of light fluid Q_e, gradually adapts to a geostrophic regime, in which the initial density difference $\Delta\rho_e$ should diffuse into a zone with an asymptotic width of $\overline{L} = \alpha L_0$. This width should be such that the maximum along-frontal velocity of the geostrophic jet stream and its flux correspond to the inflow of water to the front with a flow rate Q_e.

In reality, the frontogenetic effect in this case is incorporated into the initial existence of the flow rate Q_e, which in this particular case can be related to the deformation field.

The quasi-stationary structure of a frontal interface in cross-section for $E = 0.025$ (the same for all $\widetilde{Ro}$) is depicted in fig. 2.36 in the coordinates $\xi_{\overline{\alpha}} = \xi/\alpha$ and $\eta_{\overline{\alpha}} = \eta/\overline{\alpha}$. We can clearly see that the isopycnals, which correspond to 0.1, 0.5 and 0.9 of the total density change γ_e (thick solid lines), converge in a concentrated cluster near the surface ($\eta_{\overline{\alpha}} = 0$) in such a way that 80% of γ_e is concentrated between the values $\xi_{\overline{\alpha}} = 1.0$ and $\xi_{\overline{\alpha}} = 0.8$. We can say the isopycnal 0.9 encloses a practically homogeneous lens of lighter water. To the right of the value $\xi_{\overline{\alpha}} = 0$ (in the direction of positive values of $\xi_{\overline{\alpha}}$), the isopycnals become horizontal, so that the vertical distance between the isopycnals 0.1 and 0.9 corresponds to the thickness of the main pycnocline. Kao calls the whole surface region between the point $\xi_{\overline{\alpha}} = 0$ and the isopycnal 0.1 a frontal area (zone), and the region between the isopycnals 0.1 and 0.9 a front. In fig. 2.36, the dashed lines mark the isotachs of the along-frontal geostrophic current. The axis of this jet stream is at $\xi_{\overline{\alpha}} = -0.8$. If we accept that the jet stream is directed northward, then the lens of light water is located east of the frontal interface. Judging by the position of the isotachs, the zone of the jet stream at the surface is twice as wide as the region of the maximum horizontal density gradient. This is seen even more clearly from a comparison of the change in the density anomaly γ and the change in the elevation of the free surface along the $\xi_{\overline{\alpha}}$ axis in the frontal zone (fig. 2.37). The values of the density anomaly decrease sharply from 0.9 to 0 in a $\xi_{\overline{\alpha}}$ interval of slightly more than 0.2 in the extreme left part of the frontal zone, whereas the elevation of the free surface gradually diminishes from right to left across the entire frontal zone. The maximum velocity of the jet stream falls precisely within the region of the maximum horizontal gradient of the free surface $\Delta\eta_s$.

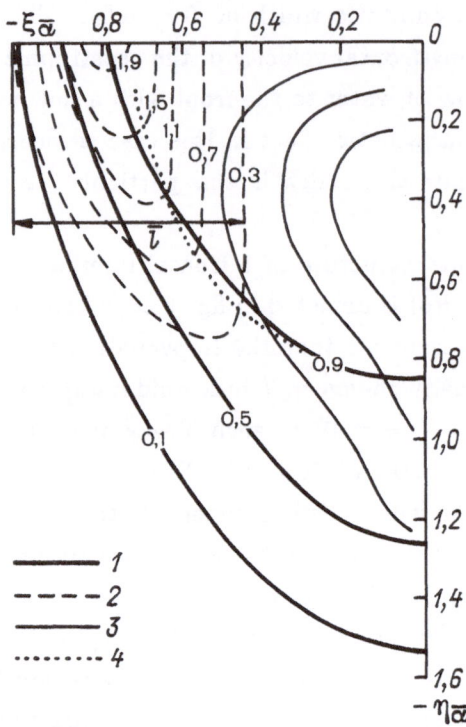

Fig. 2.36. Quasi-stationary structure of a frontal interface in cross-section with $E = 0.025$ (the same for all $\widetilde{Ro}$), based on Kao's model data [156]. 1 – isohalines; 2 – isotachs $\tilde{v}$; 3 – stream lines of the transverse circulation; 4 – $\tilde{H}/\tilde{H}_0$. (See (2.73) and (2.74)).

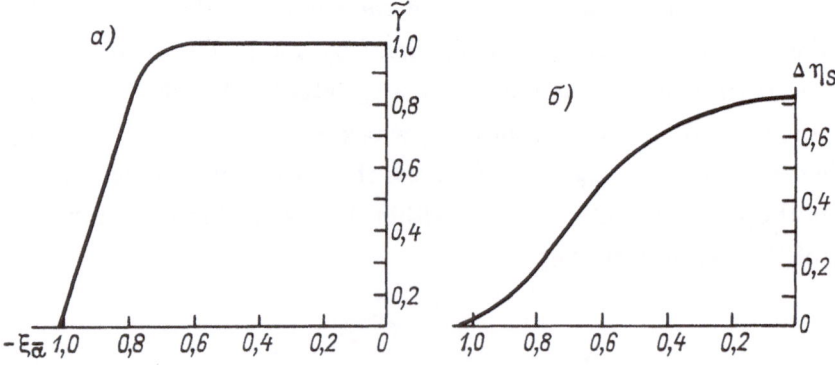

Fig. 2.37. Changes in the dimensionless density (a) and dimensionless free-surface level (b) across the frontal zone, based on Kao's model data [156].

As we can see from the position of the isotachs (fig. 2.36), the region of the frontal interface is characterized by maximum values of the vertical gradient of the geostrophic current $\partial \tilde{v} / \partial \eta_{\overline{\alpha}}$, whereas the value $0.9 \ \partial \tilde{v} / \partial \eta_{\overline{\alpha}} \simeq 0$ is observed in the lens of lighter water to the right of the isopycnal, which means that the potential vorticity here is constant. This circumstance, together with the geostrophic nature of the along-frontal flow, gives us reason to anticipate Stommel solutions of the (2.52) and (2.53) type in the lens of lighter water. Indeed, the results presented in fig. 2.36 can be approximated in the region of $0 \geq \xi_{\overline{\alpha}} \geq -0.74$ by the following expressions:

$$\tilde{v} = \tilde{v}_{max} \exp[-6(0.74 + \xi_{\overline{\alpha}})]; \tag{2.73}$$

$$\tilde{H} = \tilde{H}_0 \{1 - \exp[-6(0.74 + \xi_{\overline{\alpha}})]\}, \tag{2.74}$$

with $\tilde{H} = 0.82$ (as seen in fig. 2.36) and the ratio $\tilde{H}/\tilde{H}_0$ (dotted line in fig. 2.36) coinciding quite well with the 0.9 isopycnal. It follows from this that the total flux of the along-frontal jet stream should be equal to $0.38 \ g'(H_0)^2/f$, which is similar to the expression (2.54) obtained by Stommel [237].

The thin solid lines in fig. 2.36 depict isotachs of the cross-frontal current. In the upper part of the lens, we observe a convergent current directed towards the front. Downwelling takes place in the region of the frontal interface, and a counterflow is observed in the lower part of the lens and in the pycnocline. This transverse circulation essentially determines and sustains the quasi-steady dynamics of the frontal interface. Typical values of transverse velocities $(\tilde{u} - \tilde{u}_f)$ in relation to the fluctuating value are of the order of $1/10 \ U_\phi$ even when the front is in a quasi-steady state, whereas the downwelling speed (w) near the front amounts to $(10^{-3} - 10^{-4})v$.

Comparison with the Gulf Stream. For his numerical comparison with the Gulf Stream, Kao easily obtained the following formulae from his model:

1) the maximum velocity of a geostrophic jet stream

$$v_{max} = 2U_\phi = 2(g'Q_e)^{\frac{1}{3}}; \tag{2.75}$$

2) the slope of the frontal interface

$$\bar{s} = \frac{\bar{h}}{\bar{L}} = \frac{1.5\alpha h_0}{\alpha L_0} = 1.5\widetilde{Ro}^{-1} = 1.5f(g')^{-\frac{2}{3}}Q_e^{\frac{1}{3}}; \tag{2.76}$$

3) the width of the frontal zone

$$\overline{L} = \frac{\overline{h}}{1.5}\widetilde{Ro} = \frac{\overline{h}}{1.5}f^{-1}(g')^{\frac{2}{3}}Q_e^{-\frac{1}{3}};\tag{2.77}$$

4) the total flux of the Gulf Stream

$$Q_G = 0.253\left(\frac{\overline{h}}{1.5}\right)^2\frac{g'}{f} = 0.38\frac{g'H_0^2}{f}.\tag{2.78}$$

Table 2.3 contains the computed characteristics of the Gulf Stream for various values of Q_e and the following initial parameters: $T = 20°C$ and $S = 36.6$ ppt for the waters of the Sargasso Sea; $T = 5°C$ and $S = 35.0$ ppt for the waters of the North Atlantic; $\Delta\sigma_t = 1.6$ or $(\Delta\rho)_e/\rho_0 = 1.6 \times 10^{-3}$; $g' = 1.5 \times 10^{-2}m/s^2$ and $\overline{h}$ was taken as 1400 m when calculating $\overline{s}$ and $\overline{L}$.

Table 2.3

Computed characteristics of the Gulf Stream for various Q_e

Initial reference values				Computed characteristics of the Gulf Stream				
Q_e m^2/s	U_ϕ m/s	L_0 km	h_0 m	$\widetilde{Ro}$	$\overline{s}$	v_{max} m/s	$\overline{l}^*$ km	$\overline{L}$ km
10	0.53	5.3	19	279	1:186	1.08	182	260
25	0.72	7.2	35	205	1:136	1.44	136	194
50	0.90	9.0	55	165	1:110	1.80	107	154
75	1.03	10.3	73	142	1:95	2.06	94	134
100	1.14	11.4	88	129	1:86	2.28	85	121
125	1.23	12.3	102	120	1:80	2.46	79	113
150	1.30	13.0	115	113	1:75	2.60	74	105

* $\overline{l}$ — width of jet stream at the surface.

From the table we see that all the Q_e values taken over an order of magnitude (from 10 to 150) give us fairly realistic Gulf Stream characteristics which fit into the range of values cited by various authors. This is a broad range but this should not come as a surprise if we take into account the meandering and other known forms of variability of the Gulf Stream. On the basis of this comparison, we can conclude that the results obtained in Kao's quasi-steady model adequately describe large-scale quasi-geostrophic frontal interfaces of a climatic nature and the quasi-steady jet streams like the Gulf Stream, which are associated with them.

We have yet to discuss the physical significance of assigning Q_e as the initial condition of the problem, as well as the resulting characteristics of the given model. The fact that Stommel [237] obtained similar results from simple physical considerations without resorting to problem solving with these initial conditions indicates that he was apparently able to formulate concretely the quasi-steady problem regarding the dynamics of a large-scale front without being given the initial inflow of lighter fluid with a flow rate Q_e. This view is also confirmed by Mooers' analysis [185] which is interesting in the fact that the transverse x- and z-plane circulation in it appears as a secondary phenomenon which is generated by inertial and internal friction effects and depends only on v and γ. According to Kao and formula (2.78), the total flow rate of the Gulf Stream Q_G depends in the end on H_0, and not Q_e. However, H_0 is none other than the asymptotic value $\overline{h}$ which depends on h_0, and therefore on Q_e. Unfortunately, ref. [156] does not show how the thickness of the layer of lighter water approaches its asymptotic value. Instead, Kao gives it the value of 1400 m when calculating Gulf Stream parameters. At the same time, fig. 2.34, in which Kao has plotted a continuous profile for $\tau = 30$ and $\alpha = 2\pi$, indicates that the asymptotic thickness of the lens is $\overline{h} = 179$ m when $Q_e = 10$ m^2/s and $h_0 = 19$ m, and is 6 times greater, i.e. $\overline{h} = 1083$ m, when $Q_e = 210$ m^2/s and $h_0 = 115$ m. This means that either a different length of time with different values of Q_e is required for the thickness of the lens to reach the asymptotic value $\overline{h}$, or these asymptotic values depend on Q_e, and in that case the total flow rate Q_G also depends on Q_e. This in our opinion is the shortcoming of Kao's model in which the given flow rate Q_e is required for sustaining transverse circulation at all stages of the process. It would be more logical to expect a dependency of transverse circulation on the along-frontal component of velocity v or the flow rate Q_G, as was the case in Mooers' study [185]. However, Kao is of the opinion that the near-surface branch of the transverse circulation in the Sargasso Sea, which is directed towards the front of the Gulf Stream, is of wind origin and is equivalent to an inflow of lighter water with a flow rate Q_e. Naturally, in accordance with the model described above, the dynamics of the Gulf Stream can remain constant within a wide range of variable Q_e values. In Kao's opinion, the question belongs to the problems concerning the general circulation of oceanic waters, and goes beyond the analysis of his model.

2.8. Factors controlling the evolution of fronts

Oceanic fronts are characterized by the property that once they appear, they continue to exist for a long time due to their "self-sustaining" mechanisms, e.g. transverse convergent circulation [127, 156, 185], or turbulent entrainment [127]. It is interesting to note that descriptions of cases of frontolysis are extremely rare in the literature, whereas examples of frontogenesis or established fronts are encountered frequently. This is most probably because frontolysis begins when the above-mentioned mechanisms cease to function for some reason. Apparently, this is what primarily determines the sharpness of visible manifestations of fronts at the surface of the ocean (bands of foam, trash and slicks, rip zones, eddies, colour contrasts, etc.).

On the other hand, it would be incorrect to maintain that frontal interfaces in the ocean are very stable. On the contrary, they are known as the most variable features of ocean structure and dynamics.

The dual character of oceanic fronts discussed above requires a particularly close study of all the factors which determine their evolution and variability. At the same time, it is this area of research which is still only at the initial stage of development.

An oceanic frontal interface formed near the surface is exposed to the action of the wind, the effect of which in the simplest case manifests itself mostly through horizontal and vertical flows which intensify or weaken the convergent circulation near the front. In coastal areas changes in the wind, responsible for upwelling, more directly and even wholly determine the evolution of the fronts which are a part of the upwelling phenomenon itself (see section 3.2). The effects of the wind can be also very important in estuaries, as well as in bays, straits and marginal seas. There, tidal currents can greatly alter the pattern of fronts, forcing them to migrate or periodically disappear in the course of the tidal cycle (see subsection 3.3.2). The seasonal variation of solar heating, which leads to the formation and disappearance of the seasonal thermocline, can together with wind and tidal mixing result in the cyclic emergence and destruction of shelf fronts (see section 3.4). Seasonal changes in the fluxes of momentum, heat and moisture between the ocean and atmosphere can affect the sharpness and geographic position of climatic-scale frontal zones in the ocean [213, 214]. A mesoscale deformation field associated, for example, with a system of eddies can, in the process of their evolution or migration, undergo local intensification and weakening, which in turn cause the sharpening, relaxation or migration of frontal interfaces. Discussions of a balanced front (see section 2.5) help us to understand some of the effects of a variable deformation field. Finally, the quasi-geostrophic jet streams associated with frontal interfaces can become unstable and then wave-like disturbances (meanders) which complicate the frontal structure can begin to develop on them.

These reasons are quite sufficient to arouse our doubts about the possibility of frontal interfaces attaining some degree of persistence, and in my opinion, the quasi-steadiness of large-scale climatic frontal zones is nothing more than an illusion based on conventional oceanographic averaging of their internal variability which does not fit into any of the customary patterns on close examination.

All the above-listed factors that determine the evolution and variability of frontal phenomena in the ocean can be arbitrarily divided into two groups, 1) external factors and 2) internal factors. The first group includes the factors which are associated with variable environmental conditions, e.g. the wind or the exchange of heat and moisture between the ocean and the atmosphere, with the deformation field in which the front appeared, and with tidal phenomena. The second group should include basically the fairly specific factors which can manifest themselves in the dynamics and internal state of the frontal interface itself and the associated jet stream either quite independently, or under the influence of external factors. These are processes of barotropic and baroclinic instability, which were initially studied for application to zonal jet streams and fronts in the atmosphere. Now, similar concepts are being extensively used to study the motions of oceanic waters as well. Let us first discuss some of the external factors, namely the effects of the wind.

2.8.1. The effect of wind on near-surface fronts.

The main concepts regarding the nature of the effect of wind on oceanic fronts can be acquired by examining concrete situations in the light of what we already know from Ekman's classic theory and its more recent generalizations. In this area, we find a comparatively small number of papers which analyze numerical models or specific natural conditions. For instance, G.I. Shapiro [74] has written a paper on the effect of the wind on small-scale density fronts, which is briefly discussed in subsection 3.3.1. According to Shapiro's model, a wind directed from the open ocean perpendicularly to an immobile discharge front can completely alter the direction of the circulation in a discharge lens, transforming the currents near the front inside the lens from convergent currents into divergent ones. On the basis of the same paper, the slope of the frontal interface should change significantly, depending on the direction of the wind in relation to the direction of frontal movement.

A wind directed along the shoreline can cause the formation of along-shore jets and fronts associated with the transition from a frictional regime in shallow waters to a quasi-geostrophic regime beyond the shelf edge. In other cases, the wind can greatly alter the form of shelf fronts (see section 3.4) which separate coastal waters that are mixed to the very bottom from stratified waters beyond the shelf [100], and play an important role in their evolution, relaxation and adjustment to a geostrophic balance. Observations of shelf fronts off the coast of New England (USA) have shown that along-shore wind stress, if the latter is directed against the quasi-geostrophic along-front current, can lead to a sharp decrease in the slope of the frontal interface and to a complete destruction of the front. Csanady's simple theory [100] based on the effect of wind-generated disturbances on a geostrophic regime gives us a realistic first-order description of the behaviour of shelf fronts under the influence of the wind. Calculations based on this theory show that very strong impulses* of a longshore wind in the neighbourhood of $10^2 m^2/s$ are required to cause significant changes in the slope of the front; this corresponds to the action of a tangential wind stress of 1 Pa over a period of 28 hrs. The situation may differ in lakes where in calm conditions and with no tidal mixing the heating of the water during the spring- summer period is most intense in the shallow areas near the shore. Intense longshore winds, which are accompanied by the Ekman transport effect, can separate from the shore elongated lenses of light, heated or river-freshened water, on the edges of which discharge-type fronts are formed (see sub-section 3.3.1). Depending on the spatial scales of the phenomenon, the Earth's rotation may or may not play an important part. In Csanady's model [100], developed specially for the conditions in Lake Ontario, the typical depths near the shore $(25-30\ m$ at 3 km from the shoreline) determine the importance of the effect of the Earth's rotation, which leads to the formation of an along-front quasi-geostrophic jet with a speed of about 0.2 m/s in a lens $3-4\ km$ from the shore. The formation of such lenses requires smaller longshore wind impulses, in the neighbourhood of 2.5 m^2/s, which

---

* In this case, we have in mind the continuous (for a time interval t) action of tangential wind stress, distributed throughout a water column of unit area and per unit of mass.

is equivalent to the action of a tangential wind stress of 0.1 Pa for a period of 7 hrs.

On even larger scales, we have Roden's data [215] which are based on satellite information and show that it takes about one month with a wind-generated Ekman horizontal velocity gradient of about $2.5 \cdot 10^{-7} \ s^{-1}$ for the horizontal temperature gradient to double under the influence of the wind in climatic frontal zones with a width exceeding 100 km. This does not differ from our own data (section 2.5) on the deformation rate (D) in climatic frontal zones. Therefore, it appears that seasonal variations about the background deformation rate which maintains climatic frontal zones in a quasi-steady state, have the same order as the background, and so can induce purely seasonal fluctuations of horizontal gradients in these zones. However, the main variability within large-scale frontal zones is probably associated with the synoptic processes in the ocean itself at values of D in the neighbourhood of $10^{-5} - 10^{-6} s^{-1}$, and has scales smaller than 100 km and shorter than 1 month.

2.8.2 Wave-like instability at fronts

The balanced nature of frontal interfaces (geostrophic, cyclostrophic, isostatic) itself suggests the possibility that the main dynamic balance could be disturbed and near-frontal perturbations of a fluctuating (wave) nature could appear. However, such perturbations could later die down, or acquire a quasi-steady character, or begin to increase indefinitely in amplitude. In this case, we speak of the instability of these disturbances. Researchers would first of all like to know under what conditions these perturbations become unstable. Investigations have shown (see [198]) that two types of instability, barotropic and baroclinic, can be distinguished, depending on the energy source for the growth in the disturbances. These terms are largely conditional, and are related to the fact that, at first, theoretical investigations on the instability of perturbations were carried out on extremely simplified configurations of mean flows, i.e. 1) a flow without vertical changes in velocity, but with a horizontal velocity gradient ("barotropic" model), and 2) a flow without horizontal velocity changes, but with a vertical gradient ("baroclinic" model). Orlanski [198] correctly notes that the use of the adjectives "baroclinic" and "barotropic" to define the types of instability does not at all signify this or that configuration of the basic flow field. For example, a barotropically unstable motion does not have to be characterized by a depth-invariable velocity.

For a better understanding of the physical nature of the processes of barotropic and baroclinic instability, let us turn to the equation for the energy of a large-scale zonal geostrophic flow containing small velocity and pressure perturbations, the initial cause of which does not interest us in this case. In accordance with [30], this equation is written in the following way in the Boussinesq approximation:

$$\frac{\partial}{\partial t} \int_V \left(\frac{u'^2 + v'^2}{2} + \frac{g^2}{N^2 \rho_0^2} \frac{\rho'^2}{2} \right) dV = \int_V (-u'v') \frac{\partial U}{\partial y} dV + \int_V \left(-\frac{g^2}{N^2 \rho_0^2} v' \rho' \right) \frac{\partial \bar{\rho}}{\partial y} dV.$$

(2.79)

Here, all the values with a prime are related to the perturbations, while U and $\bar{\rho}$ characterize the distribution of velocity and density in the unperturbed zonal flow, so that

$$u = U + u'; \quad v = v'; \quad \rho = \bar{\rho} + \rho',$$

where U is the time-averaged mean zonal current, and ρ_0 in (2.79) is some constant mean density, all the deviations from which are on the whole small. N is the mean Väisälä-Brunt frequency with respect to the depth of the flow.

Integration in (2.79) is carried out over the volume V, determined arbitrarily to be a selected segment of a zonal canal with width L and depth H, in which the given current is located.

The left hand side of equation (2.79) reflects the rate of change in the energy of the perturbations, whereas the right hand side contains two terms which show the sources of this change. If the perturbation energy increases, then these two terms on the right hand side correspond to the two energy sources for this growth. The first source is related to the inhomogeneity of the velocity of the unperturbed zonal flow $\partial U/\partial y$ in the transverse direction. The growth of perturbations as a result of this energy source is called **barotropic** instability. The perturbations in this case draw their kinetic energy ($KEED$) from the kinetic energy of the time-averaged large-scale motion (KEM), i.e. in essence, we observe the action of Reynolds stresses. The second energy source for perturbations is related to the horizontal density gradient $\partial \bar{\rho}/\partial y$ across the flow. Because $N \neq 0$, the mean density gradient $\partial \bar{\rho}/\partial y$ also changes with depth. We observe a baroclinic effect, and so this type of instability is called **baroclinic** instability.

The available potential energy (APE), which is a part of the total potential energy (PE) in the zonal geostrophic flow, serves in this case as the source of kinetic energy of the growing perturbations ($KEED$). Therefore, a number of authors use $KEM \to KEED$ as a signature of barotropic instability, and $PE \to KEED$ as a signature of baroclinic instability [49].

If we base our reasoning on equation (2.79), then we have to admit that the sharper the quasi-geostrophic density frontal interface, i.e. the more $\partial\bar{p}/\partial y$ and $\partial U/\partial y$ are related to it, the greater the amount of energy (potential and kinetic) that can become available for the development of instability at the interface. However, the possibility of this occurring depends on numerous factors, including the detailed spatial structure of the velocity and density field in each specific case. Nevertheless, it seems indisputable that large values of $\partial U/\partial z$ and $\partial\bar{p}/\partial z$ at quasi-geostrophic fronts of a relatively small scale should make the latter unstable both in a barotropic, and in a baroclinic, case. The possibility of the growth of barotropically unstable perturbations in large-scale zonal currents is usually disputed on the basis that it is the synoptic-scale eddies that most commonly provide kinetic energy to the mean flow $(KEED \rightarrow KEM)$, and not vice versa. Similar claims for smaller scales are based mainly on the arbitrary transfer of the ideas of two-dimensional geostrophic turbulence to phenomena in the real ocean where, strictly speaking, two-dimensional turbulence does not take place. Therefore, it cannot be ruled out that a diametrically opposite situation holds true for fronts and jet streams, of scales significantly smaller than the scales of synoptic eddies, which could have emerged as a result of the deformation fields of these eddies. In any case, this is not quite clear from past observations, and therefore deserves special research.

Considerable mathematical difficulties are encountered when studying instability in a general case where a given flow shows both vertical and horizontal velocity gradients. There are certain advantages to studying two–layer or three–layer stratification [267, 268] which retains the effect of the baroclinicity of the mean flow, makes it possible to take the horizontal velocity gradient into account at the same time, and greatly simplifies mathematical interpretation. It should be said that a two–layer model (and also a three–layer model when the parameters are properly selected) corresponds better to the conditions observed near oceanic fronts, than do the models with continuous stratification. Quite realistic from this point of view is the two-layer model selected by Orlanski [198] as an approximation to atmospheric fronts. The results obtained by him for the frontal range of Kibel–Rossby numbers $Ki \leq 3$ and Richardson numbers $Ri \leq 5$ are applicable to oceanic fronts of the first category (see table 1.3 in section 1.3), and have already been used by Woods during a discussion of the role of fronts in the transfer of the "energy of variance" of the physical fields of an active oceanic layer through a cascade of scales [260]. These results can be summed up in the following way:

a) unstable perturbations exist at all wavelengths;

b) Rayleigh's shear instability (barotropic in nature) occurs at very low Ri values, i.e. when the difference between the densities on both sides of the front becomes almost non-existent, and the frontal interface is approximately vertical;

c) at low Kibel-Rossby numbers ($Ki \to 0$), a geostrophic baroclinic instability typical of long waves (wavenumber $k \to 0$) develops. However, if the Ri values are also small, perturbations which are unstable in the Rayleigh (barotropic) sense can also occur;

d) an increase in Ri due to a growth of the density gradient simultaneously with $Ki < 1$ combines Helmholtz instability with geostrophic baroclinic instability; appearing as a hydrostatic instability, due to the vertical velocity shear in the field of gravity;

e) geostrophic baroclinic instability prevails for small Ki values and large Ri values;

f) Helmholtz instability prevails for high Ki values and low Ri values. This corresponds to an increase in the role of the vertical velocity gradient.

On the other hand, Orlanski notes the difficulty of interpreting results with $Ri > 2$ and $Ki > 1$, which in essence means that the nature of the instability appearing at the sharpest medium- and small-scale fronts (exactly in the region where the geostrophic regime becomes ageostrophic) is not quite clear. Woods [260] classed the frontal waves with a length of $\lambda \approx 8 \ldots 10$ km which he observed in the Mediterranean Sea [259, 266] as baroclinically unstable waves of the sort discussed by Orlanski in the case of $Ki \to 0$ and $k \to 0$. Indeed, if we accept the difference in velocities ($\Delta U = U_2 - U_1$) on both sides of the Maltese front as being equal to $5 - 10$ cm/s, the value of the Kibel–Rossby number* $Ki = [(U_2 - U_1)k]/(2f)$ with a perturbation wavenumber of $k \approx 2\pi \cdot 10^{-6} cm^{-1}$ will be of the order $(1.5 - 3) \times 10^{-1}$. However, data [266] on the slope of the frontal interface $(1/300 - 1/400)$ with a density gradient across this front of approximately 0.5×10^{-3} [94] suggest a somewhat greater horizontal velocity difference of about 20 cm/s at the front, which brings the value of Ki closer to 1. The value of Ki for the Maltese front is approximately 1 if it is determined as $Ki = \Delta U/(fL_\phi)$, as was the case in [191] for the California Undercurrent. In this case, with $\Delta U = 20$ cm/s, $L_\phi = 2$ km (the width of the frontal zone with the sharpest horizontal temperature and salinity gradients [266]) and $f = 0.8 \cdot 10^{-4}$, $Ki = 1.2$. With the same data, the value of the Richardson frontal number $Ri = gH(\rho_1 - \rho_2)/\rho[(U_2 - U_1)^2]$ is approximately 12. Therefore, one cannot exclude the fact that fronts of the same type and scales

* As determined in [198].

as the Maltese front fall precisely into the parameter range ($Ri > 2$ and $Ki > 1$) where the results of an analysis such as [198] cannot be fully interpreted. On the other hand, Orlanski's analysis does not exclude the existence in this parameter range of unstable quasi–horizontal wave motions due to horizontal velocity shear, i.e. **barotropic instability**. Equally interesting in this respect is Flagg and Beardsley's study [115] of the near-shore front which separates the slope and shelf waters off the Atlantic coast of North America. The scales of this front are similar to the scales of the above-mentioned Maltese front and, in accordance with [269], fall into the Kibel–Rossby number range of $0.02 - 0.13$ and Richardson number range of $20 - 120$. As the authors of the study have shown, the wave-like perturbations observed at the front cannot be attributed to baroclinic instability for realistic values of the bottom slope which has a stabilizing effect on the front. With a flat bottom, baroclinically unstable waves can develop at the front. However, their growth can be prevented by the migration of the front in the direction of the continental slope where the slope of the bottom increases sharply. In this case, extremely high Ri values should prevent the development of Rayleigh (barotropic) instability even with a flat bottom. However, it should be said that there has always been a tendency in the literature to underestimate the value of the horizontal velocity difference $\Delta U = U_2 - U_1$ at fronts. When assessing the frontal Ri number, most authors accept $\Delta U \simeq 5 - 10 cm/s$, whereas a number of records indicate that the horizontal velocity difference can be so great that small vessels go off course when crossing oceanic fronts. No conclusive opinions can be formed on this question due to the lack of reliable measurements with a good spatial resolution of the velocity field near fronts. However, if we accept that ΔU at mesoscale fronts is only 3 times greater than the value usually accepted ($15 - 30 \ cm/s$), then the frontal Ri number with the same density gradients should be an order of magnitude smaller, which makes the barotropic instability of mesoscale fronts quite probable.

The above judgements stem from the fact that baroclinic instability* ($PE \rightarrow KEED$) is accepted as the only or main process of instability in the majority of papers on the instability of oceanic jet streams and fronts that have been published over the past ten years. In the opinion of D.G. Seidov (personal communication), this is due largely to the use of two-layer models which automatically determine the baroclinic (internal) Rossby radius of deformation through the thickness of the layers. In turn, this radius predetermines the characteristic magnitude (and the wave number k) of the eddies, the nature of which is largely determined by the

---

* The exceptions so far are [244] and [268].

value of the Kibel-Rossby number which depends on k. As a rule, the values of k in this case are small, which accords with the case where $Ki \to 0$ and $k \to 0$, for which, according to [198], purely baroclinically unstable perturbations develop. The opinion that barotropic instability of oceanic jet streams and fronts can be just as real and important as baroclinic instability has been expressed with increasing certainty over the past $2 - 3$ years. For example, D.G. Seidov's numerical eddy-resolving model of oceanic currents [49] shows that processes of the $PE \to KEED$ and $KEM \to KEED$ type are equally important for the energetics of synoptic-scale processes, since the maximum energy fluxes for these transfers are positive and characterized by similar values of the order $(2 - 4) \times 10^{-7} \; W/m^3$. It is interesting to note that some parts of the region modelled are in this case characterized by baroclinic instability, while others are characterized by barotropic instability.

Before concluding this section, we should look at the criteria for barotropic and baroclinic instability. According to Kuo [165], unstable disturbances can occur in a barotropic system if the gradient of absolute vorticity $(\beta - \partial^2 U/\partial y^2)$ changes its sign at least once across the width of the background zonal flow. Here, β is the latitudinal gradient of the Coriolis parameter, U is the zonal mean flow, and the y-axis is perpendicular to the flow axis. For the majority of fronts with a small width or a direction differing from zonal, β can be regarded as small in comparison with $\partial^2 U/\partial y^2$, and the possibility of barotropic instability can be judged by the behaviour of $\partial^2 U/\partial y^2$. In the case of a two-layer model in which the thickness of the upper layer h is much smaller than the thickness of the lower layer, a geostrophic jet stream with a width L in the upper layer is barotropically unstable if the gradient of its potential vorticity

$$\frac{\partial^2 U}{\partial y^2} - \frac{f}{H_m} \frac{\partial h}{\partial y} = \frac{\partial^2 U}{\partial y^2} - \frac{f^2 U}{g' H_m} \qquad (2.80)$$

changes its sign within the range $0 \le y \le L$ [235]. Here, $U(y) = (g'/f)(\partial \overline{h}/\partial y)$ is a geostrophic approximation in which $g' = g(\rho_2 - \rho_1)/\rho_1$ is the reduced gravitational acceleration, f the Coriolis parameter, and $h = \overline{h}(y) + h'(x, y, t)$ the total thickness of the upper layer. In turn, H_m is the mean thickness of the upper layer, $H_m = \frac{1}{L}\int_0^L h(y)dy$, over the width of the background flow L. Analysis of the equation (2.80) shows that for some given $U(y)$ the value of H_m will always be small, and the expression (2.80) will be negative throughout the range $0 \le y \le 1$. For large values of H_m, the term $\partial^2 U/\partial y^2$ will be the determining one in (2.80), as in the case where there is no rotation. Under these conditions, a symmetrical barotropic jet stream with no lateral boundaries is unstable. Therefore, all other conditions being equal, the stability of a quasi-geostrophic jet stream in a two-layer system depends on H_m. Consequently, the criterion of flow stability in a two-layer system will depend only on one dimensionless parameter, $F_1 = L/\lambda_1$, where $\lambda_1 = (g'H_m)^{\frac{1}{2}}/f$ is the baroclinic Rossby radius of deformation based on the thickness of the upper layer, and so $F_1^2 = f^2L^2/(g'H_m)$. The numerical value of the criterion depends on the width, form and depth of the near-frontal jet stream and the density difference $\Delta\rho = (\rho_2 - \rho_1)/\rho_1$ associated with it. For F_1 values below a certain critical level, the flow should be barotropically unstable [235].

Stern gives the following criterion for the baroclinic instability of long-wave geostrophic perturbations of a jet stream in a two-layer system where the thicknesses of both layers are comparable [235]:

$$F_1 F_2 \geq \pi^2/2. \tag{2.81}$$

Here $F_2 = L/\lambda_2$ and $\lambda_2 = [g'(H - H_m)]^{\frac{1}{2}}/f$ where H is the total depth of the basin.

In Stern's opinion [235], the dependence of both criteria (for baroclinic and barotropic instability) on the parameter F_1 is not accidental, but reflects the fact that in the two-layer system with the given configuration, the ratio of available potential energy to available kinetic energy APE/AKE is proportional to the value $f^2L^2/(g'H_m)$. With fairly small Kibel-Rossby numbers $U/(fL)$, Stern found that

$$\frac{APE}{AKE} \simeq \frac{f^2L^2}{3g'H_m} \simeq \frac{L^2}{3\lambda_1^2}. \tag{2.82}$$

It follows from (2.82) that in broad geostrophic zonal flows ($L \gg \lambda_1$), the available potential energy (APE), from which baroclinically unstable perturbations can draw energy, should prevail. At the same time, the available kinetic energy (AKE) should prevail in narrow near-frontal flows ($L \ll \lambda_1$).

According to Wright's results [267, 268], the necessary condition for baroclinic instability in a three-layer system is a change in the sign of the potential vorticity gradient either in one of the layers, or during the transition from one layer to another. In Wright's model, the equations for the gradients of potential vorticity in three layers contain the parameters F_1, F_2 and F_3, which are similar to the parameters F_1 and F_2 derived by Stern in the two-layer model. Therefore, the physical significance of the results obtained by Wright can be easily understood on the basis of Stern's more simple reasoning [235].

Of course, the above general information is not enough for us to form in each specific case an opinion regarding the stability of frontal interfaces and the nature of the perturbations which could emerge at them and grow. A specific example that requires detailed diagnosis, which cannot yet be carried out because of a lack of field data, is given in section 4.3. Special investigations in each specific case, which include the design of adequate models, as well as the general development of eddy-resolving and front-resolving numerical models of oceanic currents, promise to shed new light on instability and eddy formation at fronts in the near future. Programs of ship and satellite observations of oceanic fronts should be organized and coordinated so as to give theoreticians and specialists in numerical modelling the necessary initial data and material for making predictions.

2.9. On the two important functions of Ekman boundary layers

As we have already seen in sections 2.7 and 2.8, it is especially important to take into account the effects of turbulent friction when studying the dynamics of the quasi-steady state and the evolution of frontal interfaces. We would therefore like to devote a little more attention to this problem.

So far, the clearest ideas regarding the role of turbulent friction have been obtained from field data from the frontal interfaces of coastal upwelling [185, 186, 187]. In this case, the wind-generated Ekman transport in the near-surface frictional layer, which is directed along the normal from the shore towards the open sea, causes the submergence of cold upwelled waters under the frontal interface (see fig. 2.38), and contributes to the appearance of divergence and upwelling at some distance from the frontal interface on the side of the open sea. This leads to the appearance of a so-called "two-cell" circulation which also includes the internal Ekman layers that form on both sides of the frontal interface due to intensive downwelling, and an along-frontal current with a significant transverse gradient of velocity. We can say that the cross-shelf circulation established in this case, which is schematically depicted in fig. 2.38 on the basis of data from [187] with some modifications by the author, is determined in the quasi-steady case by the combined effect of the Ekman transport in the near-surface, near-bottom and internal (near-frontal) boundary layers. The interesting and commercially important characteristics of the development of near-frontal ecosystems are related to this type of circulation (see sections 5.1 and 5.2).

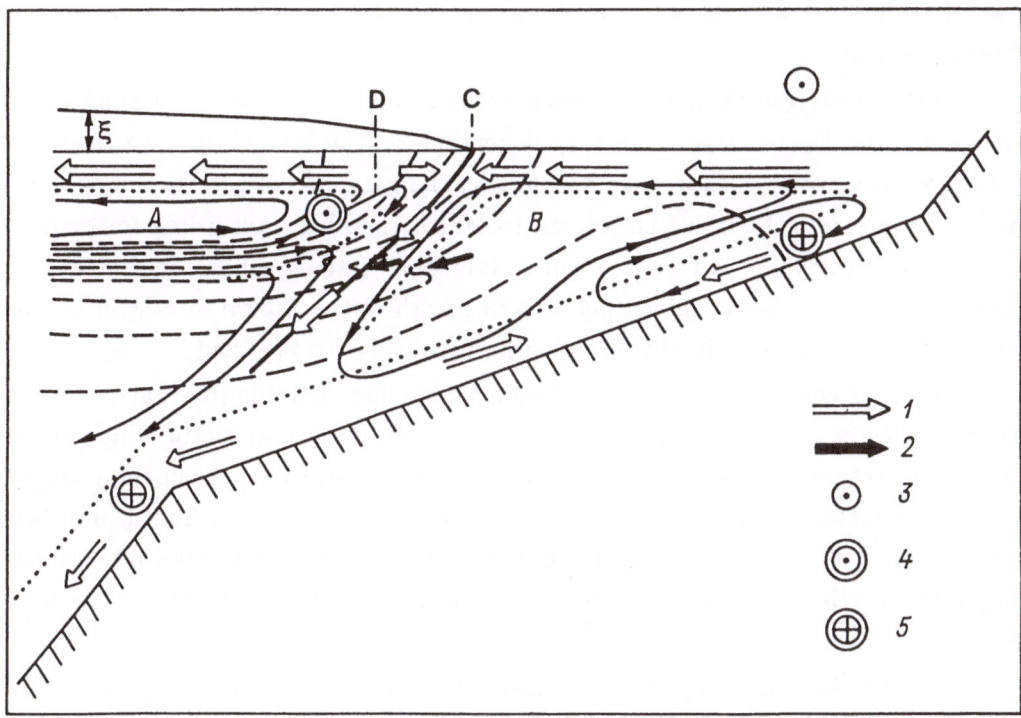

Fig. 2.38. A diagram of "two-cell" circulation in a coastal upwelling. The main cells of circulation in the vertical plane perpendicular to the shore are marked A and B. The thin solid lines with arrows indicate the general direction of flow. The dotted lines mark the borders of the Ekman boundary layers. The dashed lines mark the isotherms. The bold sloped line indicates the frontal interface. The letters D and C mark the positions of divergence and convergence respectively. Other notations: 1 – direction of flux in the Ekman layers; 2 – direction of intrusion of transformed upwelled waters across the frontal interface; 3 – wind directed out of the page; 4 – axis of the jet stream directed out of the page; 5 – axis of the jet stream directed into the page; ξ – elevation of the sea surface above the unperturbed surface.

As indicated by Kao [156, 158], Ekman boundary layers near the surface can also emerge in the absence of wind, due to a high velocity gradient at the front in the near-surface layer. The vorticity that develops in this case can be balanced only by viscous forces. On the basis of the fact that frontal interfaces are only slightly sloped toward the horizontal ($tan\alpha = 0.01 \cdots 0.001$ [71]), the effects of turbulent viscosity and turbulent heat and mass transfer are best taken into account by means of the coefficient (or coefficients) of vertical turbulent exchange. However, the reader must have already noticed in section 2.7 that Kao [156] preferred not to specify the directions of turbulent exchange in his papers on the dynamics of a quasi-steady frontal interface.

As the investigations and analyses of some authors have shown, the thicknesses of the internal Ekman layers can vary depending on the coefficient of vertical turbulent viscosity K_z which in turn depends on the gradients of geostrophic velocity and density across the front, i.e. on the frontal Ri number. According to the data of [144], this thickness can reach 15 m in typical situations, while the typical flow rate in the Ekman layers (given per unit of length of the frontal interface) can be of the order of 0.3 m^2/s in the direction perpendicular to the front.

Thus, we come to our first conclusion by stating that the internal boundary (Ekman) layers adjacent to frontal interfaces are an important factor which determines the nature of near-frontal circulation. The Ekman transport in these layers, which is directed normal to the front, helps to maintain a sharp frontal interface when combined with other dynamic and thermodynamic effects (the influence of nonlinear inertial components, turbulent entrainment, density increase during mixing).

However, the study of physics of internal Ekman layers is not only important because of their importance to near-frontal circulation. The second important function of the Ekman layers is associated with their contribution to the evolution of thermohaline finestructure, the intrusive elements of which are especially characteristic of frontal zones [62]. One can assume that in real, unsteady and inhomogeneous conditions, Ekman transport is one of the main agents of effective cross-frontal heat and mass exchange and horizontal exchange of momentum. In this case, and also bearing in mind the slope of any frontal interface, we can say that the coefficients K_ℓ of horizontal mixing inevitably depend on the coefficients K_z of vertical mixing. Perhaps this is precisely why Mooers [185] found sufficient reason to disregard horizontal turbulent mass and momentum exchange when studying the fundamental dynamics of fronts. It should be said that this relationship became quite clear in a study of the effect of the density increase during mixing near fronts (see section 4.5).

Stommel and Fedorov [238] were the first to note the significant role of internal Ekman layers in the dynamics of finestructure density inhomogeneities in the ocean. Among other things, they showed that the flow of fluid within a homogeneous lamina located in a baroclinic zone is governed by a geostrophic balance, whereas frictional Ekman boundary layers through which momentum is exchanged between motions inside and outside the lamina form at its boundary with the surrounding stratified medium. It is precisely these layers that determine the effective diffusion of the lamina in the cross-frontal direction. A complete analysis of the evolution of a large-scale intrusion and an isolated mixed patch in a uniformly stratified fluid, with the Earth's rotation and the role of the Ekman boundary layers taken into account, is presented in a paper by Zhurbas and Kuz'mina [20] which is recommended to readers with a special interest in this question.

Chapter 3

CHARACTERISTIC FEATURES OF OCEANIC FRONTS

3.1 Eddies and fronts in the ocean

One of the most important advances in the study of oceanic fronts was the realization of the fact that oceanic eddies play an important frontogenetic role of the sort which has been known for a long time in the atmosphere. The reason oceanographers lag behind in this matter is obviously related to the fact that, as we have mentioned in section 2.4, there is no complete similarity between conditions and manifestations of frontogenesis in the ocean and in the atmosphere. Another important advance was the comprehension of the role of eddies and fronts in the transfer of the energy of water motion, and in the variability of hydrographic fields, over a range of scales in the ocean [262]. The descriptive side of the interaction of oceanic eddies and fronts is discussed below.

Both from the dynamical and descriptive points of view, large meanders at the main oceanic fronts, such as the Gulf Stream front, bear the greatest similarity to atmospheric cyclones (fig. 3.1). In a meander of this type, one can observe the analogue of the atmospheric "warm sector" (this is the "cold sector" in the ocean), and the configuration of a wave-like deformed front is analogous to a combination of a warm and a cold front in a developed atmospheric cyclone. In addition to this similarity, both the meanders of oceanic fronts and atmospheric cyclones migrate along the front at which they arise.

Unlike frontal meanders, synoptic eddies of the open ocean (which were studied in the "Polygon-70", MODE-1 and POLYMODE experiments) and "rings" or frontal eddies [164] completely separated from the Gulf Stream or the Kuroshio Current, as well as small but intensive cyclonic eddies $20 - 50$ km in diameter which develop at frontal interfaces or near them [11, 50, 208, 231] (see also sections 3.1.2, 4.1 and 4.3) possess more or less independent dynamics and their own strong deformation field, the effect of which should influence the characteristics of the medium in which they migrate. Basically, we can speak of two sides of the problem, a) the frontogenetic role of eddies, and b) the eddy-forming role of fronts. This section discusses field data related to both sides of the question.

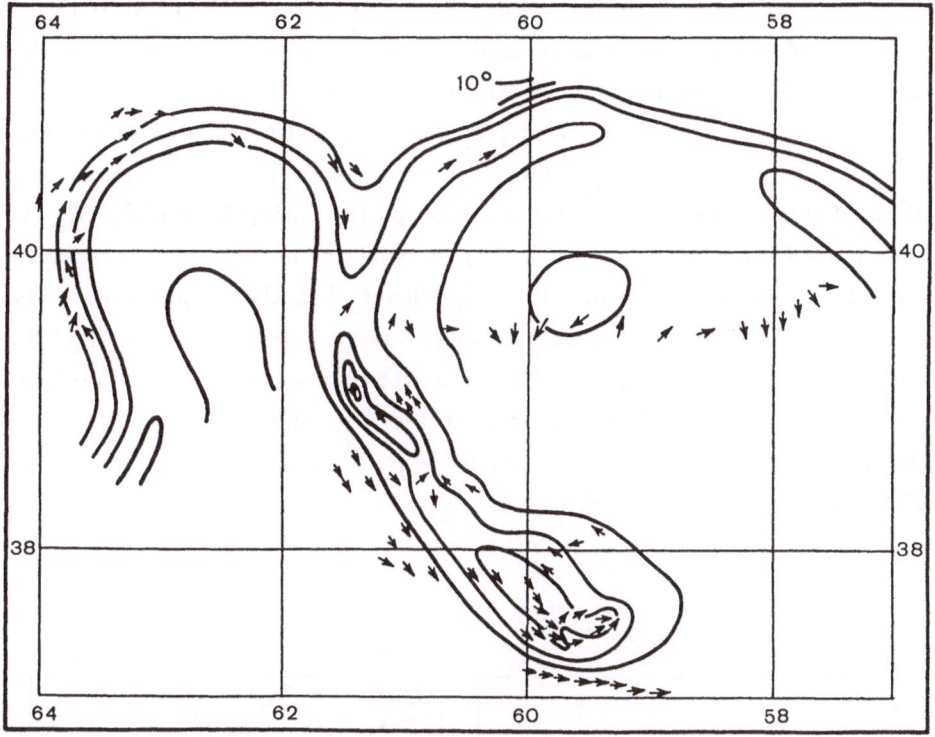

Fig. 3.1. The large cyclonic meander "Edgar" at the main front of the Gulf Stream as seen in the temperature field of the surface layer of the ocean, based on data from ref. [122]. The meander is in the process of separation and transformation into a cyclonic ("cold") ring. The isotherms run through 5.5°C. The arrows show the directions of the currents, based on measurements taken with a geomagnetic electrokinetograph from a moving vessel.

The 7-month series of data obtained in the Soviet experiment "Polygon-70" is the first fairly complete collection of hydrographic data which point to a close relationship between the synoptic eddies of the open ocean and oceanic fronts [33, 72]. Throughout the entire experiment, the author conducted an extensive program of measuring the thermohaline finestructure with the help of an AIST probe in the upper 500 – 600-metre layer of the ocean. At first, it seemed strange that the region of the most clearly defined finestructure was constantly shifting westward from the eastern side of the polygon. Later, the explanation for this strange phenomenon of "a migrating finestructure" came when we detected a synoptic-scale anticyclonic eddy shifting westward across the polygon in the velocity field [33], and a sharp thermohaline front in the temperature and salinity fields [72]. Both the eddy, the front and the finestructure proved to be very closely interrelated, i.e. the front was generated by the deforming action of a moving eddy on the local hydrographic fields, finestructure was actively formed at the frontal interface, and together as a whole they shifted westward with the phase velocity of the eddy $(3 - 6\ cm/s)$. The combined figures from [33] and [72] depicting the time of eddy detection (3 July 1970) (fig. 3.2, a and b) show, with some additional notation, the relative positions of the front and the boundary between the anticyclonic eddy (A) and the cyclonic eddy formation following it (C), as well as the averaged current vectors near the front for 3 July 1970. It is significant that the anticyclonic eddy was traced, from the vectors of current velocity, down to $1000 - 1500\ m$ (fig. 3.2, a), whereas a cyclonic horizontal velocity shear was observed in the near-surface layer $(25\ m)$ near the front, as was to be expected (fig. 3.2, b). A similar situation was later observed in the Sargasso Sea by M.N. Koshlyakov et al. [32] during the POLYMODE experiment in 1978.

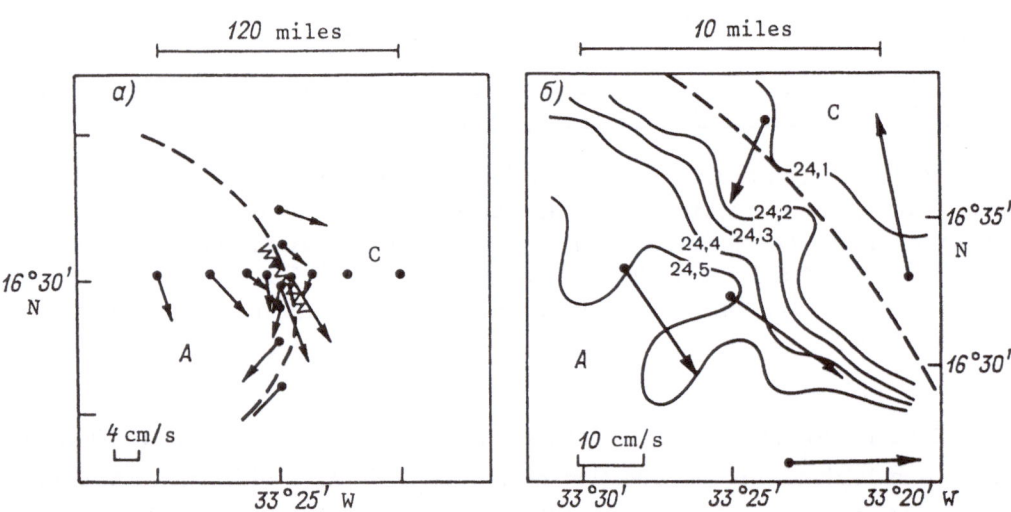

Fig. 3.2. a – daily average current vectors at the 1000-metre level at a polygon in the Tropical Atlantic on 3 July 1970, based on the data of instrumental measurements. The dashed line marks the rear boundary of the anticyclonic eddy (A) passing across the polygon from east to west. The zigzag line shows the position of the frontal interface in in the upper layer of the ocean (for greater detail see fig. 3.2 b). The scale of the vectors is indicated in the lower left-hand corner of the diagram; b – isotherms in the region of the front at the isopycnal surface $\sigma_t = 25.00$ $(50 - 60\ m)$ on 3 July 1970. Arrows indicate the daily average current vectors at the $25\ m$ level at the buoy stations closest to the front. The cyclonic vorticity of the near-frontal currents is clearly visible.

The physical nature of the observed relationships was not immediately understood. At first it was necessary to understand how the geostrophic eddy developing in the fairly deep main thermocline ($100 - 1000 \; m$) could affect the dynamics of the near-surface layer in which fronts most frequently form.

3.1.1. Frontogenesis in synoptic eddies

In 1976, Stommel and Kozyol [239] expressed the view that a mesoscale oceanic eddy localized in the main thermocline, below the vertically mixed layer exposed to the effect of the atmosphere, causes horizontal advection of waters with different temperature in this zonally inhomogeneous layer. As a result of the advection in the quasi-homogeneous layer the sharpening of temperature inhomogeneities and the formation of fronts occurs on the periphery of the eddy.

A zonally-varying temperature field in the surface layer of the ocean is just an exceptional case which corresponds to average climatic conditions. Basically, an eddy migrating in the thermocline can get into a region of the ocean where there are randomly oriented local temperature and salinity gradients and where frontal zones or previously formed local fronts can exist in the surface layer. The resulting deformation of local gradients of scalar properties should depend on the combination of their orientation in the surface layer of the ocean, the direction of the eddy motion and its orbital velocities. This type of generalized approach to the problem has never been taken by anyone, though the best approximation to it can be found in ref. [47]. This paper is devoted to the creation of a mathematical model of the evolution of a temperature field in the surface mixed layer of the ocean with mesoscale disturbances of the velocity and temperature field present in the underlying thermocline. In fact, the study is based on the same idea expressed earlier by Stommel and Kozyol [239]. However, the authors [47] successfully utilize an integral model of the surface layer of the ocean, in which diffusive processes are included in the zone of maximum density gradient. This provides us with far more accurate information from two-layer modelling of the thermal structure of the surface layer. For the sake of simplicity, the unperturbed temperature field in the surface layer is taken to be zonally homogeneous with a horizontal temperature gradient of approximately $0.007°C/km$ in the meridional direction. The background temperature distribution in the thermocline is taken to be horizontally homogenous in all directions. The temperature distribution everywhere fully determines the field of density. The motion in the main thermocline below the homogeneous layer is regarded as completely isopycnic and baroclinic. The task boils down to numerical

solution of the system of non-linear differential equations for the complete flow and the temperature field in the surface layer, in response to the assumed dome-shaped disturbance of the temperature field in the thermocline representing an individual cyclonic eddy. Those interested in the formal aspects of the model will find all the necessary details in [47]. Only the results obtained will be discussed here. The process of rearrangement of the temperature field takes 6 − 7 days. In the case of a stationary disturbance, the structure of the deformed temperature field is characterized by the appearance of warm and cold water "tongues" on the left and right peripheries of the eddy (fig. 3.3). These are formed as a result of the meridional advection of waters with a different temperature. Thermal fronts which separate the region of the disturbance from unperturbed motion form on the edges of the tongues. For an eddy moving westward, the symmetry of the pattern is disrupted. The cold tongue in front of the eddy narrows, while the warm tongue in the rear expands. The fronts sharpen significantly, and shift to the south and north respectively (fig. 3.4). Behind the eddy we observe a wake resembling a fairly broad region of approximately homogeneous temperature.

Fig. 3.3. Temperature maps of the ocean surface in the case of a stationary cyclonic eddy. Figures in top corner − day of integration. Diagram taken from [47].

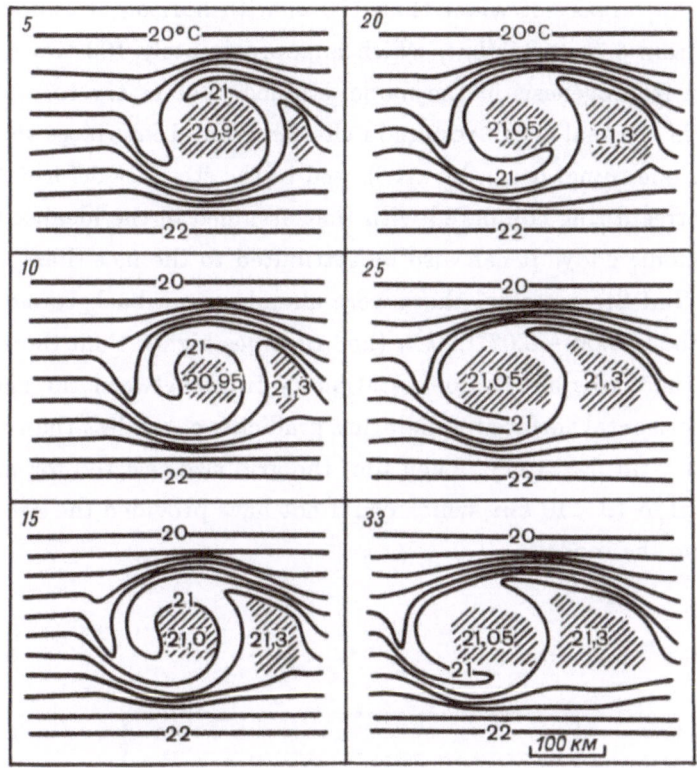

Fig. 3.4. Temperature maps of the ocean surface in the case of a cyclonic eddy moving westward with a speed of 5 km/day. Figures in upper left-hand corner – day of integration. Diagram taken from [47].

It should be said that the numerically modelled general patterns of the near-surface deformation of the temperature field due to the action, from below, of an isolated eddy (fig. 3.3 and 3.4) do not have very much in common with the widely known patterns of frontal systems in atmospheric eddy formations. On the other hand, the qualitative similarity to those observed in the ocean in recent years by a number of authors [17, 21, 32, 224, 253] is really quite significant (see also figs. 2.8, 3.5 and 3.6).

The calculations of Nelepo, Kuftarkov and Kosnyrev [47] have shown that a cyclonic eddy with a diameter of 125 km which is moving westward with a speed of about 5 km/day results in a 30-fold local sharpening of the horizontal temperature gradient in a frontal zone measuring about 10 km in width (from 0.007 to 0.2°C/km). In the frontal zones of a warm Gulf Stream ring, we observed (see sub-section 3.3.3) areas not wider than 100 m with horizontal temperature gradients ranging from 5 to 20°C/km, which is approximately 100 times greater than the amount of frontogenesis in the model studied. Let us try to understand why frontogenesis in the real ocean results in sharper frontal interfaces than model calculations. For one thing, it can be attributed to the stronger deformation field of a typical Gulf Stream ring compared with that assumed in the idealised model of an individual cyclonic eddy. It can also be attributed to the fact that, in the vicinity of the warm Gulf Stream ring, there were already mean background temperature gradients of about $0.06 - 0.07°C/km$ (i.e. 10 times higher than those given in the model), and there was also a quasi-stationary front between the shelf and slope waters with horizontal surface temperature gradients of not less than 4°C/km [188] (see also front C in fig. 3.17). Then too, the grid size selected for model calculations was equal to 10×10 km, which could not have provided the necessary spatial resolution from the model.

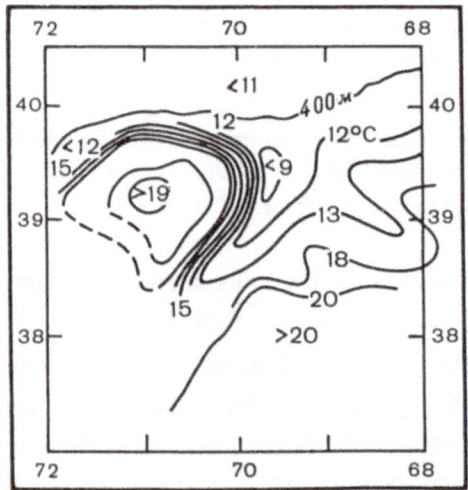

Fig. 3.5. Flow of cold water along the eastern periphery of an anticyclonic ("warm") ring north of the Gulf Stream, based on the results of an aerial survey carried out by Saunders [224] with the help of an IR radiometer and aerial expendable bathythermographs (AXBT) on 4 December 1969.

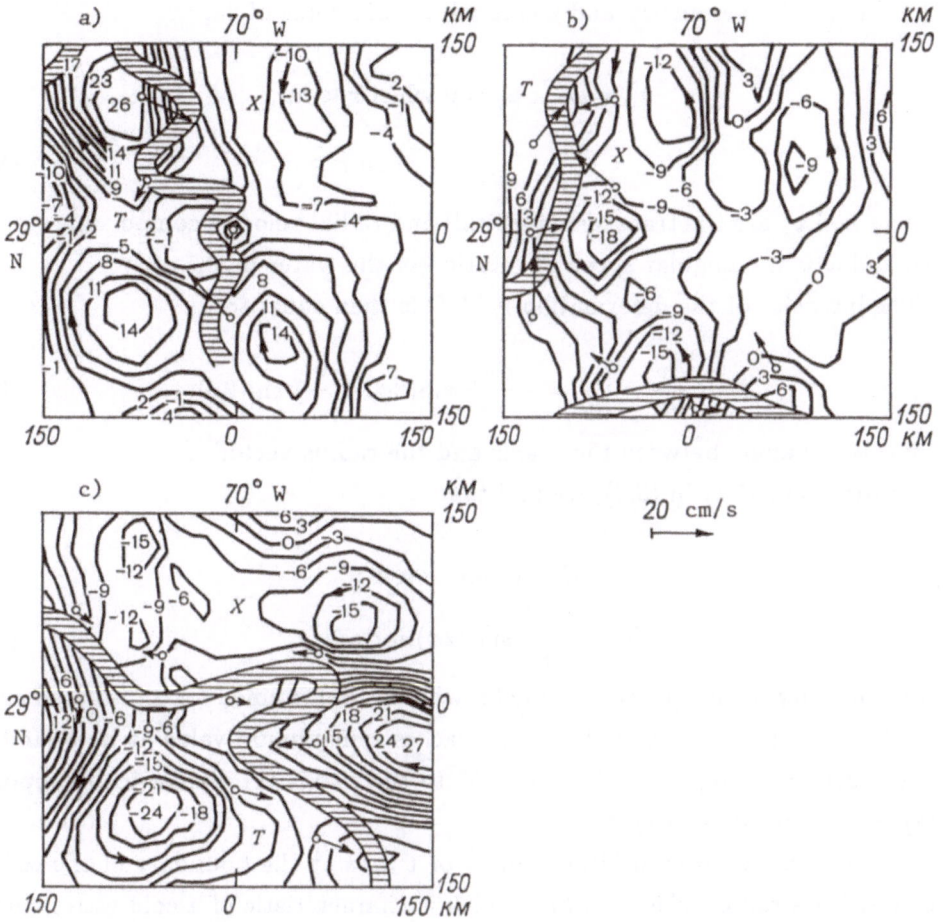

Fig. 3.6. Relative positions of fronts (hatched bands) and eddies (streamlines of the eddy component of the velocity field) at the polygon of the POLYMODE experiment (in $10^7 cm^2/s$), based on [32]. a – 21 February 1978, 100 m depth; b – 13 March 1978, 100 m depth; c – 31 March 1978, 25 m depth. Arrows (with the scale indicated to the right) show the vectors of the frontal component of velocity at the time of the survey, based on data from direct measurements at buoy stations.

This type of rough spatial resolution is also unsuitable for field studies of oceanic fronts [188].

In connection with this, I would like to give a rough estimate of the intensity of the deformation field D (see sections 2.5 and 2.6) for an isolated eddy. The fact that the orbital velocity diminishes right outside the core of the eddy can also produce high values of D. To illustrate this, N.P. Kuz'mina [38] examined a simplified system for the velocity distribution in an eddy, utilizing Rankine's model of an eddy from

[239] with circular symmetry and a core with a diameter of 2a, i.e.

$$u_\phi = \omega r; \quad u_r = 0 \text{ when } r < a;$$

$$u_\phi = \omega \frac{a^2}{r}; \quad u_r = 0 \text{ when } r \geq a, \tag{3.1}$$

where u_ϕ and u_r are the transverse (orbital) and radial velocity components respectively, and ω is the angular speed of rotation of the water in the eddy.

The intensity of the deformation field D is then equal to

$$D = \frac{\partial u}{\partial x} - \frac{\partial v}{\partial y} = -\frac{\partial u_\phi}{\partial r} \sin(2\phi) + \frac{u_\phi}{r} \sin(2\phi), \tag{3.2}$$

where ϕ is the angle between the x-axis and the radius vector r.

Substituting (3.1) in (3.2), we find that

$$D = 0 \quad \text{for} \quad r < a:$$

$$D = 2\omega \frac{a^2}{r^2} \sin(2\phi) \text{ for } r \geq a, \tag{3.3}$$

i.e. the intensity of the deformation field, which is determined for an isolated eddy, is equal to zero in the core of the eddy, reaches a maximum value at its boundary in zones located at angles of 45° and 225° to the x-axis (in diametrically opposite areas), and diminishes as $1/r^2$.

For example, with an orbital velocity of 1 m/s at the boundary of the core of an eddy with a radius of $a = 50$ km, which is characteristic of a cold water ring of the Gulf Stream, one can expect maximum values of $D_{max} \simeq 4 \cdot 10^{-5} s^{-1}$, which is significantly higher than the typical values mentioned in sections 2.5 and 2.6 and in [178].

At the present time, we have quite a quantity of experimental field data which indicate that oceanic fronts are related to quasi-geostrophic disturbances [17, 21, 32, 66, 253]. Beckerle [86] was one of the first to point out the predictable nature of frontal migration in the open ocean, and he even attempted, and quite successfully, to predict the moment of crossing of an oceanic front, identifying the velocity and direction of its migration with the phase velocity and direction of propagation of baroclinic Rossby waves. The migration of frontal systems westward with the eddies has been observed many times by the author, and noted by a number of other researchers, the velocity of propagation being in the neighbourhood of $3-4$ km/day in one case [32], and presumably not less than $5-7$ km/day in another [17, 66]. In practically all cases, the position of the fronts was peripheral with respect to the centres of the eddies. We find it interesting to study M.N. Koshlyakov's [32] time-series patterns of the positions of the frontal zone and eddy features at the polygon of the POLYMODE experiment, which are depicted in fig. 3.6. We can clearly see the advective effects of the eddy motion, which manifest themselves in the formation of tongue-like bends (meanders) of the frontal zone. Very sharp intrusive vertical stratification of waters often accompanies frontogenesis in synoptic eddies [32, 66]. This will be discussed in greater detail in sections 3.3.3 and 4.2. It would be very interesting to single out the geostrophic component of frontal currents from the total current velocity vectors measured in the surface layer of the ocean [32]. This component may be of the order of $10^{-1} m/s$, and greatly conceals the eddy nature of the motion in this layer.

As we can see from the examples given in subsection 3.3.3, an orbital eddy motion not only leads to frontogenesis, but can also advect already formed frontal interfaces (or their individual parts) over large distances. The appearance of this in the surface layer may not have been initially related to the frontogenetic effect of the eddy.

Summing up the results of our observations, we can say that a) as a rule, frontal interfaces in the surface layer of the ocean are formed at the peripheries of eddies; b) the frontal systems that appear migrate together with the eddies, undergoing an evolution related to the changes in the configuration and intensity of the eddy deformation field, as well as to the change in direction of the eddy motion; c) frontal geostrophic currents in the surface layer of the ocean can differ significantly from the orbital eddy motions of the deeper layers, as a result of which the general nature of the water motion in the upper layer turns out to be far more complex than the deep circulation.

It cannot be ruled out that eddy frontogenesis in the surface layer, which is accompanied by the development of independent frontal currents and by the formation and destruction of intrusive thermohaline finestructure, plays an important dissipative role with respect to the kinetic energy of quasi-geostrophic eddy disturbances in the thermocline.

3.1.2. Eddy generation at fronts

In subsection 3.1.1, we discussed various aspects of the frontogenetic role of oceanic eddies. This, so to speak, is only one side of the coin. The other lies in the fact that frontal interfaces themselves generate eddies, due to their own instability (see subsection 2.8.2). This process, which has been studied quite thoroughly for the atmosphere (see section 2.4), is still quite vague, even descriptively, as applied to the ocean. Only in the past $5-6$ years has some progress been made in this area as a result of the wider use of satellite data.

In our opinion, this progress is especially important in that it concerns medium- and small-scale eddies and fronts, since the processes of instability and eddy formation in large-scale latitudinal frontal zones of the ocean have been under discussion (by analogy with the atmosphere) in the literature since the middle of this century. Those interested in a detailed modern phenomenological description of eddy formation at a large-scale front can refer to Newton's paper [194] which presents the most complete physical picture of this process. Briefly, it can be described in the following way. The formation of eddies at such strong zonal geostrophic fronts as the Gulf Stream front is dynamically analogous to the formation of waves at atmospheric jet streams, and begins with the formation of a meander. The formation of a meander is usually associated with the asymmetry of the flow structure. When a cold tongue forms at the Gulf Stream front, the flow of cold waters directed southward (into the tongue) on its western side exceeds the outflow of cold waters northward (out of the tongue) on its eastern side. We observe the "filling up" of the meander with cold slope water, in the process of which the meander increases in size. The "filling up" usually takes about one month, and is accompanied by the downwelling of cold waters on the western side and around the tip of the cold tongue with velocities up to $10^{-1}cm/s$. According to Newton, the total volume of the sinking waters in this case amounts to $9 \cdot 10^6 m^3/s$. After "filling up", the meander closes as a result of the downwelling of cold waters in its neck, separates from the main current and turns into a cyclonic frontal eddy (cold ring), the subsequent lifetime of which may vary from several months to $1-2$ years. During this period, the cyclonic rings located south of the Gulf Stream usually migrate westward or to the southwest with velocities in the neighbourhood of $3-7\ cm/s$, interacting and often merging again with the Gulf Stream.

The formation of anticyclonic (warm) "rings" north of the Gulf Stream occurs in practically the same way. The literature contains descriptions of this process, based on highly detailed observations [224]. The merging of anticyclonic "rings" with the Gulf Stream after some time of independent existence occurs more frequently than in the case of cyclonic "rings". This is associated with the general configuration of the coastline and with the position of the main flow of the Gulf Stream in relation to the coast. The nearing of the southern periphery of anticyclonic "rings" to the main flow of the Gulf Stream, which precedes the merging, is accompanied by significant complication of the structure of the frontal zone and the appearance of numerous alternating frontal interfaces and a zone with a high horizontal velocity shear which forms as a result.

At the same time, special mention should be made of the obvious fact that cyclonic eddy formation at the Gulf Stream and Kuroshio fronts apparently takes place more energetically than anticyclonic eddy formation. The total number of cyclonic rings $(9 - 11)$ observed simultaneously south of the Gulf Stream usually exceeds the number of anticyclonic rings observed north of it (~ 3) [168, 212]. This fact by itself is not indisputable evidence of the greater intensity of cyclogenesis, since, as we know, anticyclonic rings merge with the Gulf Stream more frequently, and therefore have a shorter "lifetime". However, some observers [11, 161] have noted more elongated and distinct forms of cyclonic meanders, as well as a greater compactness of cyclonic frontal eddies (rings) in comparison with anticyclonic ones in the region of the frontal zone of the Kuroshio-Kurile Current. Whereas both types of rings have approximately the same size (a diameter in the neighbourhood of $100 - 150\ km$, less commonly up to $200 - 250\ km$) in the frontal zone of the Gulf Stream, the cold cores of cyclonic rings are definitely more compact and sharply defined than the warm cores of anticyclonic rings [66]. Consequently, on the basis of our observations, the orbital velocities of the first reach $2 - 3$ knots, while those of the latter rarely exceed 1 knot. It is interesting to note that the calculations based on D.G. Seidov's numerical eddy-resolving models (personal communication) demonstrate the greater intensity of cyclonic frontal eddies as compared with anticyclonic ones, and the greater sharpness of thermal fronts associated specifically with cyclonic meanders and eddies. The relationship with the general cyclonic vorticity of oceanic fronts is quite obvious here [66].

One of the most interesting facts established recently is the formation of small-scale cyclonic eddies at frontal interfaces of very diverse scales. As far back as 1940, Spilhaus [231] expressed the view, on the basis of very inadequate observations, that at least three scale categories of eddies existed in the frontal zone of the Gulf Stream, i.e. 1) large-scale frontal eddies* (diameter $100 - 200\ km$); 2) mesoscale eddies (diameter from $25 - 30$ to $50\ km$); 3) "parasitic" eddies (diameter $< 5\ km$) which develop on the peripheries of mesoscale eddies. Later Ye.I. Baranov [5], on the basis of the observations of Spilhaus and a number of other authors [231, 250], as well as Soviet data which he failed to interpret in detail, confirmed the existence of two types of eddies similar in scale to Spilhaus' first two categories in the frontal

---

* Later called "rings".

zone of the Gulf Stream. With respect to eddies of the second type, Baranov maintained that they were encountered in the main stream of the Gulf Stream, and had a cyclonic sign left of the flow axis, and an anticyclonic sign to the right of it. At the same time, he assumed that these eddies emerge because of the energy of "shear friction".† More recently, direct evidence for the existence of eddies of this scale in western boundary currents was obtained with the help of IR surveys from satellites and by surveys from research vessels [50, 170, 171, 173, 175], and in the visible range of the spectrum as well [29] (fig. 3.7). However, only cyclonic eddies located close to the main (cyclonic) front (or directly on it) and migrating in the direction of the main stream of the current at speeds in the neighbourhood of 1 knot were observed in all of these cases (see section 4.3). This makes us wonder whether we are dealing with a general result in nature, or with the imperfection of our own observations. First-hand detection of small anticyclonic eddies together with cyclonic ones in streams of the Gulf Stream or Kuroshio would appear to confirm their common origin in the shear. However, we have not yet observed any anticyclonic eddies $25 - 50$ km in diameter on any of the known IR or visible images of the Gulf Stream and Kuroshio obtained from space. Baranov's opinion regarding their existence is based on measurements of current velocities by means of GEK on general tracks at separations of about 100 km from each other, which means that this opinion cannot be considered proven as yet.

It is with good reason that Lee [170, 171] claims that the mesoscale cyclonic eddies which he calls spin-off eddies (see fig. 3.8) owe their origin specifically to barotropic instability which develops from perturbations of a cyclonic front. He also believes that the initial perturbations of a front are produced by the effect of the wind, whereas Legeckis [175] is of the opinion that the perturbing effect is associated with bottom topography. This question has not yet been resolved unanimously. In any case, the extreme of potential vorticity required for barotropic instability will apparently occur much more easily at a sharp cyclonic front (left of the current axis), whereas the anticyclonic horizontal velocity gradient to the right of the axis is probably not nearly as sharp, is more homogeneous ($\partial^2 U/\partial y^2 \approx 0$), and is not of a frontal nature. Therefore, the criterion allowing for the development of barotropic instability [see section 2.8] may not be realized here. Perhaps this is why there are no anticyclonic spin-off eddies in the Gulf Stream and Kuroshio Current.

---

† i.e. are manifestations of barotropic instability (see section 2.8).

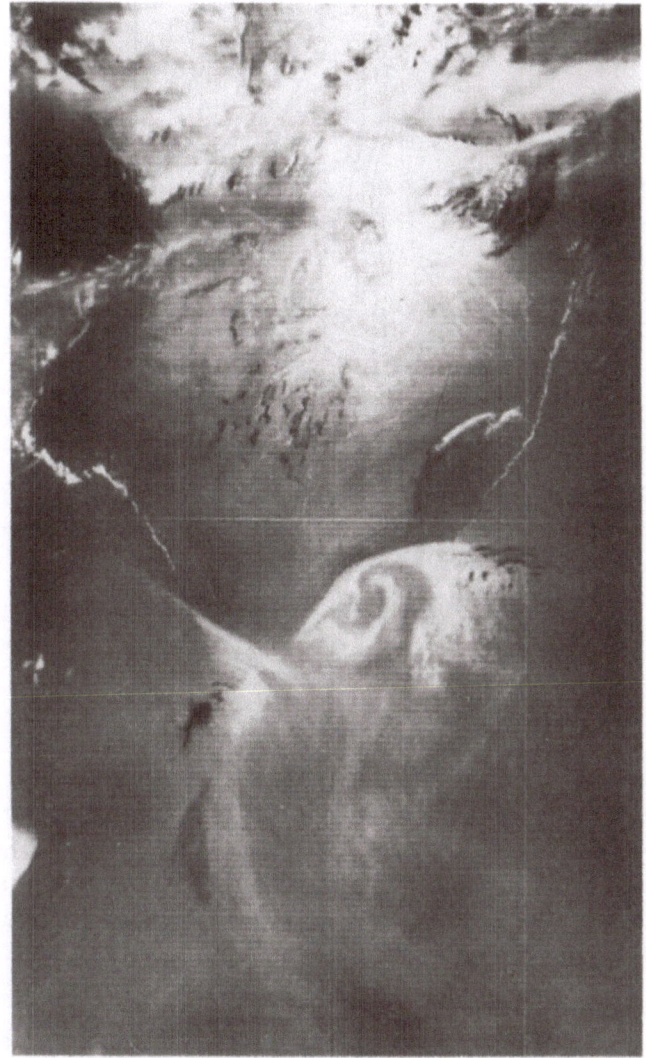

Fig. 3.7. Cyclonic eddies with a diameter of about 50−60 *km* at the boundary of the Kuroshio and Kurile current (boundary clearly marked by the chain of white clouds) east of Honshu, based on the data of images obtained from the 29th launching of the "Meteor" satellite with the help of the Scanner MCI on channel 0.7 − 1.1*μ* at 3 *h* 34 *min* Moscow Time on 6 June 1980.

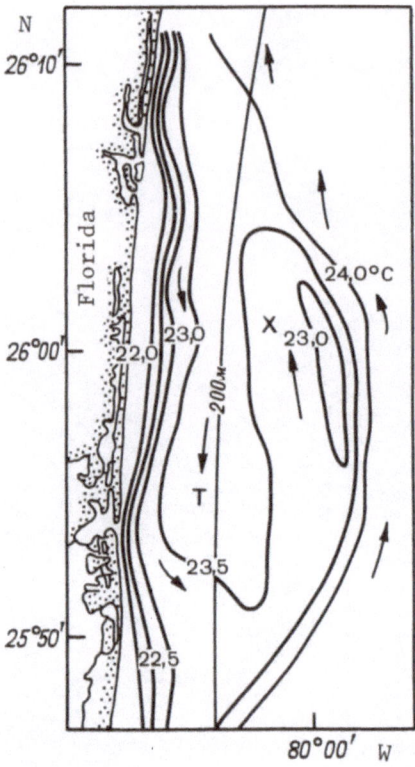

Fig. 3.8. Spin-off eddy at the boundary of the Gulf Stream off the Florida coast in the surface temperature field, based on the data of [171]. T marks the warm sector, X marks the entrainment of cold water into the main stream of the Gulf Stream, and the arrows mark the direction of water motion near the surface. The position of the 200 m isobath is shown.

It should also be said that the spin-off eddies described and studied by Lee [170] and Lee and Mayer [171] were observed by them at the western boundary of the Florida Current where the depth does not exceed 200 − 300 m. Their form, which was traced quite well by the pattern of isotherms in the temperature field, was quite unique, and gave Von Arx et al. [250] reason to compare the bands of cold slope waters entrained from that side of the front into the main stream of the Gulf Stream with "shingles" or "tiles" (see fig. 4.16). Because of this, they are referred to in our paper [50] as "entrainments" of cold water. As the latest satellite IR images have shown [195], cold slope-water intrusions of similar form move into the main stream of the Gulf Stream along the entire main cyclonic front eastward as far as 65°W and possibly further east where the depth of the ocean reaches several thousands of metres. However, it is not difficult to show that the given eddy-like disturbances of the frontal interface do not spread deeply into the ocean. The frontal interface of the Gulf Stream has a typical slope in the neighbourhood of 1/100, which with a maximum of about 10 − 20 km for the "warm sector" (marked with the letter T in fig. 3.8) signifies that the eddy structure in spin-off eddies penetrates to a depth of not more than 100 − 200 m near the front. This has been confirmed by direct measurements of the thermal structure of spin-off eddies ([50, 171]), see also figs. 4.15 and 4.17). This fact does not mean that the initial disturbances of the frontal interface in this area cannot be associated with the local peculiarities of the bottom relief. However, their subsequent existence, development and migration along the front are quite obviously not associated with the bottom relief. Indeed, even at 73°W, i.e. approximately 200 km east of Cape Hatteras, the structure of similar disturbances remains concentrated in the uppermost layer with a thickness of only 50 − 200 m (see fig. 4.17 in section 4.3). An interesting fact was the observation on an IR image of a well-formed compact cyclonic eddy with a diameter of 25 − 30 km at the eastern tip of a typical "entrainment" of cold water into a flow of the Gulf Stream [50]. This leads us to believe that the spatial structure of the given disturbances could be far more complex than conjectured by Lee [170]. This question is discussed in greater detail in section 4.3.

Small-scale cyclonic eddies are being detected more and more often at other frontal interfaces as well. At summer shelf fronts around Europe, cyclonic eddies with a diameter of 20 − 40 km are in Pingree's opinion [208] an important mechanism of cross-frontal mixing. They arise and dissipate over a period of 3 − 4 days, migrating along the front with a speed of up to 2 cm/s during this time. And in this case, satellite IR images provide us with extremely abundant material for analysis and consideration. Among other things, they give evidence of a specific "hooked" form (fig. 3.9) that is characteristic of the cold-water tongues entrained into the eddies. This is frequently observed in disturbances of a larger scale as well, for example at the Gulf Stream front [195]. Pingree [208] pays special attention to anticyclonic disturbances, indicating that they are rarely detected on satellite IR images. Nevertheless, he gives some examples of the appearance of such disturbances (off the coast of Norway and on Dogger Bank in the North Sea), which he associates with the anticyclonic horizontal shear of the current velocity and the effect of the bottom relief. At the same time, Pingree attributes the incomparably more frequent cyclonic eddies at shelf fronts to baroclinic instability of the latter.

His assessments of the scales of the phenomenon seem to agree with this point of view. If this is the case, then their physical nature differs significantly from that which is apparently characteristic of the above-described cyclonic spin-off eddies of similar size at the fronts of the Gulf Stream and Kuroshio Current.

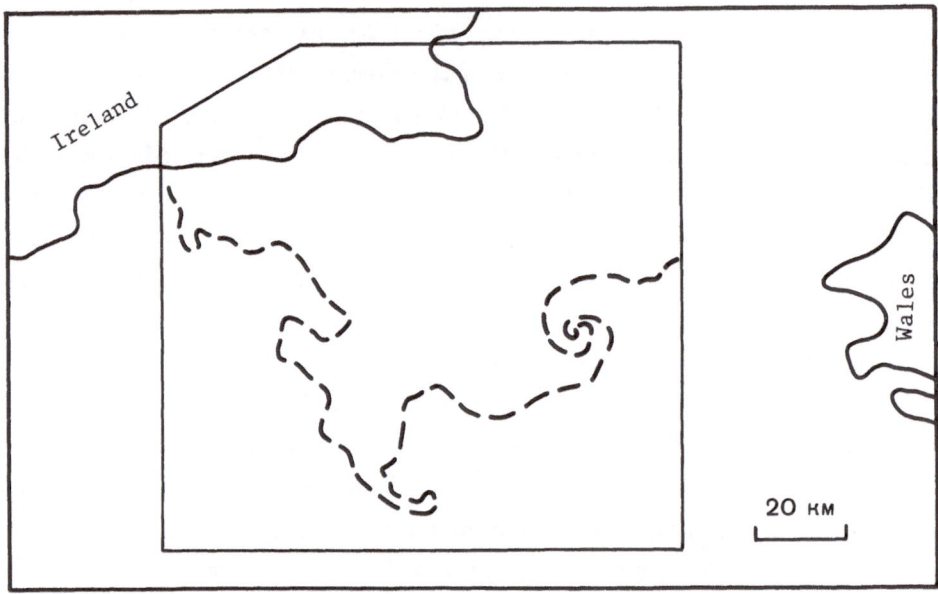

Fig. 3.9. Hooked form of the warm signatures of cyclonic eddies on IR images of the front (dashed line) in the Irish Sea (NOAA-5 satellite) on 3 September 1976, based on the data of [208]. Thin solid lines show the outer boundary of the IR image.

It cannot be ruled out that tidal currents can be the cause of small wave-like or eddy-like deformations of frontal interfaces in coastal areas. The measurement of currents with current meters in combination with satellite data gives us reason to believe that this is precisely how eddy-like meanders develop at the front of a river discharge lens* from the Po River in the Adriatic Sea [103].

---

* River discharge fronts are discussed in subsection 3.3.1.

Before concluding this section, we should mention a most interesting case of quasi-periodic large-scale anticyclonic eddy formation (with an eddy diameter of 400 km) by the Loop Current in the Gulf of Mexico, as a result of the combined effect of barotropic instability and a latitudinal change in the Coriolis acceleration [149].

3.1.3. Frontal systems of Gulf Stream rings

Basically unresearched varieties of oceanic fronts are associated with eddies ("rings") that appear as a result of the separation of meanders from the main jet streams of the ocean (e.g. the Gulf Stream). It is on the basis of these jet streams that the question regarding the dynamical analogy between atmospheric and oceanic frontogenesis was discussed in section 2.4. Therefore, it is very important to determine the degree of similarity between the frontal systems of oceanic eddies arising in jet streams, and the frontal systems of their atmospheric analogues, i.e. the cyclones and anticyclones of temperate latitudes.

The main flow of the Gulf Stream is in essence a wide ($100 - 150$ km) frontal zone, the northern part of which is a sharp temperature front with an isotherm slope of approximately $1/100$ and a horizontal temperature gradient in the thermocline in the neighbourhood of $0.2-0.5°C/km$ [224, 237]. This front is often called "cyclonic" due to the cyclonic sign of the vorticity of its velocity field. When a ring separates from the main stream of the current, the waters of the Gulf Stream together with the separated part of the frontal zone form a shell around the ring. Further development of frontal systems in separated rings may be associated with 1) the dynamics of the rings themselves, 2) their deformation fields, 3) the interaction between the initial frontal zone and the waters surrounding the ring, and 4) atmospheric effects on the ocean. The conditions for development of frontal systems are not the same due to the fact that the rings of the Gulf Stream are both cyclonic, and anticyclonic in nature.

Cyclonic ("cold") rings with a core of slope waters form on the southern side of the Gulf Stream and migrate in the homogeneous waters of the Sargasso Sea, which differ little from the waters of their shells. Anticyclonic ("warm") rings with a core from waters of the Sargasso Sea occur on the northern side of the Gulf Stream (fig. 3.10) and migrate in the region between the latter and the continental shelf of the North American continent from 55 to 75°W [168] where there are sharp local frontal interfaces between the slope and shelf waters which differ significantly in their spatial temperature and salinity gradients even outside the frontal zones.

Both types of rings move mostly in the west-southwest direction with mean velocities of $3 - 7\ km/day$, and have a lifetime of up to $1 - 2$ years, which is much greater than the characteristic time of synoptic frontogenesis in the ocean. The western movement of "rings" is their chief difference from atmospheric "analogues" which, though they move westward in relation to the upper-air jet stream, are transported eastward in this flow in relation to the Earth's surface, because of its higher velocity.

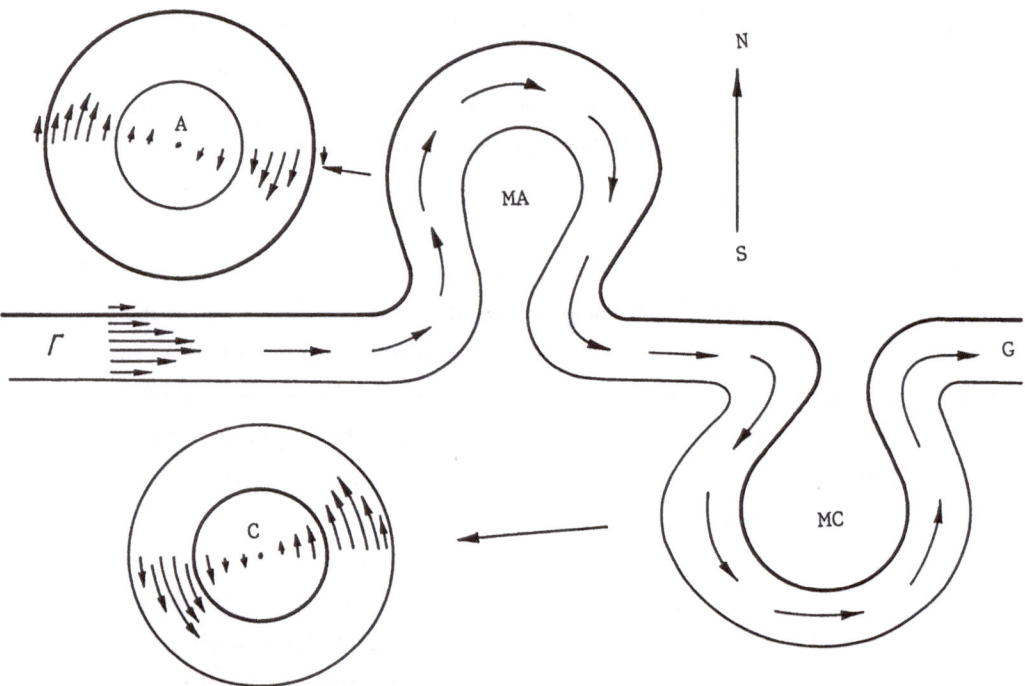

Fig. 3.10. Diagram of the separation of anticyclonic (A) and cyclonic (C) rings from the Gulf Stream (G) as a result of the formation of anticyclonic (MA) and cyclonic (MC) meanders. The bold lines indicate the position of the main cyclonic frontal interface in the upper part of the thermocline in the Gulf Stream and in the rings. The arrows indicate the direction of the motion and the distribution of velocities in the rings and in the main flow of the Gulf Stream.

A word of explanation regarding the differences in the conditions under which the main "cyclonic" front inherited from the Gulf Stream exists in the rings of both signs. In cyclonic rings, this front separates the dome of cold slope waters from the warm shell. In this case, the zone of maximum orbital velocities is found on the outside of the front. The cyclonic horizontal velocity shear across the front is naturally preserved in this case, which should help keep the front sharp. On the other hand, the equilibrium of a circular front with a small radius may differ significantly from a geostrophic one due to centrifugal effects. In anticyclonic rings, the local cyclonic vorticity of the initial front is opposite to the general anticyclonic vorticity of the water motion in the rings. In this case, the quasi-geostrophic balance of the front can be achieved only at the extreme periphery of the ring, beyond the limits of the maximum of orbital velocity, i.e. in the area where the local vorticity of the orbital motion has a cyclonic sign. Therefore, in anticyclonic rings, it is logical to expect smaller velocities of orbital motion near the front, a smaller stability of the frontal zone, or even its weakening, i.e. a decrease in the slope of the isotherms [224] and a decrease in gradients.

Frontal zone of a warm ring. We studied the frontal systems of Gulf Stream rings during the 27th cruise of the research vessel "Akademik Kurchatov" in the area of the Soviet-American POLYMODE experiment. Information on the position of the rings was provided by the Americans, and was based on satellite data. The southern frontal zone of a warm ring was detected north of the main flow of the Gulf Stream around $38°N$ and $69°W$ when our vessel was on its way from the study area to Boston at the beginning of September (5 September 1978). With practically calm weather, individual fronts in this area were clearly determined visually by slicks and the accumulation of floating Sargasso weed and trash along the line of the front. The crossing of these bands by the vessel was accompanied by an instantaneous response of the thermal sensor, being towed at a depth of 0.15 m, which recorded temperature changes in the neighbourhood of $0.5 - 2.0°C$ over a distance of only a hundred metres. The particular sharpness of the fronts at the surface of the ocean and their multiple structure were apparently due to the fact that the southern boundary of the warm ring at the surface was located above the main front of the Gulf Stream in the thermocline. What actually occurred was the merging of the frontal zones of the Gulf Stream itself and the warm ring into a single complex frontal zone, the position of which is marked by the letter A in figs. 3.11 and 3.17. The "box"-type survey carried out in this area of fronts over a square with 4-mile sides (fig. 3.12), followed by the recording of temperatures near the surface of the ocean ($z = 0.15$ m), as well as temperatures and salinities at the 3-metre level with an AIST probe in a flow-through system [65], enabled us to map the position of these fronts (fig. 3.12) and to determine their thermohaline characteristics.

In the sharpest part of the frontal zone, the overall decrease in temperature from south to north at the surface and at the 3- metre level amounted to $2°C$, and that of salinity to about 1 ppt. As a result of this, the frontal zone was characterized by a noticeable density change (about 0.14 units σ_t) due to the dominant contribution of the salinity change. The spatial variation of the zones of local temperature and salinity changes resulted in the fact that sharper density fronts with density changes up to 0.7 units σ_t were observed inside the frontal zone. The actual temperature, salinity and density gradients at the sharpest frontal interfaces amounted to $5 - 20°C/km$, $5 - 10$ ppt/km and $2 - 3$ units σ_t/km respectively.

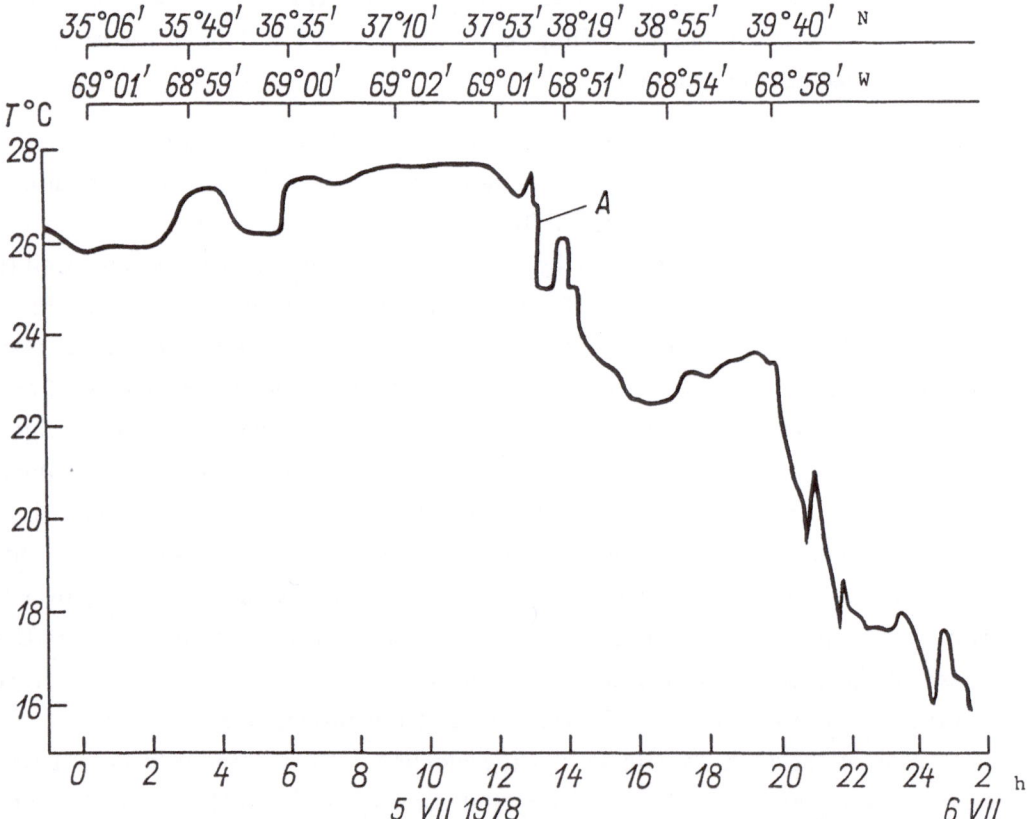

Fig. 3.11. Temperature changes in the surface layer of the ocean (depth of towed sensor $z = 0.15$ m) recorded on 5-6 September 1978 during the trip from the area of the POLYMODE experiment to Boston along $69°W$ (GMT). The letter A marks the region of the sharpest frontal interfaces associated with both the northern boundary of the Gulf Stream, and the southeastern periphery of the warm ring (see also fig. 3.12).

Fig. 3.12. Temperature distribution in the surface layer of the ocean in the frontal zone near the northern boundary of the Gulf Stream on 5 September 1978, based on the data of a towed temperature sensor ($z = 0.15m$). The dashed line depicts the course of the vessel during the crossing of the frontal zone.

The nature of the temperature and salinity changes in the frontal zone, as well as the general direction of this zone from northwest to southeast (fig. 3.12), gave us reason to believe that the detected fronts were associated with the migration of cold and freshened waters along the eastern periphery of the warm ring in the direction of the main front of the Gulf Stream. This assumption was confirmed during a second survey of the same system of fronts a week later, and from comparison of the observed pattern of fronts with the experimental frontal analysis maps for the surface of the ocean, drawn on the basis of satellite data for 16th and 23rd of August and provided by the American side. On the first of these maps, the centre of the warm ring (marked with the letter "U") was located at $39°N$ and $69°W$, and on the second one (fig. 3.13), it was slightly farther west. Based on the most probable migration direction and velocity of the ring, the return cruise of the "Akademik Kurchatov" from Boston to the POLYMODE area was made along $70°W$, which enabled us to cross the ring, most probably at the centre. We again detected frontal interfaces upon approaching the main current of the Gulf Stream (11 September 1978) near $38°12'N$. According to the temperature gradients at the 0.15 m level and the nature of the T, S-correlation at the 3 m level, these fronts were similar to those detected a week earlier. After a "box"-type reconnaissance (fig. 3.14) and determining the spatial position of the frontal zone, two sections, each measuring about 20 miles in length, were performed. The section with a south to north orientation included five stations with the AIST probe. The section with a north to south orientation consisted of five XBT profiles from a moving vessel (fig. 3.14). On the basis of data obtained with the help of an AIST probe in a flow-through system while crossing the frontal zone on the perpendicular tracks of the vessel, we compiled maps of the horizontal distribution of temperature, salinity and the nominal specific density at the 3 m depth (depicted in figs. 3.15 a, b, c respectively) and plots of the variation of the same characteristics along the route of the vessel (fig. 3.16). As we can see from these illustrations, the fronts observed in the surface layer were of an alternating, multiple, thin-jet nature. All were characterized by a positive T, S-correlation, the resulting density gradient in them being determined by the dominant contribution of the salinity gradient. The temperature, salinity and density changes in the sharpest part of the frontal zone with a width of only 2 km near the surface amounted to $\Delta T = 1.0°C$, $\Delta S = 1.6ppt$ and $\Delta\sigma_t = 0.75$ units of σ_t, while the actual temperature gradient was $15°C/km$. At two of the other weaker fronts, the changes in the same quantities amounted to $\Delta T = 0.5°C$, $\Delta S = 0.7ppt$ and $\Delta\sigma_t = 0.37$ units σ_t.

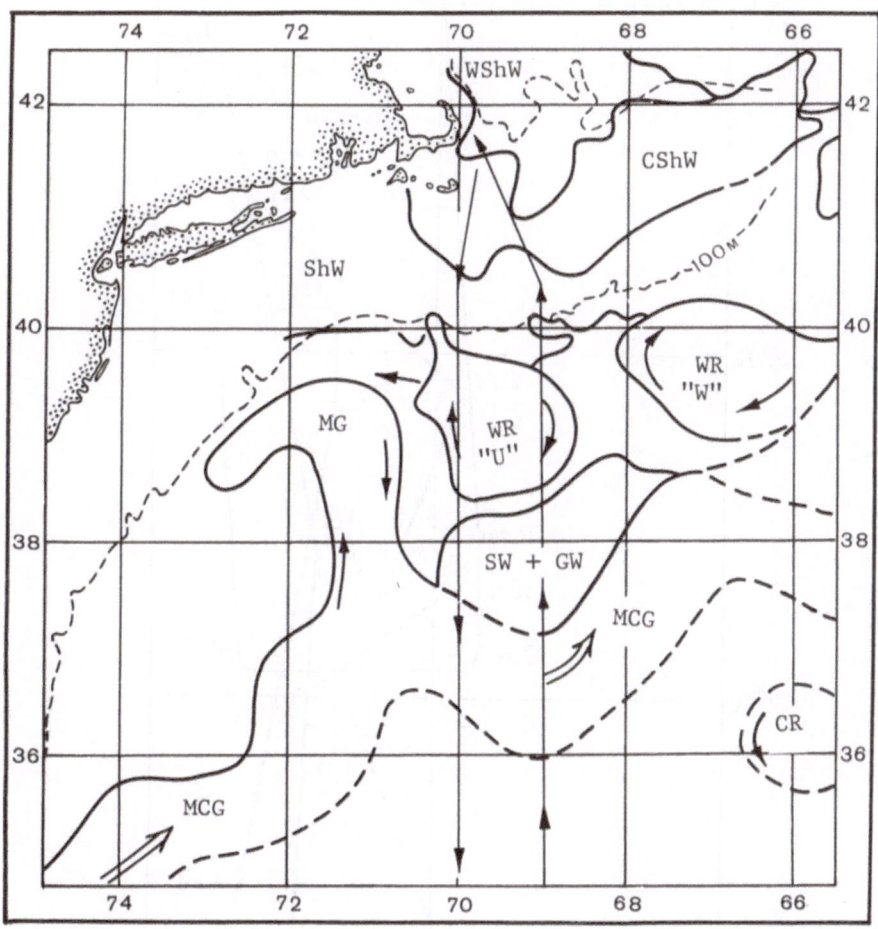

Fig. 3.13. Diagrammatic map of experimental frontal analysis of satellite data, showing the position of the main current of the Gulf Stream, its meanders and rings on 23 August 1978 (map placed at our disposal by the American side within the terms of USSR-USA cooperation in the POLYMODE experiment). *MCG* – main current of Gulf Stream; *MG* – meander of Gulf Stream; *SW* – slope waters; *GW* – Gulf Stream waters; *CR* – cold ring; *WR* – warm ring; *ShW* – shelf waters; *CShW* – cold shelf waters; *WShW* – warm shelf waters.

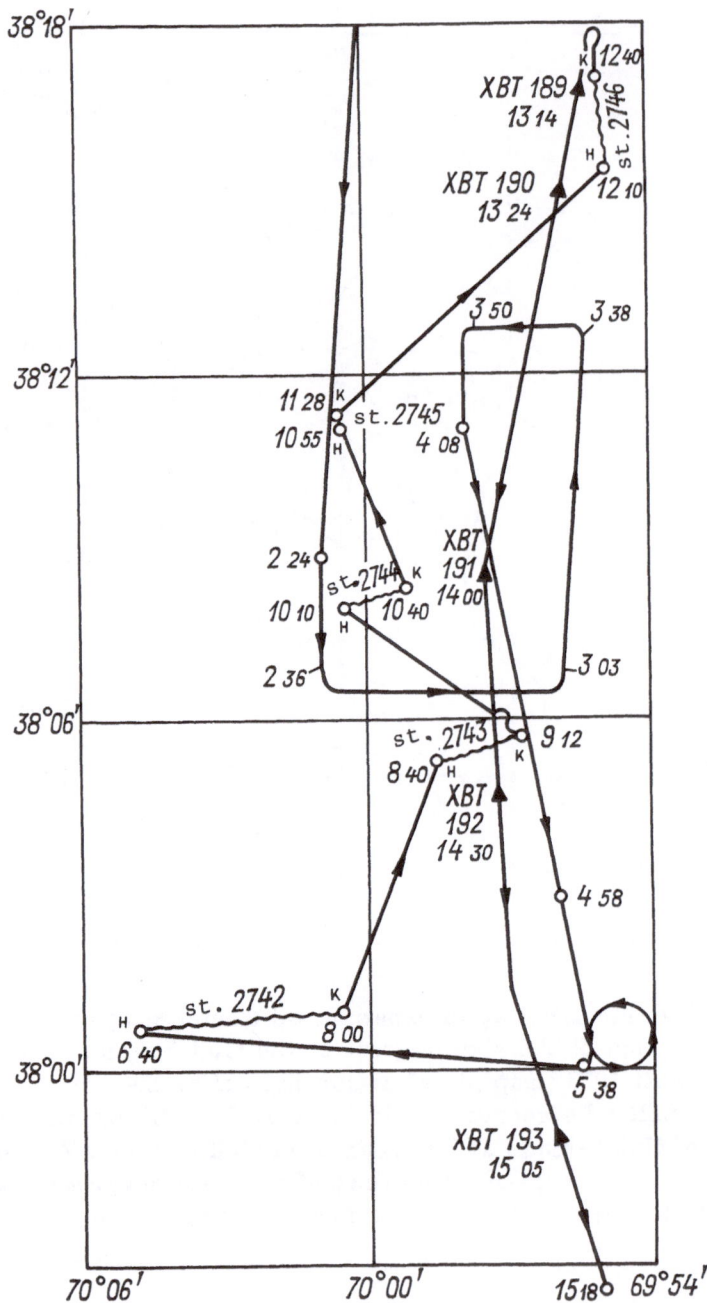

Fig. 3.14. Diagram of the track of the research vessel "Akademik Kurchatov" in the frontal zone on the southern periphery of the warm ring of the Gulf Stream on 11 September 1978 (ship time).

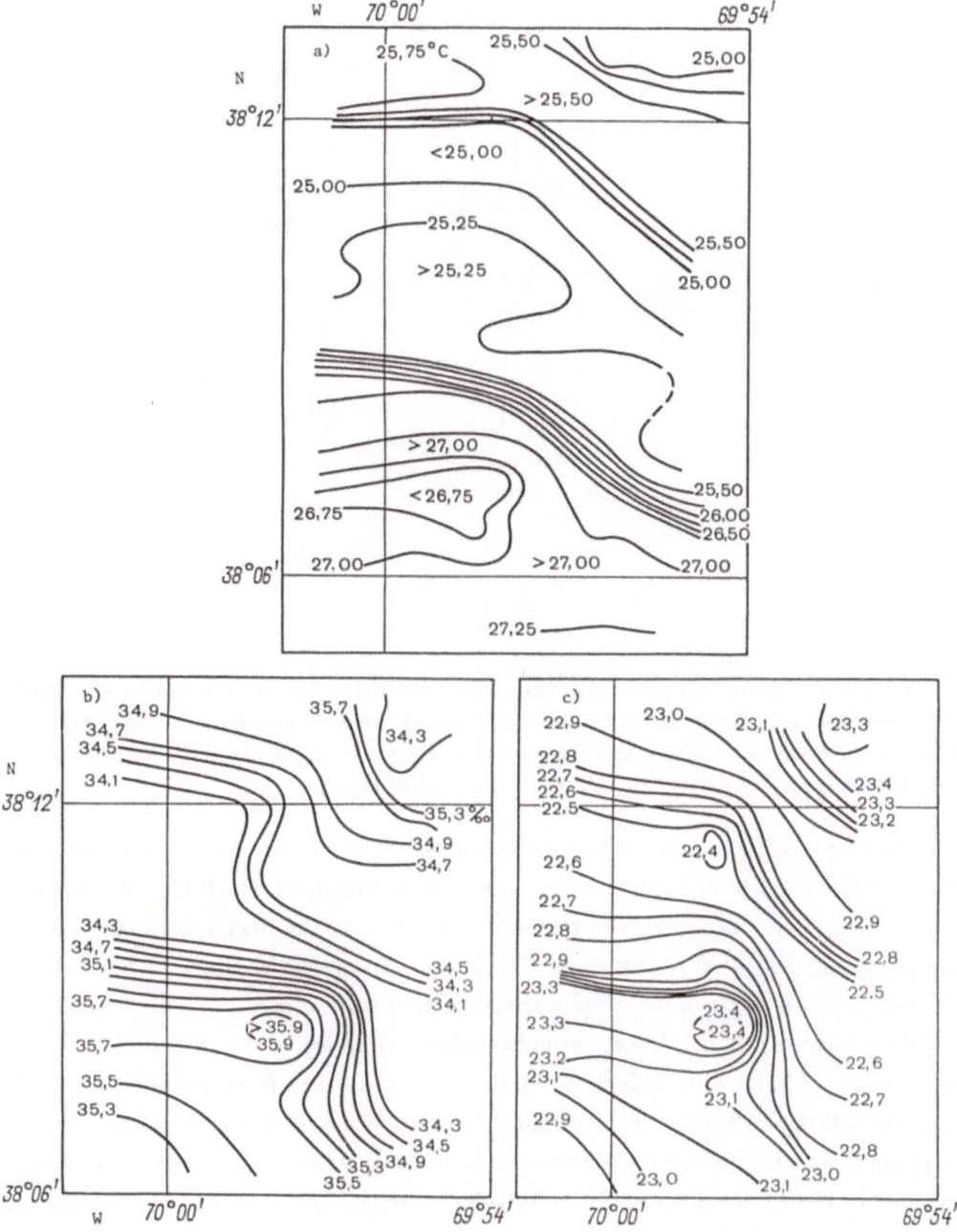

Fig. 3.15. Distribution of temperature (a), salinity (b) and density (c) at the 3 m depth in the frontal zone of the warm ring of the Gulf Stream.

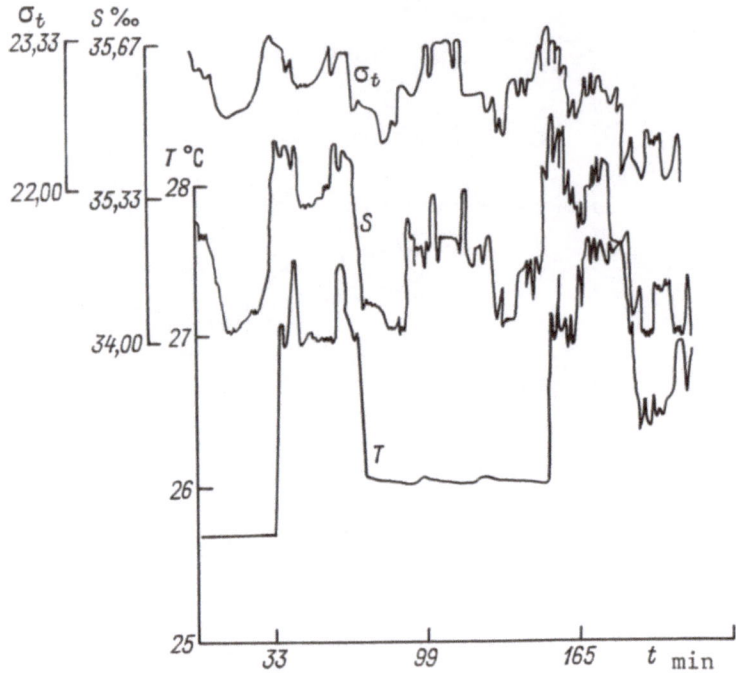

Fig. 3.16. Changes in temperature, salinity and density on the perpendicular tracks of the vessel during a survey in the frontal zone of the warm ring of the Gulf Stream, obtained with the help of an AIST probe in a flow-through system on 11 September 1978 between 01 h 47 min and 05 h 18 min (ship time).

Let us dwell in greater detail on the problem of determining the boundaries of the anticyclonic Gulf Stream ring. The frontal interface, which can be regarded as the northern boundary of the ring at the surface of the ocean, was recorded at approximately 39°27′N (point B in fig. 3.17). At the same time, a thermotrawl towed at a depth of 80 m recorded a sharp temperature increase from 13.5 to 19°C in the vicinity of 39°10′N, i.e. approximately 17 miles south of point B (lower curve in fig. 3.17). This difference in the position of the front at different levels (0.15 and 80 m) is due to the fact that the transverse dimensions of the cores of warm rings usually diminish with depth. Saunders [224] also observed a significant decrease in the dimensions of the anticyclonic ring of the Gulf Stream within the boundaries of the main front at a depth of 200 $m(120 \times 60\ km)$ as compared with its dimensions at the surface ($200 \times 100\ km$). This quite obviously shows that the rings retain the main features of the sloped front of the Gulf Stream (slope about 1/100) which at the surface of the ocean can be displaced up to $20 - 40\ km$ in the cold direction in relation to its position at the 200 m level. However, the slope of the frontal boundary of the core of the warm ring was in this case significantly smaller (2.5×10^{-3}).

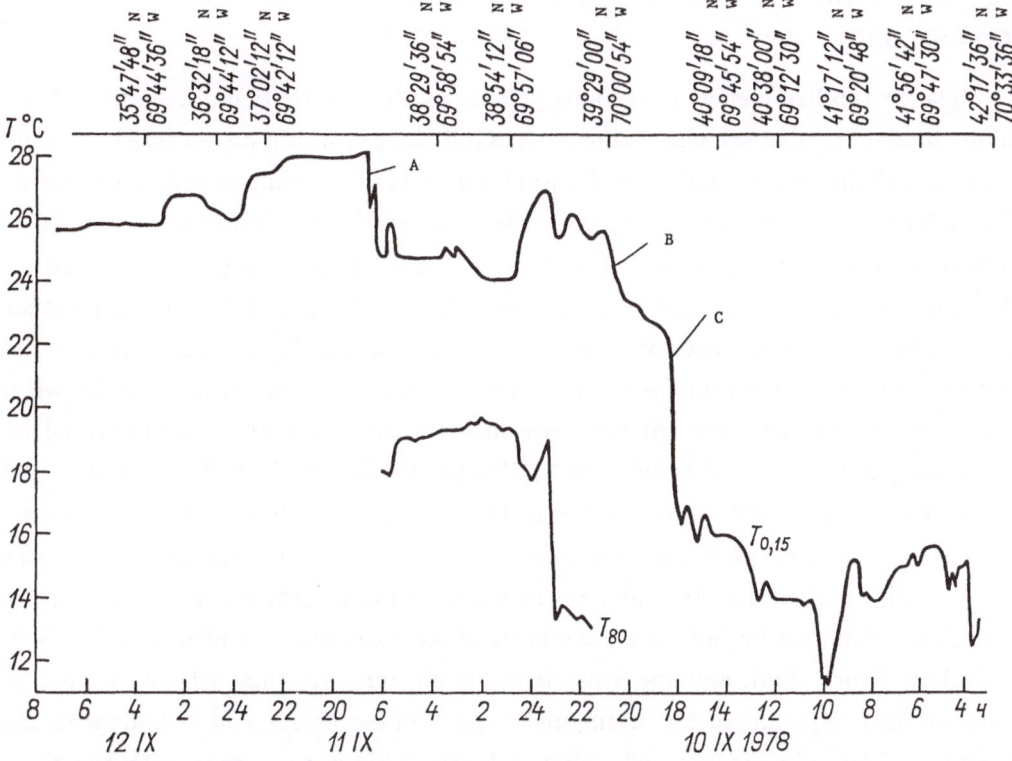

Fig. 3.17. Temperature changes near the surface ($T_{0.15}$) and at a depth of 80 m (T_{80}), recorded on 10-12 September 1978 en route from Boston to the POLY-MODE area along $70°W$. The letter A marks the region of the sharpest frontal interfaces, B - the presumed northern boundary of a ring at the surface, and C - the position of the front which separates the cold shelf waters and the warmer slope waters (time in GMT).

According to the nature of the temperature change recorded by the thermotrawl (fig. 3.17), the southern boundary of the ring was found near $38°15'N$, which in this case is in good agreement with the available satellite data (fig. 3.13). The $2 - 3°$ drop in temperature near the surface in a wide band of a part of the ring in relation to the temperature in its northern part (fig. 3.17, upper curve) was apparently due to the advective distribution in the surface layer of colder and fresher slope water that had been entrained into the circulation during the migration of the ring. The main front between the slope and shelf waters was located north of the warm ring (C in fig. 3.17).

The destabilizing effect of the anticyclonic relative vorticity on the nature of the main circular front in the warm ring of the Gulf Stream would be expected to result in at least 1) the retention of sharp frontal temperature and salinity gradients only at the extreme periphery of the ring where the local relative vorticity either diminishes to zero, or alters its sign under the effect of viscous friction, and 2) a decrease in the slope of the frontal interface. It is extremely difficult to determine whether the specific thin-jet features observed in the near-surface frontal zone are signs of this type of destabilization. Judging by the fact that the lateral distance between the northern and southern frontal zones in our case and in the case described by Saunders [224] amounted to $80 - 100$ miles ($100 - 200$ km) at the surface of the ocean, these zones really were located at the extreme periphery of the ring. As we have seen, the slope of the frontal interface on the northern side of the ring was significantly smaller than the slope of the main frontal interface of the Gulf Stream. The slopes can also be judged on the basis of sections, one of which is available in Saunders' paper [224], and the other is based on our data (fig. 3.18, a and b). In both sections, we can clearly distinguish a peak of isotherms and isohalines in the centre of the section in the $150 - 350$ m layer. This is associated with the slope waters circulating along the periphery of the ring. In our case, the width of the peak at the 200 m level does not exceed 25 km. The left-hand "slope" of the peak in fig. 3.18 a between stations 2742 and 2744 represents the northern part of the frontal zone of the Gulf Stream, while the right-hand "slope" between stations 2745 and 2746 is the southern frontal boundary of the warm ring. Here, the isotherm slope in the $150 - 300$ m layer reaches $1/100$, which is somewhat steeper than in the warm ring studied by Saunders ($1/300$). In our case, the horizontal temperature gradients at the front associated with the ring amount to $0.2 - 0.5°C/km$. Consequently, both the slope of the frontal interface, and the horizontal temperature gradients at the warm ring front in the main thermocline can be either smaller, or about the same, as at the northern boundary of the main current of the Gulf Stream. According to our data, the main front of the warm ring in the $150 - 350$ m layer had a positive T, S-correlation, the resulting density gradient being determined by temperature, and not salinity as in the near-surface fronts. Neither our own data, nor the data of Saunders indicate any presence of specific thin-jet features in the main frontal zone of the ring located below 150 m.

On the other hand, the presence of sharp inversion-intrusion jet-like features (fig. 3.18, a and b) with local temperature drops of several degrees and salinity drops of several *ppt* in the near-surface layers $(0 - 150$ $m)$ of the frontal zone of the warm rings both in the east (according to Saunders [224]), and in the south (based on our data) points to the fact that the sources of this jet-like finestructure should be located beyond the limits of the rings themselves. Consequently, the processes of formation of this structure are most likely not associated with the dynamics of the main frontal zone.

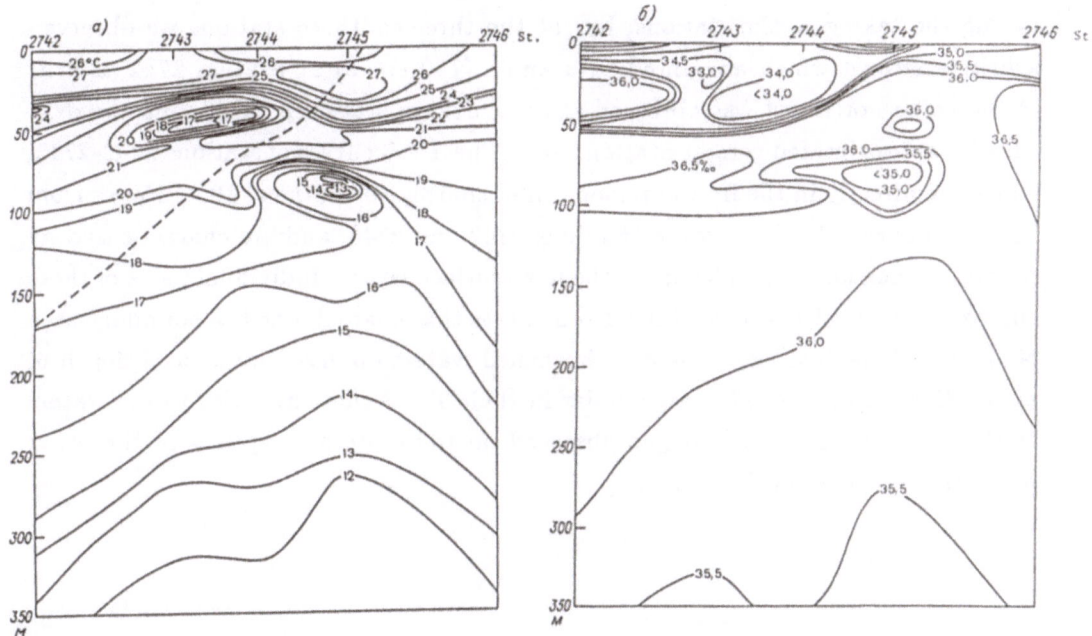

Fig. 3.18. Meridional section of temperature (a) and salinity (b) across the southern periphery of a warm ring, based on the data of vertical profiles from an AIST thermohaline probe. Position of stations Nos. 2742-2746 shown in fig. 3.14.

On the basis of the fact that we recorded salinity values of close to or even less than 34 *ppt* at individual points near the surface in the southern frontal zone (fig. 3.15 b, 3.18 b) and even 32.8 *ppt* at station No. 2745 (fig. 3.19 c), which is significantly less than the value $S = 35$ *ppt* which demarcates the shelf and slope waters in this area of the ocean [269], we can say that this region is being penetrated not only by slope waters, but also by shelf waters of low salinity.

The shelf "origin" of the waters of the lowest salinity is also confirmed by our analysis of T – S curves (fig. 3.20) based on the data of AIST probes at five stations of a section, the position of which is shown in fig. 3.14. As we can see from fig. 3.20, the T – S curves for stations executed along different sides of the frontal zone (stations 2742 and 2746) differ only in their upper part, i.e. in the region of temperatures above $20°C$, which corresponds approximately to the $40-45$ m depth. Below the surface layer ($z > 40 \cdots 45$ m), they are practically the same as the T – S curves of the waters at the northern boundary of the Gulf Stream. This permits us to assume that Station 2742 was executed at the northern boundary of the main current of the Gulf Stream, while Station 2746 was taken at the southern periphery of the warm ring. This assumption is in good agreement with the data on ship drift during the taking of the stations, i.e. at the three southern stations we observed a drift eastward which amounted to 3 knots (150 *cm/s*) at Station 2742, a drift to the north-northwest was observed at the northern station N. 2746, and the drift velocity approximated zero at Station 2745. The T – S curves of stations 2743-2745, which are directly in the frontal region, differ sharply to depths of $100-150$ m from the "boundary" T – S curves for stations 2742 and 2746, and are characterized by salinities much lower than 35 *ppt* in the near-surface layers. Individual parts of these curves coincide quite well with the T – S curves that characterize the boundary area of slope and shelf waters. The cold freshened waters observed by us at a depth of about 80 m at Station 2745 are similar in their T – S characteristics to the waters of the cold and freshened tongue observed on the eastern periphery of the other warm ring of the Gulf Stream [224].

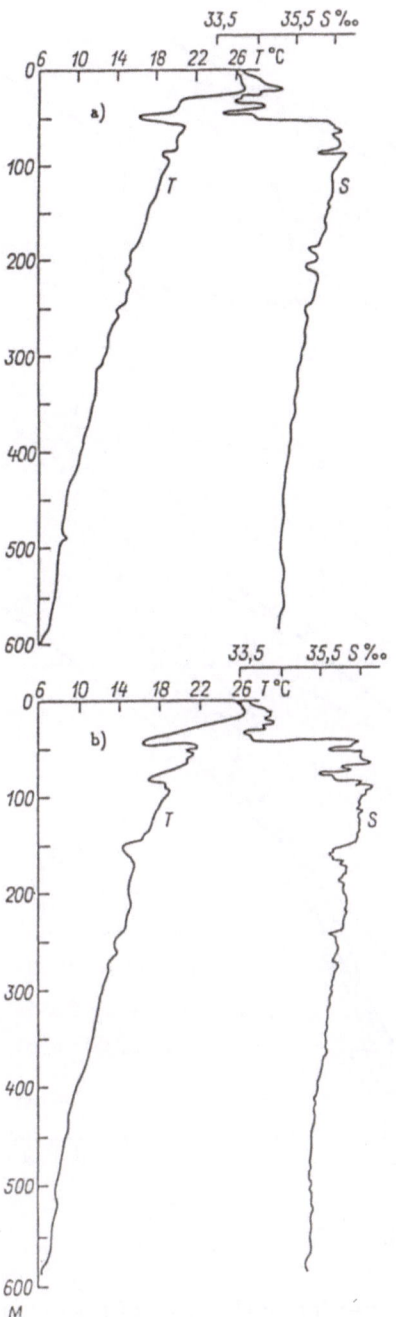

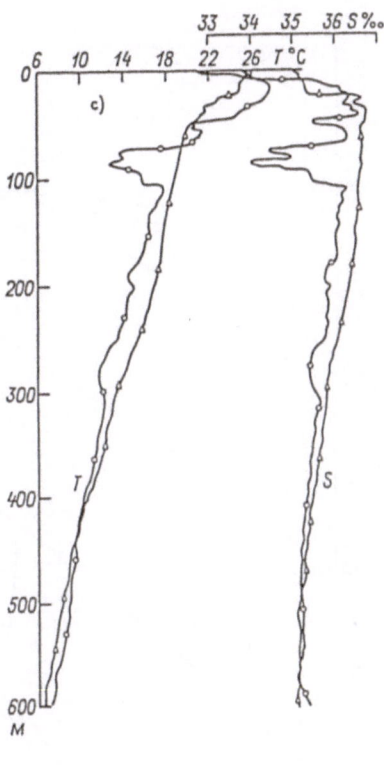

Fig. 3.19. Vertical profiles of temperature and salinity at Station No. 2743 (a) and Station No. 2744 (b); comparison of the vertical profiles of temperature and salinity at Station No. 2745 (circles) and Station No. 2746 (triangles; profile without intrusions) (c).

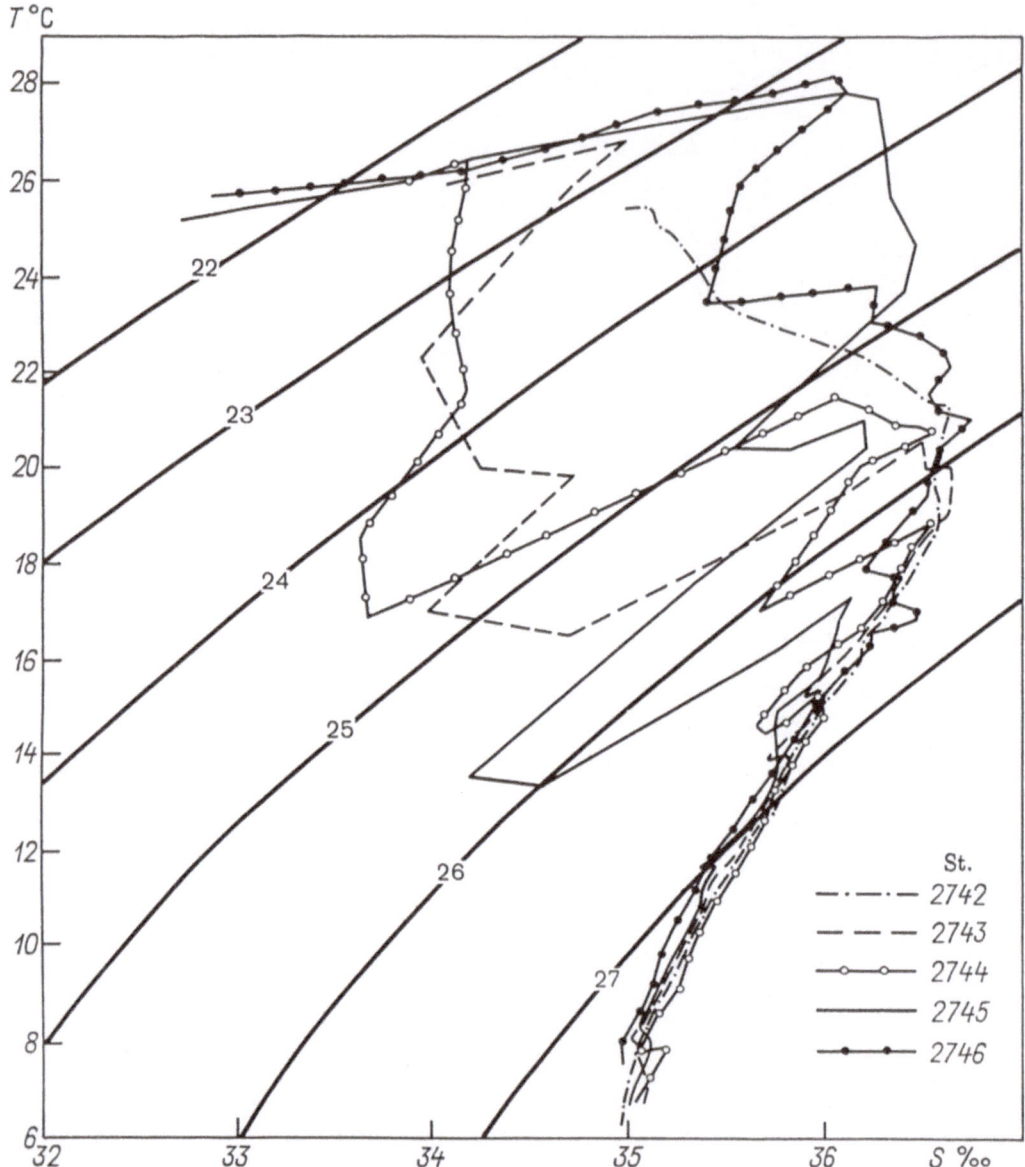

Fig. 3.20. T – S curves for stations No. 2742-2746.

The two possible mechanisms for shelf-water advection in this area of the ocean are 1) the transport of these waters from Cape Hatteras along the northern boundary of the main current of the Gulf Stream, and 2) the transport of these waters from north to south along the eastern periphery of warm rings.

Even some of the earlier studies [117, 166, 237] indicate that freshened waters with a salinity of less than 35 *ppt* were in fact spreading from the area north of Cape Hatteras along the northern periphery of the Gulf Stream. Salinities as low as the ones recorded near the surface in our case ($S = 32.8\%$) were not observed. At the same time, other surveys [189, 224] confirm the fact that surface shelf waters with a salinity of less than 33 *ppt* are transported along the eastern periphery of the anticyclonic rings of the Gulf Stream. It should be said that we have analyzed several IR images obtained from a NOAA satellite for the Gulf Stream area. On a number of them (e.g. the photo in Legeckis' paper [174]), we can clearly see the process of entrainment of shelf waters into the circulation of the anticyclonic ring and the outflow of them to the periphery. Calculations with models [47] have also confirmed that the ambient waters are entrained into the circulation, and are transported from the northern periphery to the southern one at the rear of the rings moving westward. In our case, it is precisely this mechanism that is the most probable explanation of the appearance of shelf waters in the study area.

Let us discuss in greater detail the characteristic features of the intrusion-inversion thermohaline finestructure observed by us. Apart from the surface-freshened streams at the section (fig. 3.18 a, b), we can clearly see isolated streams of cold freshened water above the main thermocline in the $0 - 150$ *m* layer. These streams are approximately 10 *km* wide, and up to $30-40$ *m* thick. The upper stream is found in the $30 - 60$ *m* layer, and the lower one in approximately the $80 - 110$ *m* layer. As we have already mentioned, similar streams of cold and freshened shelf water have been observed by Saunders [224] on the eastern periphery of a warm ring of the Gulf Stream in the surface 150 *m* layer.

The marked difference between the T – S characteristics of the shelf waters and the T – S characteristics of the slope waters in which they are distributed results in the appearance of strong temperature inversions which are clearly seen on the vertical temperature profiles based on the data of AIST probes at stations 2743-2745 in the frontal zone (fig. 3.19 a, b, c). The maximum deviations of temperature in the inversions, and the salinity deviations corresponding to them, amount to $\Delta T = -6°C$ and $\Delta S = -2.2$ *ppt* in this case, while the thicknesses of the inversion layers reach 10 *m*. The fact that the observed inversions are associated specifically with **cold** intrusions is clearly seen in fig. 3.19 c which also depicts the vertical profiles of $T(z)$ and $S(z)$ for neighbouring stations No. 2745 and No. 2746 located only 10 *km* from each other.

The lifetime of the temperature inversions determined in [142] with the help of empirical formulas for double-diffusive convection [247] amounted to only $1 - 2$

days. A similar analysis of the inversions detected by us also limits their lifetime to several days. Therefore, double-diffusive convection at the boundaries of the intrusions should result in the effective mixing of shelf and slope waters near the frontal zones.

The last point I would like to make is that the streams of shelf waters in the frontal zone are confined from the sides by sharp frontal interfaces, which is clearly seen both in the sections (fig. 3.18 a, b, and the section in [224]), and on the map of T, S and σ_t distributions at the $3m$ level (fig. 3.15 a, b, c). It is these frontal interfaces, and not the main circular front, that are characterized by maximum horizontal temperature, salinity and density gradients. We can assume from this that these fronts could exist independently of the ring, could be entrained into the circulation of the ring together with the slope and shelf waters, and then be transported by the ring in the process of its migration westward. It is quite possible that the frontogenetic action of the ring on the ambient waters, which has been confirmed by numerous calculations [47], sharpens the fronts entrained into it from the outside and at the same time maintains the entire peripheral frontal zone in the $0 - 150$ m layer in a sharp and highly divided state. At the same time, the enhanced separation of the observed jet-like intrusions of shelf waters can be the result of cross-frontal heat and mass transfer at the entrained fronts.

Frontal interfaces of a cold ring. A section across a cold ring with its centre at approximately $35°N$ and $67°W$ was performed from east to west in the course of a detailed survey of the cold core of a ring at the end of September 1978 upon the completion of the main program of the 27th cruise of the research vessel "Akademik Kurchatov".

In the section across the cold core of the cyclonic ring (fig. 3.21 a, b), we can clearly see the sharp frontal character of the walls of the core, formed by the previous "cyclonic" front of the Gulf Stream. Here at levels between 100 and 150 m, we observe horizontal temperature gradients of up to $1°C/km$ and salinity gradients of up to $0.1 - 0.2$ ppt/km with a positive T – S correlation. The density gradient across the front is determined by the dominant contribution of temperature. The slope of the walls of the core is within the limits of $1/50 - 1/100$. Near the core, we observe numerous isolated temperature inversions of an intrusive origin, which are an indication of an intensive cross-frontal heat and mass transfer associated most probably with the centrifugal ageostrophic components of the near-frontal motion. Examples of these inversions can be seen on the vertical temperature and salinity profiles for St. 2770, depicted in fig. 3.22. The circular front around the core has a radius of only about 10 miles (fig. 3.21 a, b). At this radius, the velocity of the ship drift u, which is equal to the orbital velocity of the motion in the ring near the surface, was not less than 2 knots (100 cm/s) according to our data, which makes the centrifugal term u^2/R in the equation of motion equal to $5.4 \cdot 10^{-3}$. The term corresponding to the Coriolis acceleration fu is equal to $8.9 \cdot 10^{-3}$ for the same velocity u at latitude 35°. In this case, the same order of both effects could not but affect the dynamics and structure of the ring and its main frontal interface. On the strength of this characteristic of the balance of forces, the orbital motion in the ring should be regarded as cyclostrophic, instead of geostrophic.

The T – S curves plotted by us for three stations (fig. 3.23) underline the significant differences of the characteristics of the core (St. 2769) and the peripheral waters of the ring (St. 2773). The differences in the T – S curves for stations No. 2768 and 2769, located only 8 km from each other, point to the sharpness of the main frontal interface within the temperature range of $17 - 22°C$ and salinity range of $35.6 - 36.2$ ppt. This part of the main frontal interface within the $100 - 200$ m depth range is apparently the sharpest. This is particularly evident in the field of anomalies (deviations) of temperature (ΔT), salinity (ΔS) and density ($\Delta \sigma_t$) (fig. 3.24 a, b, c) in relation to the corresponding values at the same levels at St. 2773. The values of the anomalies at the outer stations of the section, No. 2772 and 2766, are still so great that the width of the core of the ring at the 500 m level cannot be regarded as less than 150 km. The transformation of the waters of the core is best noticed at a section in the density field (fig. 3.25) where, against a background of isopycnals, the dome-shaped curvature of which is significantly weaker than that of the isotherms and isohalines (fig. 3.21), we have marked the temperature anomalies in relation to the values characteristic of the given density values in the Sargasso Sea. These anomalies are significantly smaller than those depicted in fig. 3.24 a, as the anomaly values no longer include the effect of the upwelling of water (and therefore of the isopycnals) in the central part of the ring. On the other hand, the relatively high values of these negative anomalies (up to $-0.5°C$) are noticeably drawn out to the side between the isopycnals $\sigma_t = 25.0$ and $\sigma_t = 26.0$, which indicates that the cross-frontal heat and mass transfer is the highest in this particular range of densities (or in the depth range $100 - 200$ m). It is in this depth range that the most perceptible cold intrusions with temperature inversions are located (see fig. 3.22).

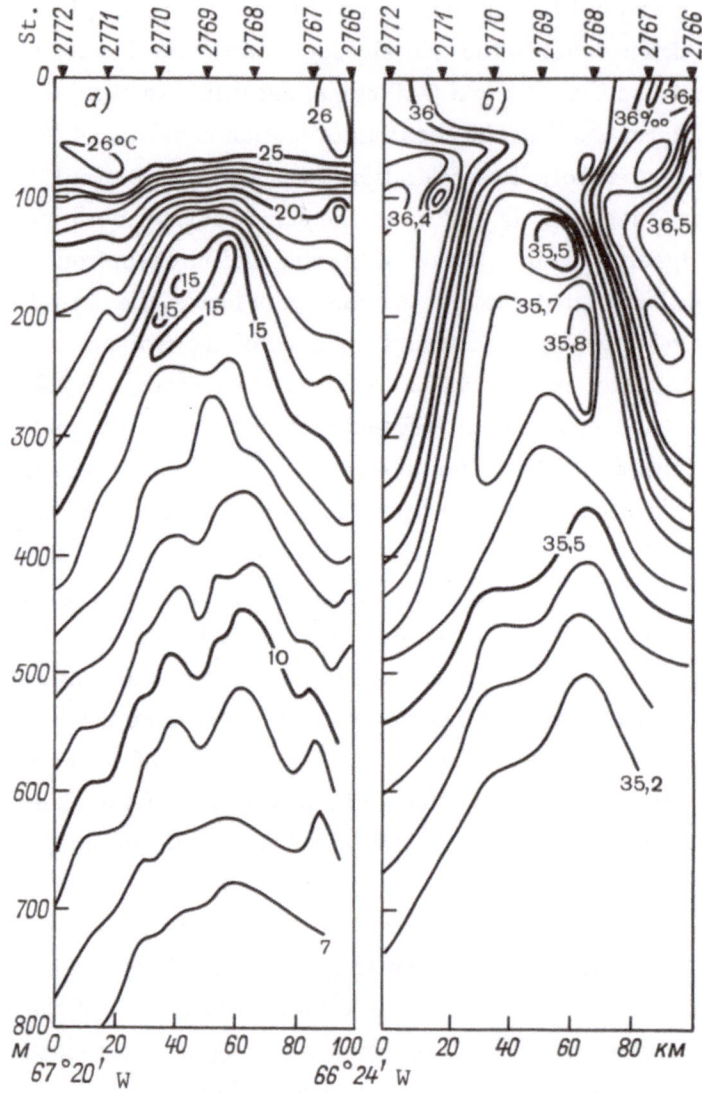

Fig. 3.21. Section across the core of a cold ring along 35°N. a – temperature, °C; b – salinity, ppt.

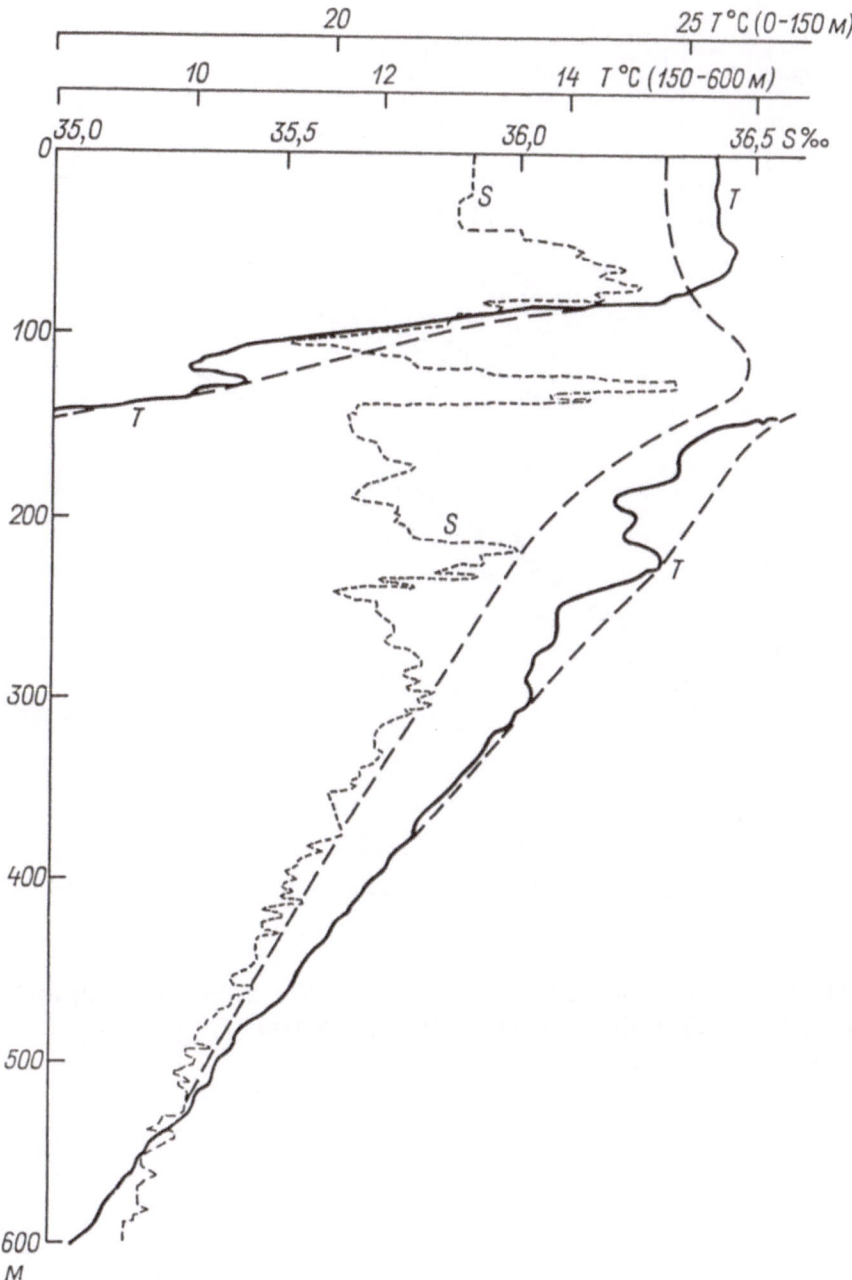

Fig. 3.22. Vertical profiles of temperature (T – solid line) and salinity (S – dotted line) at the edge of the cold core of a cyclonic ring (St. 2770, position shown in fig. 3.21). Dashed line shows the typical vertical distribution of temperature and salinity in the waters of the Sargasso Sea. We can clearly see that the observed intrusions are colder and fresher than these waters.

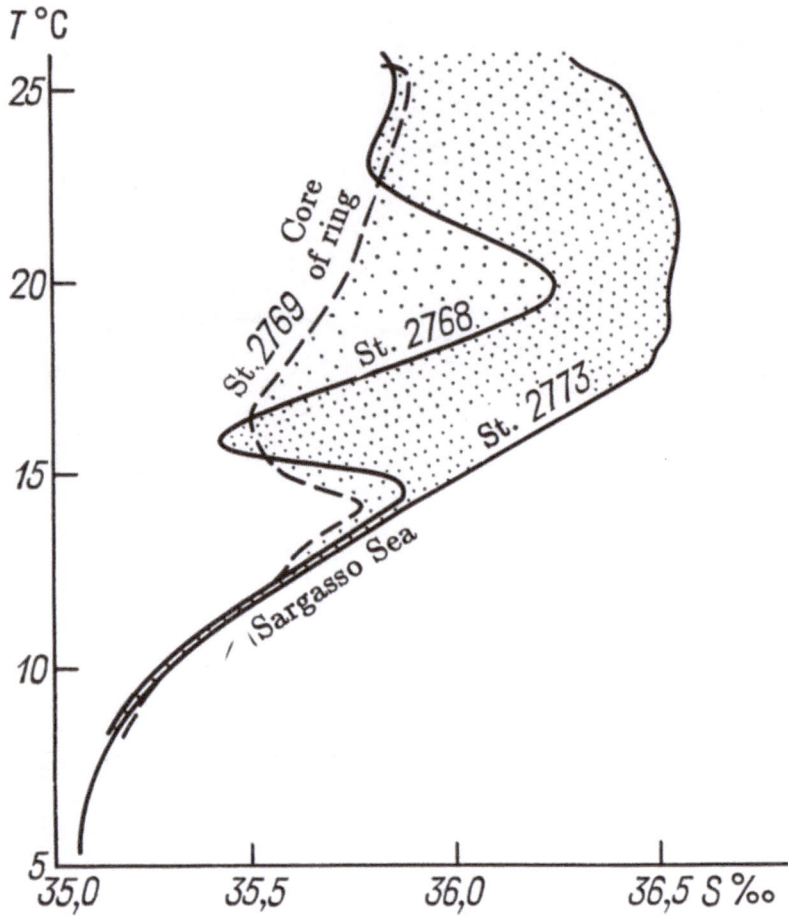

Fig. 3.23. T – S curves of stations in the centre (No. 2768 and 2769) and on the periphery (No. 2773) of the cold ring of the Gulf Stream.

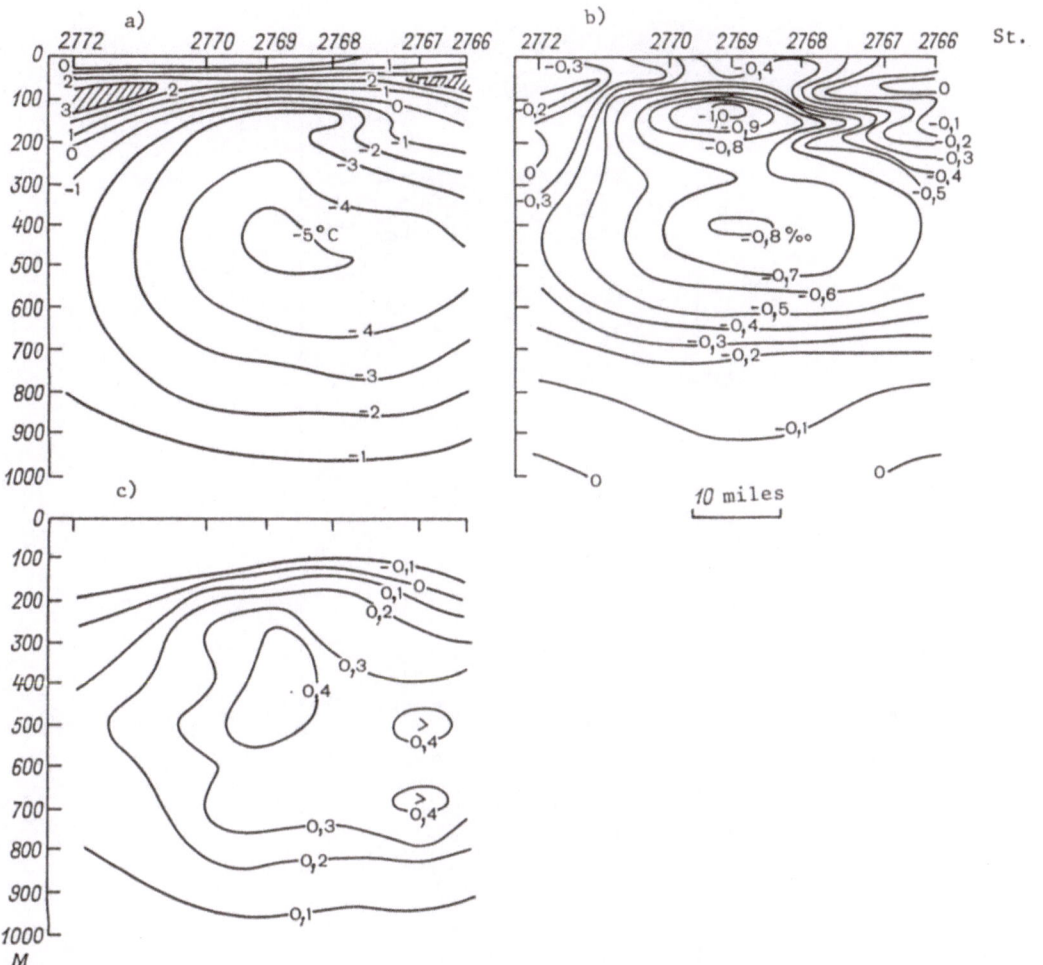

Fig. 3.24. Anomalies (compared with St. 2773) of temperature $\Delta T°C$ (a) salinity ΔS *ppt* (b) and nominal density $\Delta \sigma_t$ (c) at a section across the core of a cyclonic ring of the Gulf Stream.

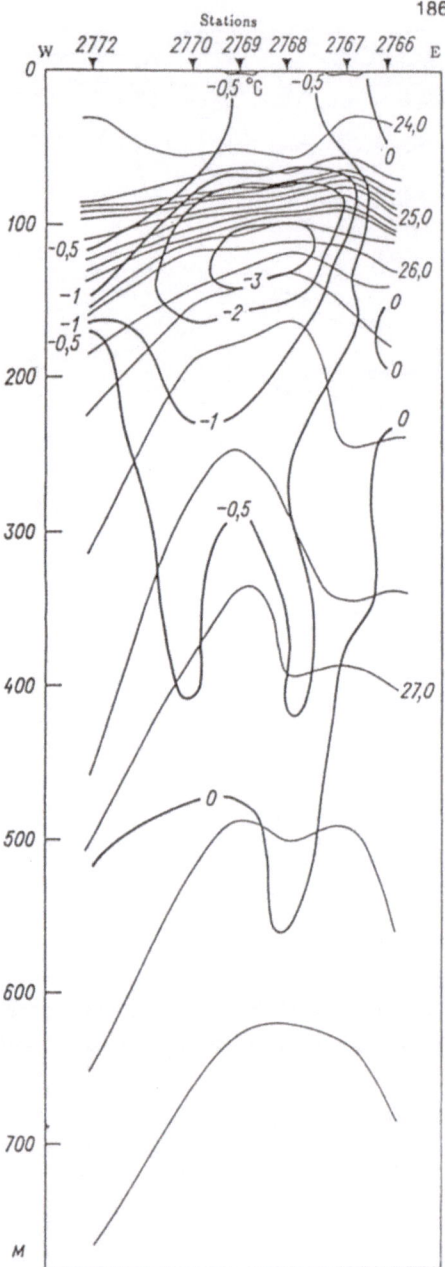

Fig. 3.25. Temperature anomalies (bold lines, °C) with respect to the waters of the Sargasso Sea at isopycnic surfaces (a section of which for a profile across the cold ring is shown by thin lines).

A comparison of the positions of cold intrusions of anticyclonic and cyclonic rings on the same T – S diagram (fig. 3.26) demonstrates their fundamentally different origin. All the points corresponding to intrusions of warm anticyclonic rings lie outside the T – S region which characterizes the waters of the rings themselves, whereas all the points of intrusions of cold cyclonic rings lie within this region. Proceeding from the position of the intrusions on the T – S diagram, we can once again substantiate the conclusion that the intrusions detected on the periphery of the warm ring are the entrained streams of shelf waters that spread from the north along the eastern boundary of the ring right up to its southern periphery. Cold intrusions of cyclonic rings are encountered much closer to the centre of the ring, and consist of transformed slope waters of the core, which penetrate the shell consisting of Gulf Stream waters possibly under the effect of centrifugal forces. Since the ageostrophic cross-frontal motions should approximate isopycnic motions, the temperature inversions associated with cross-frontal cold intrusions cannot be greater than the maximum absolute value of the isopycnic temperature anomaly in the core of the ring (i.e. not more than $3°C$, see fig. 3.25). In actuality, they should be even smaller, as the horizontal scales of cold-ring intrusions, judging by our observations, do not usually exceed 10 km. Indeed, the temperature inversions in fig. 3.22 are within the range of $0.4 - 0.9°C$.

With the help of a towed AIST thermohaline sensor, we found that the surface layer of the cold ring contained numerous frontal interfaces, not associated with the main front surrounding the core, which showed small differences, but sharp temperature gradients. Many of these fronts had a negative T – S correlation, and therefore sharp density gradients. Measurements taken at the same time did not enable us to establish the horizontal and vertical extent of these frontal interfaces. However, it is clear that they were not associated with the general deformation field of the ring, as the background large-scale horizontal temperature and salinity gradients in this area usually show a positive correlation. In summer, the less saline slope waters that are heated up near the surface apparently override the more saline waters of the shell in a thin layer from the central regions of the ring. In this case, nonuniform wind-driven mixing can give rise to shallow fronts with a negative T – S correlation [60]. Our measurements have confirmed that the surface layer temperature in the centre of the ring in September was only $0.2 - 0.3°C$ lower than on the periphery, while the salinity diminished by $0.3 - 0.4$ ppt toward the centre of the ring. As a result, the waters of the core at the surface ($\sigma_t = 23.73$) were significantly lighter than the peripheral waters ($\sigma_t = 23.94$), whereas the exact opposite was observed at a small depth.

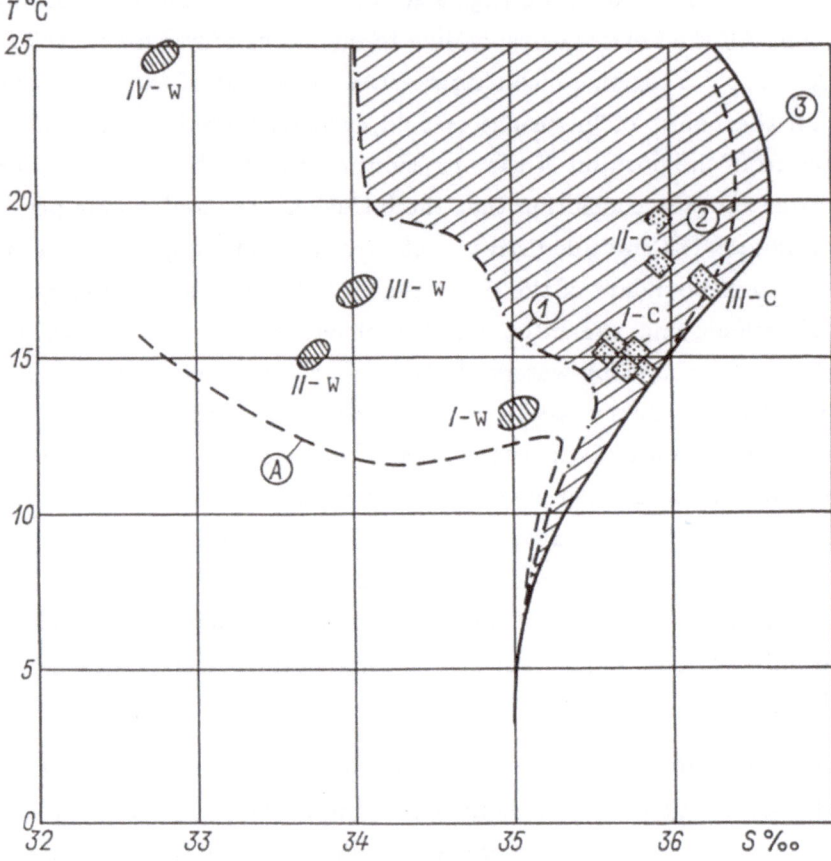

Fig. 3.26. Position of intrusion-related temperature inversions on a general T – S diagram of waters in the Gulf Stream area. *W* – inversions of a warm ring; *C* – inversions of a cold ring; *A* – shelf waters; 1 – slope waters; 2 – waters of the Gulf Stream; 3 – waters of the Sargasso Sea. *I, II, III, IV* – different groups of inversions.

3.2 Peculiarities of coastal upwelling fronts

We have already briefly discussed in this or that connection some of the characteristics of frontal zones that form in areas of coastal upwelling (see sections 2.1, 2.2, 2.9). Let us now compare some of the best known areas of coastal upwelling in order to establish the differences in the characteristics of frontal zones and fronts, which could arise as a result of different hydrographic regimes in these areas with the same frontogenetic process. Of greatest interest to us is the role of spatial variability of salinity in the formation of the main characteristics of frontal interfaces. It is remarkable that the close relationship of the positions of local current velocity maxima (based on direct measurement) with the positions of local maxima of the horizontal salinity gradient in frontal zones has already been ascertained [193], though it has not yet been explained.

Coastal upwelling recently became the subject of intense and detailed scientific research (national and international programs CUE, CUEA, CINECA, JOINT-1, etc.*), in the course of which many new questions arose. One such question arose on the basis of the author's own observations and study of available data. The author questioned whether the fact that the thermal fronts formed in a coastal frontal zone during upwelling in the presence of significant horizontal salinity gradients are characterized by considerably higher horizontal temperature gradients than in the cases where the salinity field is spatially homogeneous was coincidental or not. During a discussion of this question at a SCOR working group on coastal upwelling processes in 1975, I gave the conditions in the Oregon and Peru coastal areas as examples. The sharpness of the thermal fronts in the first area contrasts clearly with their diffuse character or even absence in the second area. This difference obviously indicates that one and the same frontogenetic process (in this case coastal upwelling) can bring about different hydrographic conditions.

In 1975, there were simply no reliable qualitative arguments to substantiate the proposed hypothesis. However, even today it is difficult to find the necessary numerical data in the literature. Only a small number of authors give information about the actual values of the horizontal temperature, salinity or density gradients observed by them at fronts in coastal upwelling zones. In many cases, we had to search through expedition reports for relevant summaries of hydrographic data and analogue records of continuous temperature and salinity measurements from a moving vessel in order to determine the approximate values of these gradients. Preliminary studies on the basis of selected American data were carried out in 1977 by a thesis student of the Moscow Physicotechnical Institute, S.V. Sergunin, under the author's supervision. Now that the necessary data have been collected and organized and a better understanding of some of the physical details of oceanic

---

* CUE – Coastal Upwelling Experiment (Oregon coast, USA, 1972-1973).

CUEA – Coastal Upwelling Ecosystem Analysis (USA, Oregon coast, Peru coast, 1973-1977).

CINECA – Cooperative Investigations in the North-East Central African region (International study under the aegis of IOC/UNESCO off the northeastern coast of Central Africa, Canary Upwelling, 1971-1974).

JOINT-1 – Joint investigations of the USA, FRG and France in the vicinity of the Canary Upwelling in 1974.

frontogenesis has been achieved on the basis of numerical models [36, 38, 178], it has become possible to find the correct physical explanation of the observed phenomenon.

The following coastal upwelling areas will be subjected to a comparative analysis below:

1) the Oregon upwelling;

2) the South African (Capetown) upwelling;

3) the Brazil upwelling (near the Cabo Frio cape);

4) the Canary (West African) upwelling;

5) the Peru upwelling;

6) the Venezuela upwelling.

Some of the hydrophysical characteristics of the first five areas are given at the end of this section in table 3.1, where they are presented in a natural sequence based on the "intensity" of the observed thermal and density fronts. In the first four areas, the fronts with their characteristic manifestations were observed firsthand by a number of researchers, and the corresponding comments or descriptions can be found in the literature. We have no direct or indirect information about fronts for the Peru upwelling area, while the available data indicate that the horizontal temperature gradients in this area are so diffuse that we can hardly speak of frontal zones or fronts even within the scope of the formal criteria proposed by us (see section 1.2). We can only say that in this area, we are actually dealing with a classic example of the superposition of certain "events" of an intensive coastal upwelling, associated with local winds, on the general background of a so-called "climatic" upwelling determined by the direction of the general ocean circulation. At the same time, however, the local and horizontal surface-layer temperature gradients, even in the most favourable season for intensive upwelling "events", are only $2 - 3$ times higher than the background (climatic) ones due to a whole series of factors (including the absence of salinity stratification). The sixth area selected by us (the Venezuela upwelling area in the Caribbean Sea) is an interesting example. It appears to have all the conditions for the formation of sharp density fronts due to the strong freshening effect of the Orinoco and Amazon rivers. However, investigations related to the upwelling in this area do not directly mention any fronts associated with the upwelling, and the overall temperature difference between the waters of the coastal upwelling and the ambient waters of the Caribbean Sea does not usually exceed $2-3°C$ in any season, which obviously eliminates the possibility of any sharp fronts forming here in the surface-layer temperature field. We shall now discuss the conditions in each of the selected areas in greater detail.

1. **Oregon upwelling.** The existence of clearly defined fronts in the Oregon upwelling zone has been noted many times in the literature [101, 150, 186], and the thermal fronts at the surface are quite discernible on satellite IR images as well [195]. The overall temperature contrast during the period of intensive upwelling here can reach $6 - 8°C$ in a coastal zone with a width of only $20 - 30$ miles. The frontal interfaces in this area are clearly defined zones of convergence [150], which has been confirmed by experiments with surface and subsurface buoys, conducted within the framework of the CUE program. Fronts in this area are accompanied by intensive intrusions of warm layers with temperature inversions, the stability of which is achieved by an increase in salinity [186, 230]. According to the data of analog measurements of temperature and salinity during traverses of fronts in the Oregon upwelling zone [101], the horizontal temperature gradients which reach $1 - 4°C$ near the surface and the salinity gradients which amount to $0.7 - 2.0$ ppt/km near the surface are typical for this area, and are higher than for all the other coastal upwelling regions. The fact that the values of these gradients during the period of intensive upwelling are always negatively correlated with each other makes the horizontal density gradients at the fronts in this area also the sharpest known in the coastal upwelling zones (see table 3.1). In our opinion, the sharpness of frontal interfaces observed in the Oregon upwelling area is due to the following factors:

1) the combination of a sharp salinity stratification with a thermal stratification of the opposite sign in the pycnocline closest to the surface;

2) the great depth of the littoral region (150 m depth only 20 km from the shore) which is conducive to the formation of geostrophic along-frontal jet streams.

The very high relative contribution of salinity to the spatial variability of density in this area is characterized by low values of the σ_T/σ_S ratio of the mean standard deviations of temperature σ_T and salinity σ_S from the mean values, with low values of the ratio, $\alpha\Delta T/(\beta\Delta S)$, of the contributions of the overall limits of temperature and salinity changes (ΔT and ΔS) in the upwelling zone to the overall change in density $\Delta\rho = \Delta\rho_T + \Delta\rho_S = -\alpha\Delta T + \beta\Delta S$ (see columns 5 and 6 in table 3.1).

The conditions of formation of an intrusive finestructure at frontal interfaces of the Oregon upwelling are highly distinctive. For example, the warm intermediate layer that frequently arises at intermediate depths below the main pycnocline is a positively correlated T – S anomaly on the background of the general negative T – S correlation of thermohaline variability [61]. The appearance of warm intrusions in this area can be attributed to the sinking "under the front", along the surface of the frontal interface, of the upwelled waters that heat up slightly while they are near the surface and as they flow out from the shore in a near-surface drift [186] (see fig. 2.38).

2. **South African (Capetown) upwelling.** The intensive thermal fronts of the South African upwelling have been described in detail by N.D. Bang [83]. The horizontal temperature gradients which typically amount to $0.5 - 1.0°C/km$ and more*at these fronts place the South African upwelling area at the same level as the Oregon area. The slope of the bottom in the Capetown area has the same order of magnitude as in the area of the Oregon coast. The presence of an intrusive thermohaline finestructure in the profiles and sections given in Bang's paper [83] makes the conditions in this frontal zone even more similar to the conditions near the Oregon coast. Basically, it would seem that the South African upwelling could be regarded as the exact analog of the Oregon or California upwellings at southern latitudes. However, a careful analysis of the data reveals significant differences which do not permit us to draw a complete analogy between them. Unlike the conditions of stratification in the surface layer along the Pacific coast of North America, the vertical temperature and salinity gradients in this layer near the Atlantic coast of South Africa have the same sign. This determines the significant **positive** correlation of the horizontal temperature changes in the South African upwelling zone. Furthermore, the waters of the Agulhas Current, which approach this area from the south, have high temperature and salinity values, unlike the river-freshened waters flowing from the north along the entire Pacific coast of North America. It is these two circumstances that make the horizontal density gradients at the fronts of the South African upwelling less intensive than at the fronts of the Oregon upwelling, despite the greater overall temperature gradient near Capetown which amounts to $8 - 10°$ in a coastal zone only $10 - 20$ miles wide. On the other hand, the positive T − S correlation does not prevent the horizontal temperature gradients at the fronts here from being just as sharp as in the Oregon upwelling area.

The total positive T − S correlation of thermohaline variability in this area greatly facilitates the formation of thermohaline intrusions with temperature inversions at the frontal interfaces, due to the available energy of "thermoclinicity" [263]. In this case, their formation apparently does not require the somewhat special con-

---

* Bang [83] notes that the local temperature changes at the upwelling front sometimes amounted to several degrees over several hundreds of metres (i.e. about $10°C/km$).

ditions (heating from the surface) which are necessary for the formation of warm intrusive intermediate layers off the coast of Oregon and California.

Another interesting characteristic of the South African upwelling is the additional zone of divergence and upwelling located $30 - 35$ miles further out to sea, i.e. at a distance of $50 - 60$ miles from shore. The depth of the ocean at this point reaches $1000\ m$, so that the formation of this zone can hardly be associated with the bottom relief. It is also difficult to associate this divergence with the Ekman outflow from shore in the near-surface layer, as was the case in the Oregon upwelling (fig. 2.38). The distance between the divergence and the coastal frontal interface is too great. Bang [83] suggests that this zone and the adjacent seaward frontal interfaces and jet streams are associated with the large-scale general circulation.

3. **Brazil upwelling** (in the vicinity of Cabo Frio). The presence of fronts in this area can be ascertained on the basis of the results of continuous measurements taken from a moving vessel with an automatic temperature and salinity recording system [152]. Our studies based on these data and the sections in [180] give typical values of the horizontal temperature gradients at these fronts as not higher than $0.5°C/km$, which places the Brazil upwelling in third place in table 3.1. In the upwelling region in spring, we observe a subsurface salinity maximum at the $50\ m$ level, which is due to the fact that coastal waters freshened by river run-off flow in from the west near the surface [180]. This situation is the cause of a negative T – S correlation in the patches of cold water and at the fronts of the shallow upwelling during this season, which is clearly seen in the temperature and salinity records published in [152] (see also fig. 2.14 in section 2.2). During the rest of the year, the T – S correlation of spatial temperature and salinity variability in the vicinity of the Brazil upwelling is a positive one. During more intensive upwelling from depths of $100 - 200\ m$, the T – S correlation in the upwelling patches becomes positive over large areas even in spring. As a result, the density contrasts at the fronts cannot reach the same values that are observed during a negative T – S correlation. The ratio $\alpha|\Delta T|/(\beta|\Delta S|)$ for the overall limits of density variability due to temperature $\alpha\Delta T$ and to salinity $\beta\Delta S$ in the Brazil upwelling zone is significantly higher than in the two areas already discussed, which points to the less significant role of salinity in the local dynamics of fronts, as compared with temperature. This applies directly to the upwelling area near the Cabo Frio cape. Somewhat farther west in the vicinity of Ilha Grande, where water freshened to 33 ppt flows out of Sepetiba Bay, we observe purely saline fronts with horizontal gradients of up to $1.0 - 1.6 ppt/km$ [183]. No temperature contrasts are observed. However, it is unlikely that these

fronts are related to the upwelling process.

4. **The Canary (West African) upwelling.** Indications of fronts at the surface of the ocean (rips, rough water) were observed in this area by navigators over 40 years ago [216, 226]. References to, and descriptions of, fronts in the zone of the Canary upwelling can also be found in the recent scientific literature [245]. On the other hand, there are also indications that no convergence zone is observed near the surface in the area studied by the JOINT-1 expedition (between 21 and 22°N) (in contrast to the area of the Oregon upwelling), which could be due to the weakness of the horizontal density gradients at the surface in the vicinity of the Canary upwelling [150]. It cannot be ruled out that the hydrographic and topographic conditions vary greatly within the upwelling zone from 23 to 9°N, and that the observed cases of rough water and rips were associated with near-bottom convergence at the shelf-break where it passes into the continental slope at depths of about $100 - 200$ m, $40 - 50$ to 100 km from the shore. This point of view has been confirmed by studying the data of Römer [216] and Schumacher [226]. All of the averaged monthly average coordinates of rough water and rip sightings given by Römer fall within the region between 9 and 12°N near the 200 m isobath (fig. 3.27). In turn, Schumacher's detailed maps which display the rough water and rip sightings show that these phenomena were hardly ever observed off the coast of Africa between 20 and 22°N. In the more southern areas (from 20 to 5°N), the points of rough water and rip sightings closest to the African coast are located at a distance of not less than 50 km from shore (see fig. 1.1 a, b), while in the Oregon upwelling zone north of Newport (about 45°N), the sharpest zone of convergence is located (because of the very deep coastal waters) within a 10 km distance from the shoreline [150]. On the basis of the available observational data and measurements (especially those pertaining to the CINECA and JOINT-1 expeditions), we can conclude that in the immediate area of the Canary upwelling ($100 - 200$ km from shore), thermal fronts are not encountered very often, and the horizontal temperature gradients at them are weak (as has also been confirmed by satellite data [174]). Only in exceptional cases do they reach $0.5°C/km$. At the edges of the patches of intensive upwelling, the horizontal temperature gradients do not usually exceed $0.1 - 0.2°C/km$.

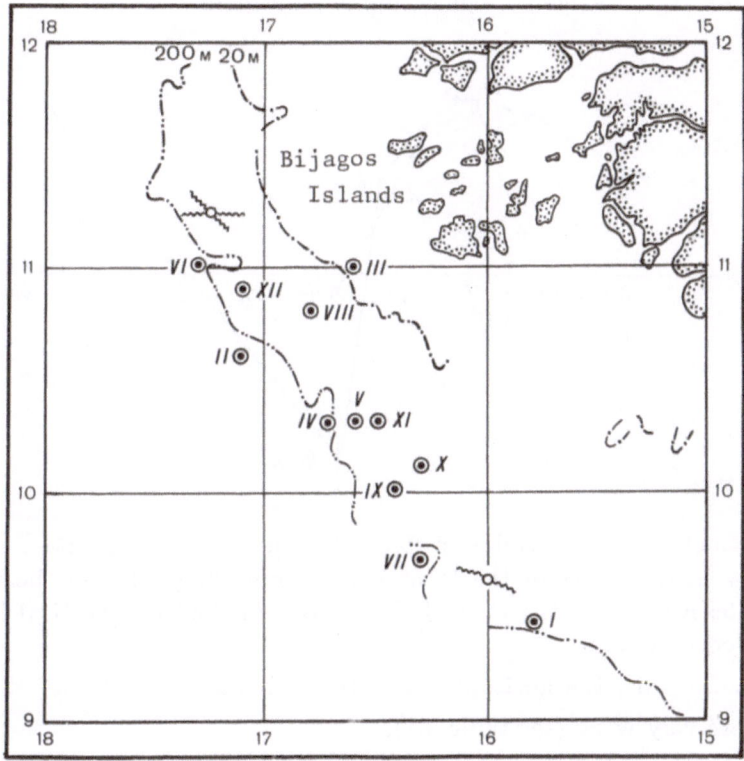

Fig. 3.27. Average monthly positions of the points at which rough water and rips are most frequently observed near the coast between 9 and 12°N, based on [226]. Wavy lines show the direction in which the rip lines have been seen to extend.

The thermohaline variability at the surface in this area is characterized by a variable sign of the T – S correlation, negative during the transition from summer to autumn, and positive the rest of the time (fig. 3.28). The vertical stratification off the coast of Africa is much weaker than off the Oregon coast, and because of this, the range of temperature variations $|\Delta T|$ in the zone of intensive upwelling does not exceed $5°C$ when $\alpha|\Delta T|/(\beta|\Delta S|) \simeq 1.5$. In our opinion, the weakness of the fronts in the zone of the Canary upwelling is associated primarily with the following factors:

1) with the prevailing positive T – S correlation for the spatial thermohaline variability over a relatively small range of temperature variations $|\Delta T| \simeq 5°C$;

2) with the relatively wide and shallow shelf, the effect of which excludes the formation of a near-surface zone of convergence near the shore, and also prevents

the formation of an along-shore geostrophic jet stream.

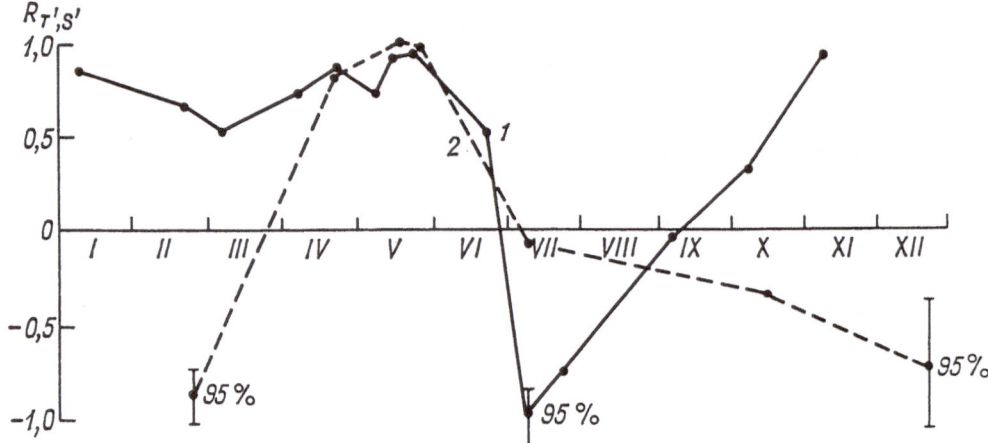

Fig. 3.28. Coefficients of correlation $R_{T',S'}$ for the spatial variability of temperature and salinity in two areas of the West African upwelling. 1 – in the area of Cap Blanc (20°55′N), 2 – in the area of Cape Roca (12°00′N). Vertical lines show a 95% confidence interval.

Because of this, the horizontal density gradients at the fronts of the Canary upwelling are very weak (see table 3.1).

It cannot be ruled out that sharper fronts are located at the boundary of the wider ($\sim$ 1000 km) zone of climatic upwelling between the $20 - 22°C$ waters surrounding the Canary Islands and the Cape Verde Islands, and the $25 - 27°C$ surface waters of the tropical Atlantic. The high concentration of rough water and rip sightings along the 25°W meridian between 20 and 5°N on Schumacher's maps is an indication of this (see fig. 1.1 a, b) [226]. This zone is apparently similar to the secondary upwelling zone observed farther out to sea from the South African coastal upwelling.

5. **Peru upwelling.** This most typical and widely known upwelling area is characterized by diffuse horizontal temperature gradients and the apparent absence of fronts. The author, who participated in a survey of this area in March 1974, when patches of intensive upwelling with a temperature of about 16°C accompanied by thick coastal accumulations of fog were found near the shore, observed that these patches were surrounded by only weak contrasts of surface layer temperature of about 5° per 100 km (fig. 3.29). At the same time, the sharpest local horizontal temperature gradients did not exceed 0.1°C/km. Satellite IR measurements in this area [174] indicate that the values of the "frontal" gradients of the temperature field of the ocean do not exceed 0.2°C/km. Numerous direct measurements of

the temperature field of the Peru upwelling during different years have shown the same. Fronts are not mentioned at all in any of the numerous descriptions of the Peru upwelling. No intrusive intermediate layers with temperature inversions in the vertical profiles have been observed either.

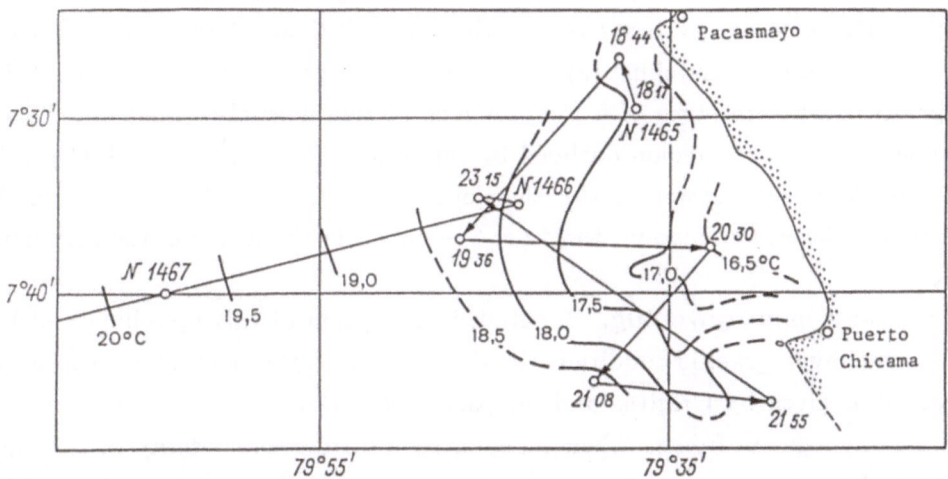

Fig. 3.29. Temperature distribution for the surface layer of the ocean near Puerto Chicama and Pacasmayo (Peru), based on data from a survey carried out by the author from the research vessel "Akademik Kurchatov" on 28 Feb 1974. Thin solid lines show the tracks of the vessel. The stations used to plot fig. 3.30 a are marked.

At the same time, we note that the slope of the bottom in the Peruvian coastal area is practically identical to that in the Oregon area. The background climatic upwelling in the Peru area is perhaps the most intensive of all known ones (with the exception of El Niño periods), and is felt within a wide band extending at least 1000 km from shore, as in the area of the Canary upwelling. The main thing that distinguishes the hydrographic regime of the Peru upwelling from conditions in other areas is the practically complete absence of salinity stratification. The complete range of salinity variations in the zone of intensive upwelling between Callao and Chicama (from 12 to 8°S) does not usually exceed $0.1 - 0.3$ ppt. During anomalous years, this range can increase to $0.3 - 0.5$ ppt in the vicinity of Talara ($4 - 5°S$). Such a small salinity variability gives the Peru upwelling area the highest values of the ratio $\alpha|\Delta T|/(\beta|\Delta S|)$ which fluctuates from 3 to 22 (table 3.1) with an overall positive T – S correlation. In our opinion, it is these conditions that are responsible for the diffuse nature of the frontal zones in the Peru upwelling area. MacVean and Woods' model [178] explains this quite clearly, indicating that this is associated with the low degree of "thermoclinicity" of local stratification. By "thermoclinicity"

MacVean and Woods mean the angle of slope of the isothermal surfaces to the isopycnals. In the area of the Peru upwelling where the density field is almost completely determined by temperature, this angle approximates zero. Because of this, the frontogenetic effect of the isopycnic motions prevailing in the water masses is also close to zero. The degrees of "thermoclinicity" of the stratification in the Peru and Oregon coastal areas, based on the data of [67] and [140], are compared in fig. 3.30. In the Peru upwelling (a), the isotherms are parallel to the isopycnals for the greater part of the area of a 130-mile long section normal to the shore. Their maximum slope in the region outlined by the dashed line amounts to $1°C$ per 50 miles. In the Oregon upwelling (b), the slopes are large everywhere, except in the near-bottom layer, and amount to $1°$ per $3 - 5$ miles in the region of the maximum value.

6. **Venezuela upwelling.** A detailed description of this upwelling and the background hydrographic conditions in the vicinity of the northeastern coast of Venezuela is given by Griffiths and Simpson [138]. However, the authors do not mention any signs of fronts. Their maps of temperature and salinity distribution at the surface of the ocean are based on the data of hydrographic stations which are located quite far from each other $(20 - 40\ km)$. Therefore, the horizontal temperature and salinity gradients shown on the maps (up to $0.2°C/km$ and up to $0.3\ ppt/km$) do not appear very high, indeed they are rather diffuse, with the overall contrasts of temperature $|\Delta T| \simeq 2 \ldots 3°C$ and salinity $|\Delta S| \simeq 8 \ldots 10\ ppt$, and a generally negative T – S correlation which characterizes this area. We have no other information about the conditions in this area, except for Wüst's classic study [271] based on archived (and far from synoptic) data and Gade's paper [123]. The data of Griffiths and Simpson give us reason to believe that the freshening of the surface waters in this area due to the run-off of the Orinoco and Amazon rivers leads to intensive heating of the near-surface layer by solar radiation. Obviously, this heating and the stable tradewinds from November to May greatly reduce temperature contrasts in the surface layer of the ocean.* Griffiths and Simpson [138] note that, according to their observations, sharper horizontal temperature gradients are observed in the deeper layers 100 m and more below the surface. This agrees with the above assumption, but still requires further verification from continuous temperature and salinity measurements at several depths from a moving vessel and from satellite IR data.

---

* This has been confirmed by R. Griffiths in a personal communication in response to my request (K.F.).

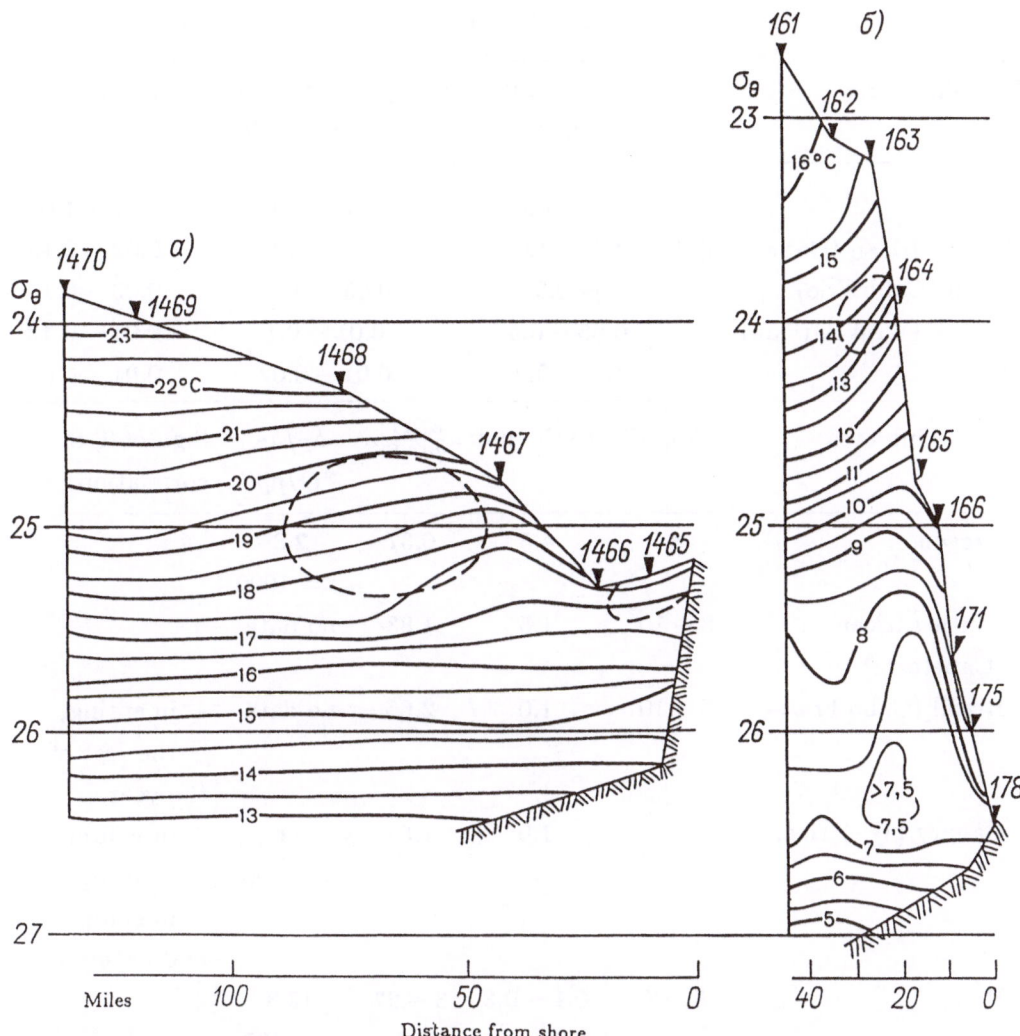

Fig. 3.30. Temperature distribution as a function of nominal potential density (σ_θ) in the vicinity of the Peru upwelling (a) based on the author's data, and in the area of the Oregon upwelling (b) based on the data from [140]. Dashed lines mark regions of maximum thermoclinicity associated with the upwelling of cold subsurface waters.

Table 3.1

Upwelling area	Typical values of $\partial T/\partial x$, $°C/km$	Typical values of $\partial S/\partial x$, ppt/km	Typical values of $\partial \sigma_t/\partial x$, km^{-1}
Oregon	$1.0 - 4.0$	$0.7 - 2.0$	$0.1 - 1.0$
South African (Capetown)	$0.5 - 1.0$	$0.1 - 0.2$	$0.02 - 0.15$
Brazil (Cabo Frio)	~ 0.5	$0.05 - 0.08$	$0.03 - 0.18$
Canary (West African)	$0.05 - 0.5$	$0.01 - 0.1$	$0.01 - 0.12$
Peru	$0.1 - 0.2$	$0.01 - 0.02$	$0.01 - 0.03$

Upwelling area	$|\Delta T|, °C$	$|\Delta S|$, ppt	$\frac{\alpha|\Delta T|}{\beta|\Delta S|}$	δ_T/δ_S $°C/ppt$	Sign of T-S correlation
Oregon	8	4.5	0.37	2.8 $n = 1000$	—
South African (Capetown)	$8 - 10$	1.5	1.83	no data	+
Brazil (Cabo Frio)	$6 - 10$	1.0	2.66	no data	− in spring, + the rest of the year
Canary (W African)	5	1.0	1.53	9.6 $n = 400$	+ in winter and spring, − in summer and autumn
Peru	$5 - 7$	$0.1 - 0.3$	$3 - 22$	13.3 $n = 400$	+

3.3. Salinity fronts originating from river
discharge into coastal areas of the ocean

We can now regard as legends the references of ancient navigational records to the fact that during the flooding of the Nile River, skippers could replenish their supply of fresh water by scooping it up directly from the sea surface near the Levantine coast of the Mediterranean Sea. The discharge of the Nile is now regulated by means of the high Aswan Dam, and records of the abundant effluxes of fresh water from the delta into the sea eastward and then to the north in a narrow band along the Levantine shore have been preserved only in the bottom sediments. Like the old navigational records, they are evidence that the waters of the Nile spread over great distances from the delta, practically without mixing with the waters of the sea or becoming saline for a long time, which in itself points to the curious aspects of dynamics and mixing at the boundaries of fresh waters flowing into the sea. These boundaries must have been very stable and sharp during the flooding of the Nile.

Despite the significant number of areas where situations like this occur and the highly interesting physical, geochemical, biological and geological consequences of such "encounters" of river and sea waters, the detailed physical conditions in the regions of intensive river discharge into the oceans and sea have not yet been documented very well. Though the literature does contain data on the volumes of discharge and the flow rates for the majority of the large rivers that flow into the ocean (see table 3.2), we do not know the characteristic dimensions of the freshwater lenses that form on the surface in coastal regions during different stages of flooding, nor do we know the relationship between these parameters and the intensity of the discharge of the river. Consequently, we still cannot predict at what distance from shore the sharpest salinity fronts should be located, and what the mean and minimum salinity of the freshened water in the lens enclosed by these fronts will be. We do know that all of these characteristics should differ for river discharge through the delta and for a typical estuarine regime (see ref. [91]). However, we also encounter intermediate regimes, the characteristics of which depend on the form of the mouth of the river, the intensity of its discharge and local tidal phenomena. If the river has an estuary, the latter can have its own **estuarine fronts** independent of those which enclose the lens of freshened waters in the open sea, and which we shall refer to as **river discharge fronts**. In large estuaries, tidal mixing can be so intensive that river discharge fronts may not be present at all beyond the estuary.

Table 3.2

River	Average annual discharge, km^3/yr	Maximum discharge, $10^3 m^3/s$	Reference and remarks
Amazon	5500	200	[134, 222]
Yenisei + Ob*	1428	210	70% of discharge in 2 months of high water [2, 3]
Congo	1190	40	[126, 134]
Orinoco	1080	65	[123]
Lena	650 − 700	100 − 120	80% of discharge in 2 months of high water [2, 3]
Yangtze	677	?	[151]
Mississippi	300 − 500	22 − 56	[134, 151]
Volga	250	24.8	[46]
Columbia	230	20	[196]
Danube	200	10.1	[46]
Pechora	130	?	[46]
Neva	80	3	[46]
Connecticut	17 − 18	1.7 − 2.5	[126]
Hudson	17	?	[90]

* Due to the closeness of their mouths in the Kara Sea, these two rivers form a single freshened lens on the sea surface.

In the end, the whole variety of regimes is determined by the correlation between the river discharge, the form and size of the mouth, the intensity of tidal mixing, the overall depth, and the area of the shelf. Under such conditions, it is very difficult to develop a sufficiently universal theory of the phenomenon, and to describe in sequence all of its aspects, including fronts. Nevertheless, the progress achieved in research over the past few years instils the hope that we will soon be able to describe successfully the main features of the structure, behaviour and dynamics of both river discharge and estuarine fronts. In this section, I would like to sum up the available field data on these fronts, and to review the main theoretical results. Let us first look at river discharge fronts.

3.3.1. Open-sea discharge fronts

The freshening effect of the great Siberian rivers the Ob, Yenisei and Lena on the marginal seas of the Arctic Basin has been known from time immemorial. Judging by the "dead water" phenomenon* observed here, the layer of freshened waters during river flooding in these seas has a thickness in the neighbourhood of $3-5\ m$. Flood outflows of such rivers as the Yenisei and Lena can amount to $70-80\%$ of their annual discharge, their discharge rates in this case reaching $100,000 - 120,000\ m^3/s$ [3]. However, the conditions in this area are greatly complicated by the melting of sea ice during the spring-summer period, due to which considerable amounts of meltwater of very low salinity are added to the flood discharge of the rivers along the entire Arctic coast of Eurasia [2]. The melting of shore ice and the winter ice cover in the 100 km wide coastal zone along the entire shore from the White Sea to the Bering Strait produces an amount of fresh water equal (in order of magnitude) to the total annual discharge of all the large rivers of this area. The presence of powerful freshening effects in the coastal zone of the Arctic seas results in a situation where the freshened waters in the summer time occupy vast areas, i.e. there are $573,000\ km^2$ within the 25 ppt isohaline in the Kara Sea alone (which receives the Ob and Yenisei rivers) [2]. The Arctic Basin being relatively closed, further mixing of these waters with seawater results in a low salinity, of the upper layers of the Arctic Ocean, which does not exceed $32 - 33\ ppt$ even under ice. However, even against this background, river discharge fronts in the salinity should develop during the spring-summer period. In view of the specific nature of

---

* An abnormally high resistance to the movement of ships, due to the internal waves generated at the interface between salt and fresh waters.

the above-mentioned conditions of their formation in this area, the river discharge fronts of the Arctic Basin should be subjected to special analysis.

River discharge fronts in the ocean arise in cases where the waters of river discharge, not having sufficiently mixed with the seawater in the mouth or estuary, form a comparatively thin layer ("lens") of fresh or considerably freshened waters on the surface of the ocean. Sharp salinity fronts enclose a lens of this type along its outer, seaward side, and directly under it there forms a vertical density gradient so sharp that the turbulent mixing across this boundary is probably significantly suppressed. The volume of freshened water in the lens corresponds only to a certain fraction of the annual discharge; at the same time, this fraction must depend significantly on the conditions of mixing of the discharge with the waters of the ocean, and may vary greatly. For example, if the drainage lens of the Orinoco R. contained the total discharge of fresh water, its area within the 20 ppt boundaries would be 12 times greater than observed in reality, i.e. $290,000\ km^2$ instead of $25,000\ km^2$. On the other hand, the river discharge lens of the Connecticut R. [126, 128] contains only 1/2000 of the annual discharge. According to observations [123, 128, 151] the $\overline{h}/\ell$ ratio of the average thickness of a lens ($\overline{h}$) to its width along a normal to the shore (ℓ) is a highly constant value for all rivers, and is of the order of 10^{-4}, which makes the lens similar in its proportions to a sheet of writing paper. In this case, it is natural to expect that the main processes of mixing of the lens with the ambient waters of the ocean can be described in terms of "lateral" or horizontal turbulent diffusion in the $x-y$ plane. On the strength of the high hydrostatic stability of the pycnocline, it would seem that vertical mixing should not play a significant role. However, this is still debatable. Earlier papers, for example Takano's [242], were characterized by a clear tendency to take only horizontal diffusion into account, whereas later studies [84, 91, 127, 196] emphasized the importance of vertical turbulent exchange. To mention one among others, Garvine's model [127] of small-scale river discharge type fronts (see later) arouses our interest by its plausibility. This is to a considerable degree associated with this author's assumption that the fresh water is turbulently entrained vertically downward across the frontal interface into the layers of saline ocean water flowing under the lens. It is precisely this direction of entrainment that can explain both the retention of a high degree of freshness in the lens, the nature of the circulation of the fresh water in a vertical plane normal to the front, and the sharpness of the front itself.

We can say that, in a stationary case, the losses of fresh water from the lens due to vertical entrainment are exactly compensated for by the constant addition of water to the lens from the river discharge. The total amount of water leaving the lens because of entrainment should be directly proportional to the area of the frontal interface and, therefore, the perimeter of the external front of the lens b_f and its thickness $\bar{h}$. On account of the thinness of the lens $(\bar{h}/\ell \simeq 10^{-4})$, one can picture the process of water loss from the lens due to entrainment as horizontal diffusion across the entire frontal boundary b_f in length and facing seaward. In this case, the coefficient of diffusion should be a constant value, since it parameterizes one and the same process in all the cases. The overall horizontal dimensions of a patch spreading under the influence of an excess horizontal density gradient should then be related to the discharge of the river q and the difference between the mean values of water salinity in the lens $(\bar{S})$ and beyond it (S_0). Therefore, on the basis of dimensional considerations, we can say that

$$K_\ell b_f \sim qd, \tag{3.4}$$

where $d = S_0/(S_0 - \bar{S})$ is the coefficient of dilution of the river discharge which inevitably takes place due to tidal and wind-wave mixing. The right-hand side in (3.4) represents the river discharge diluted to the average salinity of the water in the lens. The left side in (3.4) is proportional to the discharge of the entrained freshened water integrated along the entire length b_f of the drainage front. In order to convert (3.4) to an equation, we must determine the dimensionless coefficient of proportionality R which depends on the thickness of the lens $\bar{h}$, since the area of the frontal interface across which the waters are entrained depends not only on b_f, but on $\bar{h}$ as well. Then, (3.4) becomes

$$RK_\ell b_f = qd. \tag{3.5}$$

Let us determine R by means of the well-known expression for the time of relaxation of inhomogeneities through diffusion,

$$t = \frac{\ell^2}{K_\ell} \tag{3.6}$$

Here, t is the characteristic time of relaxation of the lens when it is not being replenished by the river discharge. Basically, in a steady state, t can be equated to the time of renewal of the entire fresh water content of the lens due to river discharge, i.e.

$$t = \frac{Q_\pi}{q},$$

where Q_π is the total content of fresh water in the lens, and

$$Q_\pi = Q\frac{S_0 - \overline{S}}{S_0} = \frac{Q}{d} = \frac{A\overline{h}}{d},$$

where $Q = A\overline{h}$ is the total volume of the lens and $A = b_f \ell$ is its area. Consequently,

$$t = \frac{A\overline{h}}{qd} = \frac{b_f \ell \overline{h}}{qd}, \tag{3.7}$$

By equating (3.6) and (3.7), we get

$$K_\ell = \frac{qd\ell^2}{A\overline{h}} = \frac{qd\ell}{b_f \overline{h}}. \tag{3.8}$$

Comparing (3.8) and (3.5), we find that

$$R = \overline{h}/\ell. \tag{3.9}$$

We shall now utilize the available published data on the characteristics of the river discharge lenses of the Connecticut, Mississippi and Orinoco rivers [123, 126, 128, 151] to determine the order of magnitude of K_ℓ, its constancy and, consequently, the accuracy of the relationships (3.5) and (3.8). For the sake of convenience, the necessary data have been tabulated in (3.3) where we can find the discharge values for the rivers and the scales characteristic of the lenses, as cited by the authors or taken from their tables or maps (lines 1-7), as well as the calculated derived values of d, R, Q, Q_π and t (lines 8-12) and the relative (dimensionless) characteristics $\ell_i, t_i, \overline{h}_i, d_i$ and q_i (lines 13-17), normalized by division by the mean* values of the corresponding variables for the Connecticut River ($\ell_c, t_c, \overline{h}_c, d_c$ and q_c).

As we can see from line 18 of table 3.3, the substitution of the known values of $q, \overline{h}, \ell, A$ and d in equation (3.8) gives us very similar values of K_ℓ for all three rivers, the mean value being equal to $2 \cdot 10^3 m^2/s$. This circumstance enables us to determine quite accurately "the replenishment time of the lens" t using only the known diameter of the lens ℓ from (3.6). It is especially interesting to note that the replenishment time of the lenses t for all the given rivers differs significantly, and it is directly proportional to the discharge of the river q (see line 12 in table 3.3). On the basis of the fact that the main parameters of the freshened discharge lenses of

---

* Of all the values cited in [126, 128, 131].

$$t_i = \ell_i^2 \qquad (3.10)$$

is correct, and enables us to calculate t, if only ℓ is known, by the formula

$$t = t_c \ell_i^2 \qquad (3.11)$$

which yields satisfactory results (line 22). However, I would like to be able to predict on the basis of the known value of the river discharge q at least one of the main characteristics of the lens, e.g. ℓ, b_f or $\overline{h}$. This can be done by using the equation (3.8) from which we can obtain the value of b_f if we assume roughly that $R = \overline{h}/\ell \simeq 10^{-4}$. For all the rivers in table 3.3, these values appear to be quite close to the real ones. Of the known correlations used for this purpose, Bondar's formulas [89] are also suitable. However, they were obtained under the assumption that absolutely no mixing took place between the river water and seawater, and so they are applicable only to the area directly adjacent to the mouth. Better results are provided by our empirical relationship

$$\overline{h} = \overline{h}_c (q_i d_i)^{\frac{2}{3}}, \qquad (3.12)$$

which satisfies the wide range of q and $\overline{h}$ values given in table 3.3.

Table 3.3

Line No.	Notation**	Unit of measurement	Connecticut*	Mississippi [151]	Orinoco [123]
		Observations			
1	$q \times 10^{-3}$	m^3/s	1.68	22.6	65.0
2	$A_{20} \times 10^{-6}$	m^2	41.8	0.9×10^4	2.5×10^4
3	$\overline{h}_{20}$	m	1.0	4.5	10
4	$l_{20} \times 10^{-3}$	m	5.4	50	75
5	$b_{f(20)} \times 10^{-3}$	m	8.3	200	350
6	S_0	ppt	25	36	35
7	$\overline{S}$	ppt	10	10	10
		Derived values			
8	d	—	1.67	1.38	1.40
9	R	—	1.85×10^{-4}	0.9×10^{-4}	1.33×10^{-4}
10	$Q \times 10^{-6}$	m^3	41.8	40.5×10^3	250×10^3
11	$Q_\pi \times 10^{-6}$	m^3	25.1	29.2×10^3	178×10^3
12	$t = \frac{Q_\pi}{q}$	h	4.15	358	759
		Dimensionless characteristics			
13	l_i	—	1	9.3	13.9
14	t_i	—	1	86.4	182.9
15	h_i	—	1	4.5	10
16	d_i	—	1	0.83	0.84
17	q_i	—	1	13.4	38.7
		Estimates			
18	$K_\ell \times 10^{-3}(3.8)$	m^2/s	2	1.9	2.1
19	$b_f \times 10^{-3}(3.8)$	m	7.6	174	343
20	$\overline{h}(3.12)$	m	1	5	10
21	$l \times 10^{-3}(3.9)$	m	$5 - 10$	$25 - 50$	$50 - 100$
22	$t(3.11)$	h	4.15	359	802

* Averages of the values given in [126, 128, 131].

** Subscript "20" denotes the boundaries taken within the 20 *ppt* isohaline.

Knowing $\bar{h}$, we can obtain the approximate value of ℓ from (3.9). It is interesting to note that $\bar{h}$ values varying only slightly within about $1\ m$ are obtained from (3.12) for different values of the Connecticut discharge (table 3.4), and with an increase in discharge they increase to the extent where, as Garvine assumed [126], they balance the limits of lens expansion A_{20} when a maximum area of $60-70\ km^2$ is reached. The deviations from observational data are insignificant for the Mississippi and Orinoco rivers. The discharge for the Amazon R., which according to [222] amounts to $200 \times 10^3 m^3/s$, is found to give $\bar{h}_{20} \simeq 21.5m$ from (3.12), which is actually impossible, as seaward of the delta there is a very wide ($100\ km$) and shallow (about $12-15\ m$) shelf on which intensive mixing of river waters and seawater takes place. Waters with a salinity of $\bar{S} \leq 20ppt$ reach the bottom in a zone measuring $70-100\ km$ in width (fig. 3.31 a, b), and the 20 ppt isohaline then rises obliquely to the surface in a zone $30-40\ km$ wide. The total width of the lens within the 20 ppt isohaline at the surface varies from 120 to 150 km between low and high discharge [134].

Having obtained the value of $\bar{h}_{20}$ from the known q, we can determine approximately the value of ℓ_{20}, assuming that $\bar{h}_{20}/\ell_{20} \simeq (1-2) \times 10^{-4}$. This value gives us the approximate distance from the shore to the river discharge salinity fronts, since, as indicated in [123, 128, 151], the maximum salinity gradients in all the given cases practically coincide with the position of the isohaline surface $S = 20\ ppt$ (fig. 3.32). Knowing $q, \bar{h}_{20}, \ell_{20}, d$ and K_ℓ, we can use (3.8) to determine A, and (3.11) or (3.7) to determine t.

Tidal mixing is apparently an important factor to consider when it is quite obvious, as in the case of the Connecticut drainage lens where the area A_{20} and diameter ℓ_{20} vary perceptibly throughout the tidal cycle (12 h 24 m) [128]. It was because of this that the average parameters of the run-off and drainage lens of this river had to be used in previous estimates. However, for large rivers like the Mississippi, and especially the Orinoco, the fluctuations in discharge caused by the tidal current in the mouth constitute only a very small part of the discharge. In this case, tidal mixing is not clearly defined, and so is sufficiently taken into account empirically through d and K_ℓ.

Table 3.4

Characteristics of the river discharge lens of the Connecticut River (initial data from [126])

Line No.	Notation	Unit of measurement	18/IV 1972 h.t.	21/IV 1972 l.t.	21/IV 1972 h.t.	1/V 1972 h.t.	1/V 1972 l.t.	12/VI 1972 h.t.	16/VI 1972 l.t.	30/III 1973 l.t.	13/IV 1973 l.t.	Average
1	$q * \times 10^{-3}$	m^3/s	1.44	3.13	2.26	1.20	2.20	0.47	1.29	1.55	1.56	1.68
2	$A_{20} \times 10^{-6}$	m^2	19.7	48.7	55.4	52.7	77.4	0.8	23.0	47.5	51.0	41.8
3	$\bar{h}_{20}$	m										1
4	$l_{20} \times 10^{-3}$	m	?	7	5	5	8	1	?	6	6	5.4
5	$b_{f(20)} \times 10^{-3}$	m	?	7	11	11	10	1	?	8	10	8.3
6	S_0	ppl					25					25
7	$\bar{S}$	ppl					10					10
8	d	—					1.67					1.67
9	R	—					1.85×10^{-4}					1.85×10^{-4}
10	q_i	—	0.86	1.86	1.35	0.71	1.31	0.28	0.77	0.92	0.93	1
11	l_i	—	?	1.30	0.93	0.93	1.48	0.19	?	1.11	1.11	1
12	$\bar{h}(3.12)$	m	0.9	1.5	1.2	0.8	1.2	0.4	0.8	1.0	1.0	1.0
13	$l \times 10^{-3}(3.9)$	m	4.9	8.1	6.5	4.3	6.5	2.2	4.3	5.4	5.4	5.4
14	$t(3.11)$ **	h	?	7.0	3.9	3.9	6.1	0.8	?	4.6	4.6	4.0
15	$t(3.6)$	h	3.3	9.1	5.9	2.6	5.9	0.7	2.6	4.1	4.1	4.3
16	$b_f \times 10^{-3}(3.8)$	m	6.5	14.0	10.2	5.4	9.9	2.1	5.8	7.0	7.0	7.5
17	$b_f = (T)A/l_{comp.}$ $\times 10^{-3}$	m	4.0	6.0	8.5	12.2	11.9	0.4	5.3	6.8	7.3	6.9

h.t. — high tide, l.t. — low tide

* Mean value for one-half of a tidal cycle (6 h 12 min) with correction for the tidal current (net river discharge).

** $t = 4.15 \, l_i^2(h)$.

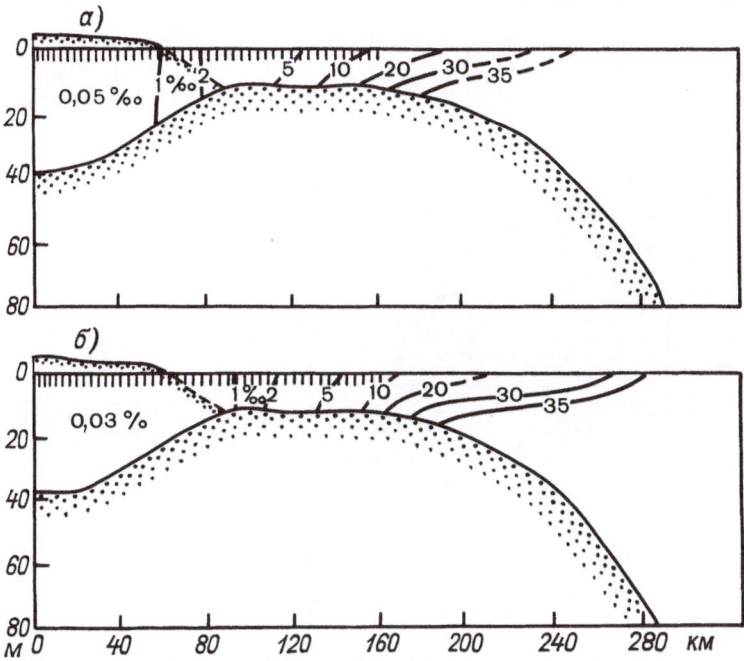

Fig. 3.31. Vertical section of the salinity field across the river discharge lens of the Amazon R., based on [134], during minimum discharge (a) and during high discharge (b).

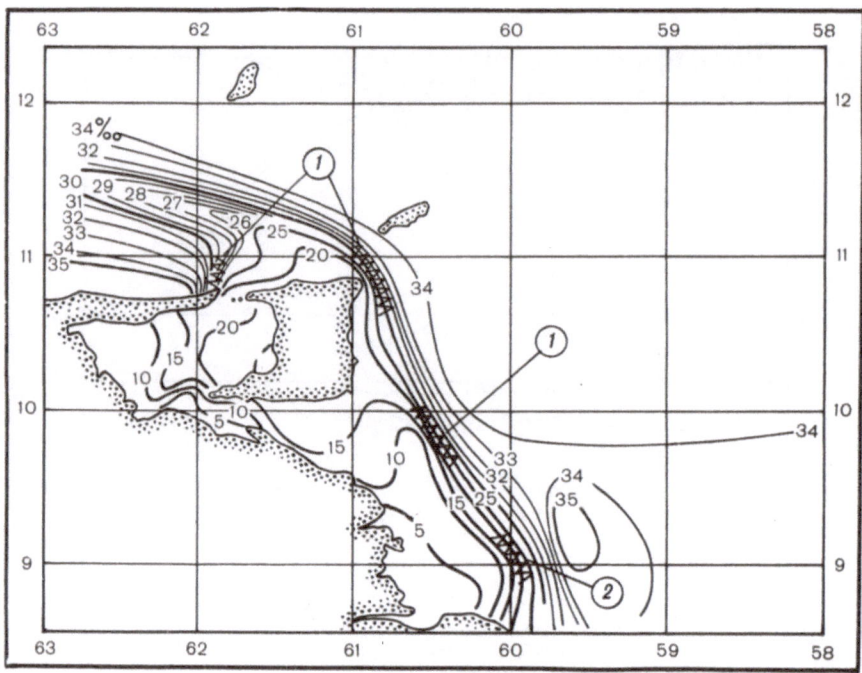

Fig. 3.32. Salinity distribution at the ocean surface in the river discharge lens of the Orinoco R., based on [123]. The zigzag line marks the regions of the most significant frontal salinity gradients, which amount to more than $1ppt/km$ (1) and $0.5ppt/km$ (2).

Let us now examine the characteristics and dynamics of the river discharge fronts that confine the freshened lens. The section across the river discharge lens of the Connecticut River (fig. 3.33 a, b, c) along the normal to the main front is displayed in the salinity field (a) and has a sharp, almost vertical line for the interface (ϕ), into which the $20 - 24$ ppt isohalines have converged. At a depth of about 3 m, the frontal interface breaks off, and then occupies a practically horizontal plane. The density distribution (c) is practically a replica of the salinity distribution. The temperature contrasts between the lens and the ambient water are insignificant (b), and in the region of the main frontal interface (ϕ) are slightly diffused by comparison with the salinity contrasts. At a more detailed section across the frontal interface, which manifests itself at the surface as a colour change (from brownish gray to bluish green), we can clearly see (fig. 3.34 a) that the main salinity change of 11 ppt (from 12 to 23 ppt) falls within a zone which is only about 150 m wide, with a maximum salinity gradient amounting to 4 ppt, over 20 m (or $0.2 ppt/m$).* The density change in the same zone (fig. 3.34 c) amounted to $9\sigma_t$ units, with the temperature change of $1.5°C$ at temperatures of about $5 - 6°$ contributing little to the density change due to salinity. Such sharp horizontal salinity and density contrasts, which are $1 - 2$ orders of magnitude greater than the horizontal salinity gradients at fronts in the open ocean, are apparently characteristic of all river discharge fronts. Becker and Prahm-Rodewald [85] report that in the Heligoland Bay of the North Sea, the freshened lens of the Elbe and Weser rivers is bounded by discharge fronts with horizontal jumps of $0.2 - 0.4$ ppt over 100 m. Salinity gradients of up to $0.8 - 1.0$ ppt per 100 m characterize the river discharge fronts of the Hudson River in New York Bay [90, 91]. Even comparatively rough measurements in the vicinity of the Orinoco run-off, with stations 15 miles apart, [123] show horizontal salinity gradients of up to $0.5 - 1.0$ ppt per km (see fig. 3.32). Actually, local salinity changes of the order of several ppt, occurring in narrow frontal zones up to $100 - 200$ m wide, can also be expected here.

Further evidence for this comes from the sharp colour contrasts observed by numerous researchers in the river discharge lens of the Amazon.

---

* Even sharper contrasts and gradients have been recorded at estuarine fronts (see subsection 3.3.2 below). For example, the salinity changes at the surface in Delaware Bay amounted to 4 ppt/m [166]

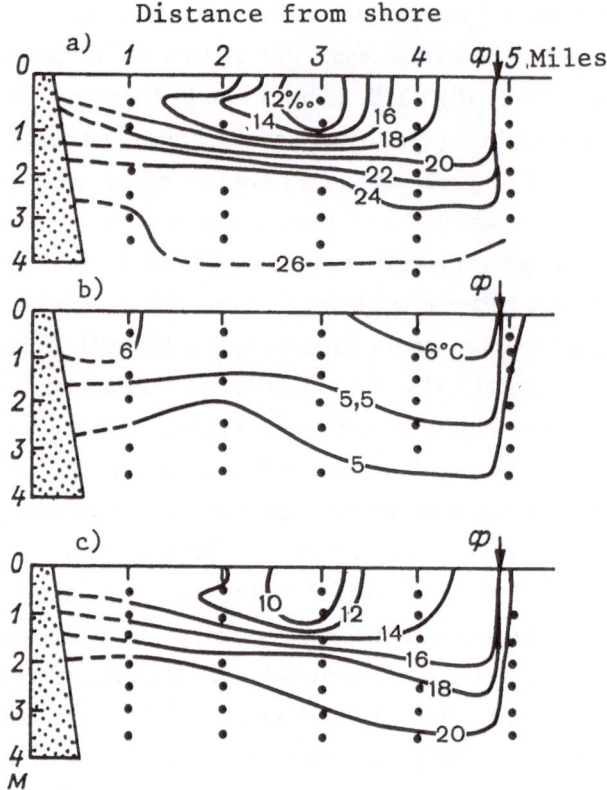

Fig. 3.33. Distribution of salinity (a), temperture (b) and nominal specific density (c) in the drainage lens of the Connecticut R., based on Garvine's data [126]. The letter ϕ marks the frontal interface.

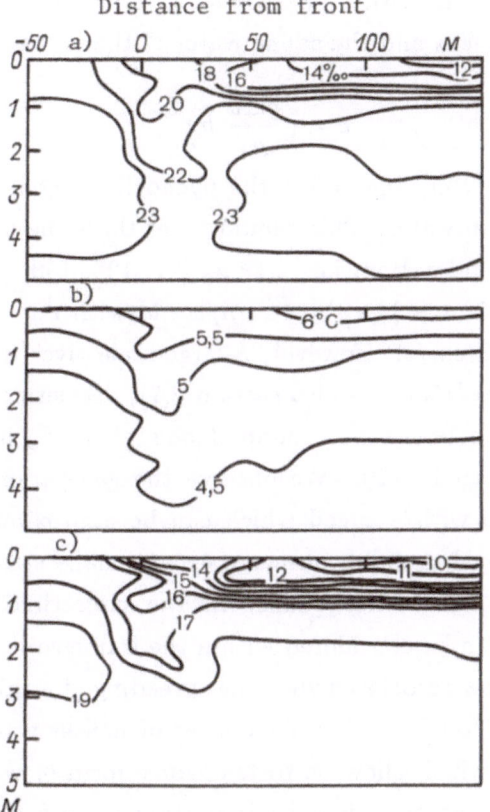

Fig. 3.34. Details of the distribution of salinity (a), temperature (b) and nominal specific density (c) near the frontal interface of the drainage lens of the Connecticut R., based on Garvine's data [126].

Such contrasts permit us to accept 100 m as the characteristic transverse scale L_ϕ for the discharge fronts of all rivers, large and small. This, in turn, gives a frontal Kibel-Rossby number

$$Ki = \frac{u_\phi}{fL_\phi}, \qquad (3.13)$$

where u_ϕ is the characteristic speed of advance of the river discharge front with respect to the underlying saline water (or the velocity of the salt water flow under the discharge front) in the neighbourhood of 30 (when $u_\phi = 30 \; cm/s$ and $f = 10^{-4}s^{-1}$).

The best means of determining u_ϕ comes from consideration of the dynamics of long, free internal waves with a phase velocity c, at the surface of the interface between the freshened lens and the saline water of the ocean, equal to

$$c = \left(\frac{g\Delta\rho}{\rho_0}\,\overline{h}\right)^{\frac{1}{2}}, \tag{3.14}$$

where $\Delta\rho$ is the density change across the pycnocline under the lens, while ρ_0 is the density of the ocean water. This should be of the same order of magnitude as the velocity of frontal migration, i.e. $c \approx u_\phi$ [91, 128, 158]. With $\Delta\rho = 2 \cdot 10^{-2}$ and $\overline{h} = 1m$ (Connecticut R.), $u_\phi = 44\ cm/s$ and with $\Delta\rho = 10^{-2}$ and $\overline{h} = 10m$ (Orinoco R.), $u_\phi = 98\ cm/s$. However, a large-scale river discharge front, as we have already seen from [156] (see also section 2.7), decelerates its forward motion rather quickly, the baroclinic zone expands (the scale of L_ϕ increases) and the Ki number can diminish to $1 - 10$. We observe the generation of an along-frontal geostrophic jet stream with a speed which can be approximately equated to u_ϕ. For example, Alexander Humboldt observed exceptionally strong jet streams at the edge of the Orinoco river discharge lens. At the same time, we believe that the geometry of the lens can be considered within the framework of a schematic model for a viscous 2-layer flow associated with the spreading of a thin layer of light water isostatically "floating" on a much deeper layer of denser water. The question of boundary conditions, which allow us to take some form of vertical and horizontal turbulent viscosity into account, is quite important when formulating the problem.

In the literature, there are two widely known physical models for the slope of the river discharge lenses in estuarine regions of the ocean. These are Takano's model [242] and Garvine's model [127]. The first model utilizes the conventional 1950's method for the integrated flow, which makes it possible to obtain an analytical solution of the system of linearized Navier-Stokes equations, taking into account only horizontal turbulent friction and the effect of the Earth's rotation on the motion of water from river discharges into the ocean. Garvine's model also examines a freshwater lens isostatically floating on the surface of water of oceanic salinity. In the initial equations of motion, the nonlinear advective terms and the vertical shear stresses are retained, whereas the effect of the Earth's rotation and lateral friction is ignored. In addition to this, Garvine also introduced the effect of turbulent entrainment at the surface of the interface of two layers moving toward each other with a speed $u_\phi = c$, where c is determined by the equation (3.14).

Having studied Takano's model, as applied to the drainage lens of the Hudson River, Bowman [90] found that the results come closest to field data when $K_\ell = 10^4 m^2/s$. The effect of the Earth's rotation is in this case responsible for the moderate deflection of the river discharge lens to the right of the direction of the run-off. However, it is difficult to say whether Takano's model really does correspond to field conditions in this sense, as the geometry of New York Bay allows for only a right-hand deflection of the outflow. The shore here prevents any deflection to the left. However, we know that some of the rivers of the northern hemisphere are characterized by cyclonic ("left-hand") deflection of the outflow, instead of anticyclonic deflection. This may be related to the constant along-shore current which is not taken into account by Takano's model [242]. For example, observations of the Connecticut run-off [128] showed a deflection of the river discharge lens to the left of the mouth. Deflection to the left is also observed in the run-off of the Orinoco, the river discharge lens of which extends for 350 km northwest of the mouth [123]. As we have already mentioned, the effect of the Earth's rotation in this case influences mainly the dynamics of the river discharge front, and not the geometry of the lens.

We can also say that Takano's model, because it ignored vertical turbulent viscosity, requires higher values of K_ℓ, even higher than those obtained earlier in this section when analyzing the typical dimensions of drainage lenses for various intensities of river discharge.

Because of its integral character, Takano's model cannot describe the drainage fronts themselves. The value of the lens thickness according to this model should gradually diminish to zero as we move farther from the mouth; furthermore, a transverse section of the river drainage lens along the axis of motion is not very similar to that observed in reality. As far as we can judge on the basis of the available data [90, 91, 126, 131], the thickness h under natural conditions is approximately constant over the entire area of the lens, and begins to diminish sharply only near the outer front. According to the data of Garvine and Monk [131], the region with a significant interfacial slope in the Connecticut lens was only about 50 m wide (see fig. 3.33 and 3.34).

Obviously, horizontal turbulent diffusion ("lateral friction") is of importance only when solving problems related to the maximum dimensions of river discharge lenses.

Garvine's model [127] and Bowman's later analysis [91] have shown that it is more important to take into account vertical turbulent viscosity and the turbulent entrainment at the interface between light and heavy fluids when describing the dynamics of the river discharge fronts themselves.

Garvine's model calculations [127] have shown that the distribution of current velocities near the frontal interface is closest of all to that observed in reality when

fresh water is entrained from the top downward into the saline water. The calculated distribution of specific density in this case also proved to be similar to that
observed in nature; at the same time the region with a significant interfacial slope
was found to be not wider than 40 m, while the maximum thickness of the lens
was approximately 2 m, which agreed fully with field data. The entrainment of
water from the river discharge lens downward across the surface of the interface is
also confirmed by a comparison of the observed integral flux in the lens with the
computed value. In all cases, the value of the integral flow $U = \int_0^h u\,dz$, obtained
from observations, proved to be negative (from -0.1 to $-0.4m^2/s$), which is possible only with turbulent entrainment from the top downward. Calculations using
the model give us typical U values of $-0.1m^2/s$, which is in good agreement with
field data. From the calculations we see that the sharpness of the river discharge
front really does increase with entrainment from the top downward. This intensifies
near-frontal convergence and, therefore, increases the velocity of downwelling near
the front. Garvine's calculations with entrainment from top to bottom show that
the downwelling speed amounts to approximately 18 cm/s. Vertical velocities of
this order should not surprise anyone, as the difference of the horizontal velocities
across the front in this case amounts to 80 cm/s over a distance of only $50-100$ m.

Garvine's assumption regarding the importance to river discharge fronts of the
turbulent entrainment of fresher water from the river discharge lens into the underlying saline waters is fundamental to our understanding of the characteristics
of motion within the river discharge lens and in its immediate vicinity. From observations on the river discharge lens of the Connecticut R. [128], we can say that
the main transport of water in the lens, which arises as a result of river outflow, is
directed mostly to the southeast, whereas the transport of water near the main river
discharge front farther south is directed to the southwest, toward the front, i.e. normal to the main transport. This additional transport can be explained only within
the framework of the above-described dynamics of the front with the contribution of
turbulent entrainment directed downward. Basically, it follows that the dynamics
of a narrow near-frontal region can and should be examined separately from the
dynamics of the whole river discharge lens, and that totally different hypotheses
can be applied when solving these different problems.

Another important result of Garvine's work [127] was that, according to the
theory, a river discharge front should advance with a speed determined by (3.14)
with respect to the underlying fluid. However, in a stationary system of coordinates

associated with the Earth or, in this particular case, with the shore, the expansion of the river discharge lens and the migration of the river discharge front are controlled, as we have seen earlier, by the discharge of the river. Basically, the position of the river discharge front becomes fixed in a stationary case. In this case, the downwelling and convergence at the front will become the mechanism that entrains the ambient saline water under the front. Obviously, Garvine and Monk's observations [131] confirm the existence of this type of flow under the quasi-stationary front. However, in one of his later studies, Garvine [128] stresses the independence of the field of motion in the ambient water in Long Island Sound from the motion of the lens itself and the water in it. However, we should take into consideration that strong tidal currents were observed in the Sound in this case. In Garvine's opinion [128], the importance of the parameter determined by (3.14) also lies in the fact that, through comparison with it, one can determine the nature of the flow in the river discharge lens. If the water in the lens spreads with a speed exceeding the phase velocity c, then long internal waves cannot reach the source of freshened water (the mouth of the river) and affect the initial flow. Such a speed of the discharge current is called supercritical, as the densimetric Froude number

$$Fr = \frac{u^2}{c^2} = \frac{u^2 \rho_0}{gh\Delta\rho} \tag{3.15}$$

exceeds 1 in this case, and the Richardson number becomes less than 1.

Observations in the discharge lenses of numerous rivers have shown that discharge currents are usually supercritical, but that the stability of the interface of the river discharge lens is not affected in this case.

Further generalization of Garvine's theoretical analysis was recently carried out by Kao et al. [158] with a time-dependent numerical model, and by G.I. Shapiro [74] with a simple analytical model for a transient density front exposed to tangential wind stress. The numerical model by Kao et al. [158] includes the Coriolis effect and, therefore, applies to fronts of a larger scale, but with the same intrusive density structure as for river discharge fronts (e.g. fronts between shelf and slope waters off the eastern coast of North America). Such fronts are characterized by small Kibel-Rossby numbers ($Ki \sim 10^{-1}$) and their stationarity is achieved through a quasi-geostrophic along-frontal current. The results of Kao et al. have shown that this requires a time of the order of $10/f$. On the other hand, it appears that the effect of the Earth's rotation weakens near-frontal convergence and the downwelling of light water near the front. The model of Kao et al. also makes it possible to study the effect of wind and Ekman drift on near-frontal circulation and on the slope of the frontal interface.

From the very beginning, Shapiro [74] ignored the Earth's rotation, the inertial terms and "lateral friction", retaining the effect of time dependence only in the equation of continuity and taking w as equal to the change in the thickness of the layer of light water. The introduction of the tangential wind stress tangent τ as the boundary condition at the free surface linearly adds to the circulation of light water in the surface lens a wind-driven circulation which satisfies the well-known relations of the theory of wind-driven currents in a small inland sea [76]. The formulation of the problem enables us to prescribe a continuous influx of light water into the lens as river run-off. However, the initially imposed conditions of the insignificance of the densimetric Froude number ($Fr \ll 1$) limits the application of this theory to the phase of development of river discharge fronts where the lens has already formed and further inflow of water into the lens has ceased. In the case where $\tau = 0$, the solutions derived by Shapiro are similar to Garvine's results [127], but the special emphasis placed on the role of the wind and the disregard for turbulent entrainment at the frontal interface do not allow us to derive anything new from these solutions about the dynamics of the river discharge fronts themselves during the active phase of their existence. The calculations carried out by Shapiro are difficult to apply to such fronts, due to the insignificance of the assumed density contrast $\gamma = 10^{-3}$ and the requirement of a low Froude number.

3.3.2. Estuarine fronts

The hydrographic conditions of coastal waters, lagoons, fiords, bays and estuaries are so unique that this problem has been discussed in special extensive references in journals and monographs. The water motion in estuaries is characterized by many of the tendencies typical of river flows. On the other hand, however, estuarine waters are highly stratified, which is not at all characteristic of typical river flows but, on the contrary, makes the dynamics of estuarine waters similar to the dynamics of oceanic waters. In this comparatively small section, it would be impossible to describe in detail the specific conditions under which estuarine frontogenesis develops and the evolution of the most peculiar frontal interfaces takes place. Therefore, the reader who has a special interest in the physical oceanography and hydrodynamics of estuaries will not find most of the information required by him in this section, but should refer to the widely known classic studies by Bowden, Cameron, Pritchard, Rattray, etc. The main purpose of this section is to show the properties which govern the behaviour of the boundaries between fresh and salt water in flow-through basins. The limitations and comparative shallowness of these, together with the water exchange with the open ocean, determine the pattern of internal water circulation.

As Klemas rightly indicates [162], the classic pattern of estuarine circulation, evolved in the course of 10 − 20 years of research, was based on the study of a simplified two-dimensional flow pattern in the vertical and longitudinal plane of an estuary, and on the concept of a gradual change of the horizontal thermodynamic characteristics. Observations of the past few years and especially the results of re- mote sensing surveys combined with detailed measurements from research vessels have prompted us to review this classic concept. It has become clear that extremely sharp longitudinal and transverse velocity and density gradients, which are related to frontal systems and play an important role in the dynamics of estuarine circula- tion, occur in estuaries.

The frequently observed manifestations of fronts (bands of foam and trash, changes in the colour and transparency of the water), as well as the horizontal velocity and density gradients associated with fronts, are much sharper in estuaries than in the open ocean. According to recent measurements, for example, the frontal salinity gradients in Delaware Bay [167] amount to 4 ppt/m.

It is observations of estuarine fronts, carried out from small vessels and with greater detail than is possible in the open ocean, that revealed interesting charac- teristics of the position of bands of foam and trash, and of the change in water colour (and transparency) near fronts. The bands of foam turned out to be a sign of flow convergence at a front which at the surface restricts the maximum extent of the freshwater lens. The band of trash and the line of colour contrast are usu- ally displaced relative to the band of foam in the direction of the freshened waters, as shown in fig. 3.35. An interesting aerial photograph of a coastal front with a displaced band of foam is shown, with a clearly defined brightness contrast, in ref. [162].

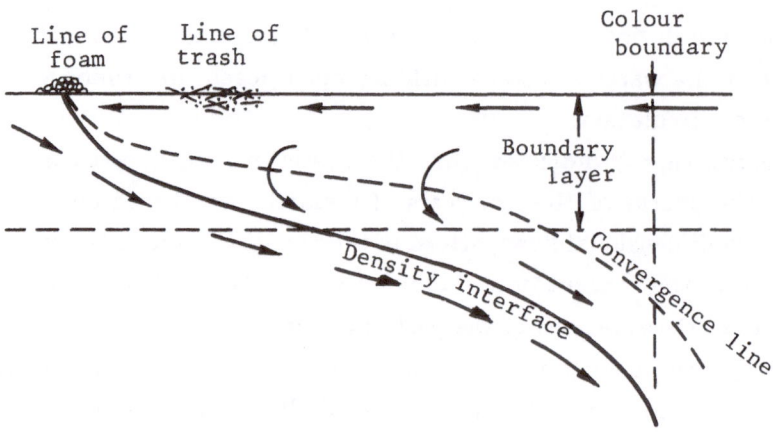

Fig. 3.35. Relative distribution of foam and trash accumulations, and of the colour boundary near an estuarine front, after Klemas [162].

As a rule, two types of frontal interfaces can be observed in estuaries, type I – frontal interfaces associated with a "salt wedge" and occupying a position perpendicular to the axis of the estuary, and type II – frontal interfaces related to the interaction of tidal currents with the bottom topography of the estuary and located along the axes of the characteristic features of the bottom topography (mainly along the axis of the estuary). The classic "salt wedge" and frontal interfaces of the first type are more characteristic of estuaries resembling a canal. The discharge of the river water in this case should be comparable to the maximum inflow of seawater into the estuary during high tide, and only slightly greater than it. In such estuaries, the seawater in the lower layer penetrates far upstream, and forms a near-bottom "wedge" with a sloped upper boundary (see diagram in fig. 3.36 a). The tip of the wedge forms a sharp near-bottom front inside the estuary, whereas the front near the surface may be located outside the mouth of the estuary and may represent a river discharge front which bounds the freshwater or freshened lens at the surface of the sea or ocean (see subsection 3.3.1). The frontal interface of the wedge and its near-bottom front may migrate over great distances along the estuary from one phase of the tidal cycle to the other. If the estuary happens to have a substantial widening near its mouth then the near-surface fronts associated with the "salt wedge" can also be observed within the lower part of the estuary. Typical examples of estuaries with a salt wedge are the estuaries of the Thames and Mersey rivers, or Knight Inlet in British Columbia (Canada).

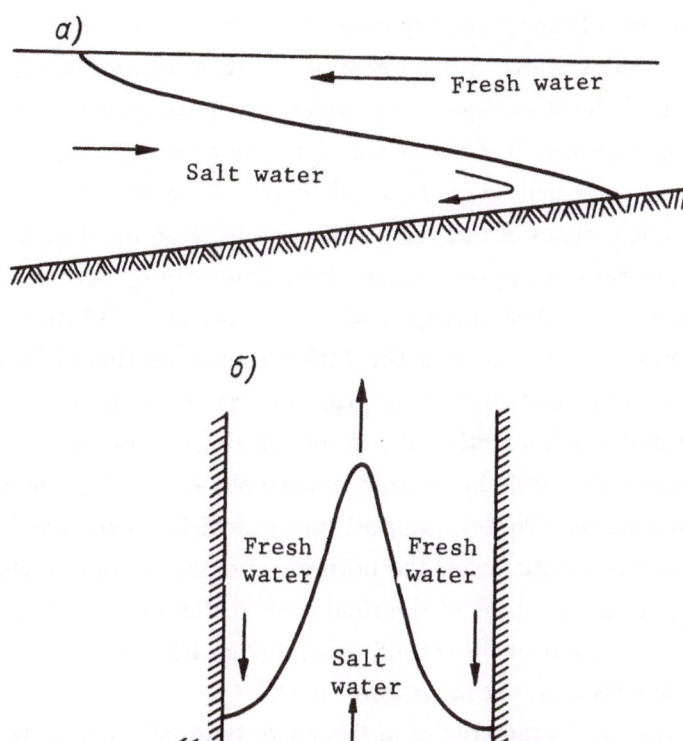

Fig. 3.36. Salt wedge in the vertical axial plane of an estuary (a) and in the horizontal plane at the mouth of a small river (b).

Frontal interfaces of the second type are encountered more frequently in shallow estuaries of considerable width, with a complex rugged bottom topography. They can also arise when the discharge of the river is comparatively weak, when the salt waters during high tide force their way into the estuary, along the axis of the deepest part of the river bed, over the whole depth, forming a V-shaped wedge in the horizontal plane instead of the vertical plane (fig. 3.36 b). This type of wedge moves quickly upstream, being constrained from the sides by very sharp fronts. A film showing the movement of this type of wedge upstream and its retreat during low tide was shown by J. Simpson at the 14th General Meeting of SCOR in 1978. However, even when the distribution of river and sea water in an estuary remains basically two-layered, the interaction of currents moving in different directions (river discharge and flood tide) with the bottom topography can lead to the formation of strong local convergences (fronts), the positions of which, on the one hand, clearly correlate with the characteristics of the bottom topography, and on the other, can vary significantly from one phase of the tidal cycle to the next. Delaware Bay [162] especially abounds in fronts of the second type, though Klemas indicates that fronts of the first type are also formed in its lower part.

Table 3.5 gives the characteristics of these two types of fronts in Delaware Bay, based on the observations of Klemas.

As observations have shown, along-shore jets, which are in many ways similar to the coastal jets described by Csanady [100], can develop in estuaries. However, they differ from the latter in that the main driving forces in estuaries are the density and pressure gradients associated with the inflow of fresh water, and not the tangential wind stress as shown in Csanady's models. However, strong winds can greatly alter the pattern of currents and fronts in estuaries.

If an estuary is deep (e.g. the Gulf of St. Lawrence), then the along-shore jets and the frontal interfaces associated with them acquire a quasi-geostrophic character. An interesting example of this is the Gaspé Current in the Gulf of St. Lawrence [243, 244], with which a characteristic salinity front (fig. 3.37) and an along-frontal surface cold-water band, clearly depicted on satellite IR images, are associated. The shore has a stabilizing effect on the Gaspé Current. Farther out from shore, the jet becomes unstable, and substantial wave-like disturbances [244] which cause significant frontal movement are formed on it.

Table 3.5

Characteristics of estuarine fronts, after Klemas [162]

Characteristic	Type I Fronts associated with a "wedge"	Type II Fronts of interaction with the bottom relief
Location	Lower part of bay	Upper and lower part of bay
Direction	Perpendicular to axis of river flow	Parallel to axis of river flow
Speed of frontal migration, cm/s	10–60	5–20
Limits of frontal displacements, km	10	± 0.3
Speeds of convergent currents, cm/s	2–20	5–40
Velocity change across the front, cm/s	1–5	5–20
Change in depth of disappearance of white disc, m	1.0–2.2	0.4–1.6
Change in colour	Moderate	Intense
Concentration of foam and oil	Moderate	High
Concentration of trash and colour	Moderate	High
Refraction and suppression of waves	Moderate	Strong
Temperature change, $^\circ C$	0–2	0–3
Salinity change, ppt	1–4	0.5–3

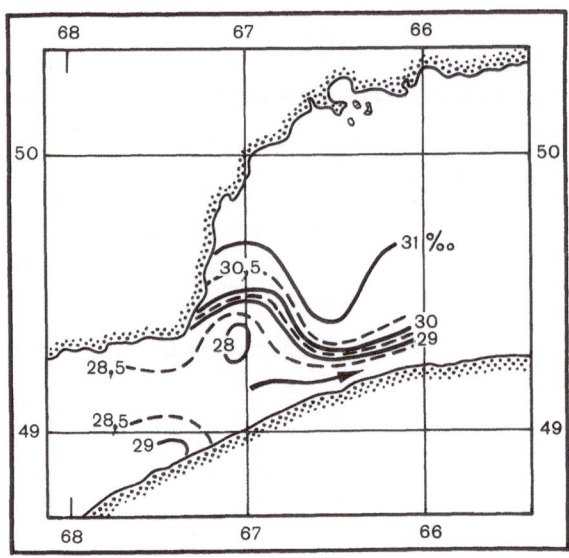

Fig. 3.37. Salinity distribution (in *ppt*) in the Gulf of St. Lawrence (based on [243]).

The drawn out trough-like form of numerous estuaries makes it much easier to formulate model problems. In the case of a "salt wedge", the problem of theoretical modelling lies in predicting the position of the interface between the run-off and sea water on the axis of the estuary, the relation between the slopes of the free surface and the interface, and consequently, the relation between the thickness of the layers and the velocity of the current in each section of the estuary. The modelling difficulties here may be due to the presence of abrupt inhomogeneities, sills and banks in the bottom topography along the axis of the estuary. Significant nonlinear effects may be associated with them. As the recent investigations of Farmer and Smith have shown [110, 111], it is precisely these nonlinear effects that are extremely intense and create near the sill in Knight Inlet (British Columbia, Canada) the most intensive wave-like pycnocline disturbances due to waves behind an obstacle. A hydraulic jump, an internal bore and a series of solitons arise during different phases of the tidal cycle. These disturbances are accompanied by surface effects [110], horizontal density gradients and interfacial slopes, making them very similar to frontal phenomena (see section 3.5). In the case of frontal interfaces of the second type, the problem is complicated by the fact that the complex bottom topography in the transverse section of the estuary creates conditions for the formation of transverse interfacial slopes and randomly oriented frontal interfaces during the intrusion of oceanic waters. The currents in the lower layer in this case may be due to the tides, while those in the upper layer may be due not only to the discharge, but also, as we have already mentioned, to the effect of the wind.

The literature on theoretical modelling of currents and fronts in estuaries is quite extensive. Different sections of Officer's monograph [196], as well as the paper by Pearson and Winter [202] can be recommended to the reader.

3.4. Coastal fronts with tidal mixing

In coastal regions of the ocean where the relatively shallow shelf occupies large areas, dissipation of the kinetic energy of the tides in shallow water causes the formation of sharp frontal boundaries between well-mixed shelf water and the stratified water of the deeper areas adjacent to the shelf. At moderate latitudes, these frontal interfaces are of a seasonal nature, as it is only in summer that the waters of the open ocean acquire a sharp thermal stratification in the active layer due to intensive solar heating. The transverse structure of this type of frontal interface is depicted in fig. 3.38. Fronts of this nature first attracted attention in the Irish Sea in 1971, due to the formation of intensive thermohaline finestructure at the fronts [227]. Somewhat later (in 1974), Simpson and Hunter [228] proposed a rather effective physical criterion for predicting the position of such fronts in the coastal zone.

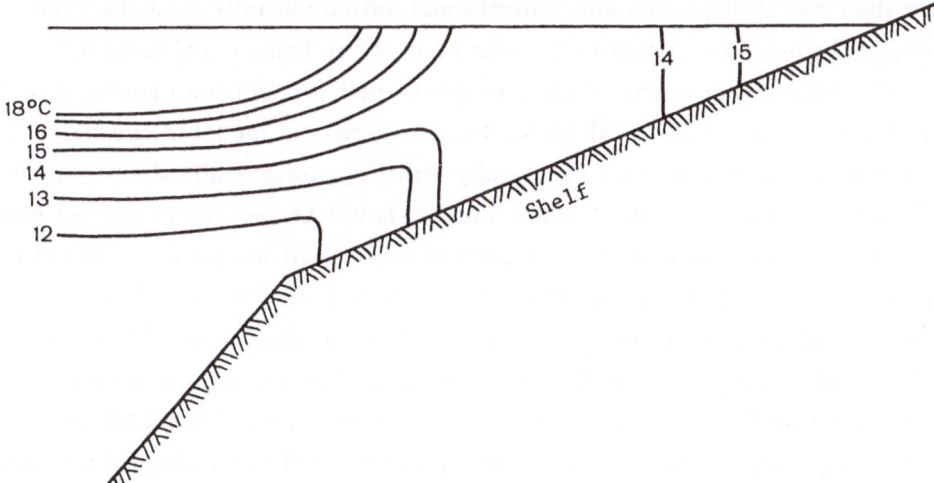

Fig. 3.38. Diagram of the position of isotherms in a summer shelf front that forms due to mixing in shallow waters.

This criterion represents the ratio R of the rate of production of the potential energy required for maintaining complete vertical homogeneity of the upper layer, in opposition to the continuous stratifying effect of the solar heat flux Q through the free surface, to the energy dissipation rate of tidal flows due to near-bottom friction, i.e.

$$R = \frac{g\alpha Qh/(2c_p\rho)}{C_D|u|^3}, \tag{3.16}$$

where h is the depth of the sea, c_p – the specific heat at constant pressure, C_D – the coefficient of friction, and $|u|$ – the amplitude of the tidal flow.

If we regard Q and C_D as constant values on the average, then (3.16) takes on the simple form of

$$R \sim \frac{h}{|u|^3}, \tag{3.17}$$

which is known as the Simpson-Hunter criterion. When calculating it, the mean spring amplitude of tidal flows is taken for $|u|$. It should be said that at approximately the same time, independently of Simpson and Hunter, the same criterion was also obtained by Fearnhead [113] who also examined wind mixing in addition to tidal mixing. Pingree and Griffiths [210] later suggested that the value $[(h/C_D|u|^3]^{-1}$, which is measured in m^2/s^3, can be interpreted as the specific (per unit of mass) rate of dissipation of tidal energy in a column of water with thickness h (see also [211]).

In their calculations, the above-mentioned authors usually used the value $s = lg[h/C_D|u|^3]$ where the coefficient C_D was accepted as being equal to 0.0025. Refs. [207] and [210] contain a map of the European shelf, which shows contours of the parameter $s = lg[h/C_D|u|^3]$, calculated from a numerical model of the tides. The regions with $s > 2$ correspond to thermally stratified areas, while the regions with $s < 1$ correspond to well-mixed ones. Fronts should be located in regions where $s \simeq 1.5$, as has been confirmed from numerous satellite IR images presented in refs. [210], [211] and [229]. Basically, the criterion (3.16), or Fearnhead's variation of it, can be used to predict changes in the positions of shelf fronts from month to month, as well as for predicting the time of their appearance and disappearance, as has been done by Pingree et al. [207, 210]. It has also been shown [207, 229] that regions of high concentrations of chlorophyll-a (fig. 3.39) and biological organisms are in some way associated with shelf fronts. The position of these regions can also be predicted on the same basis.

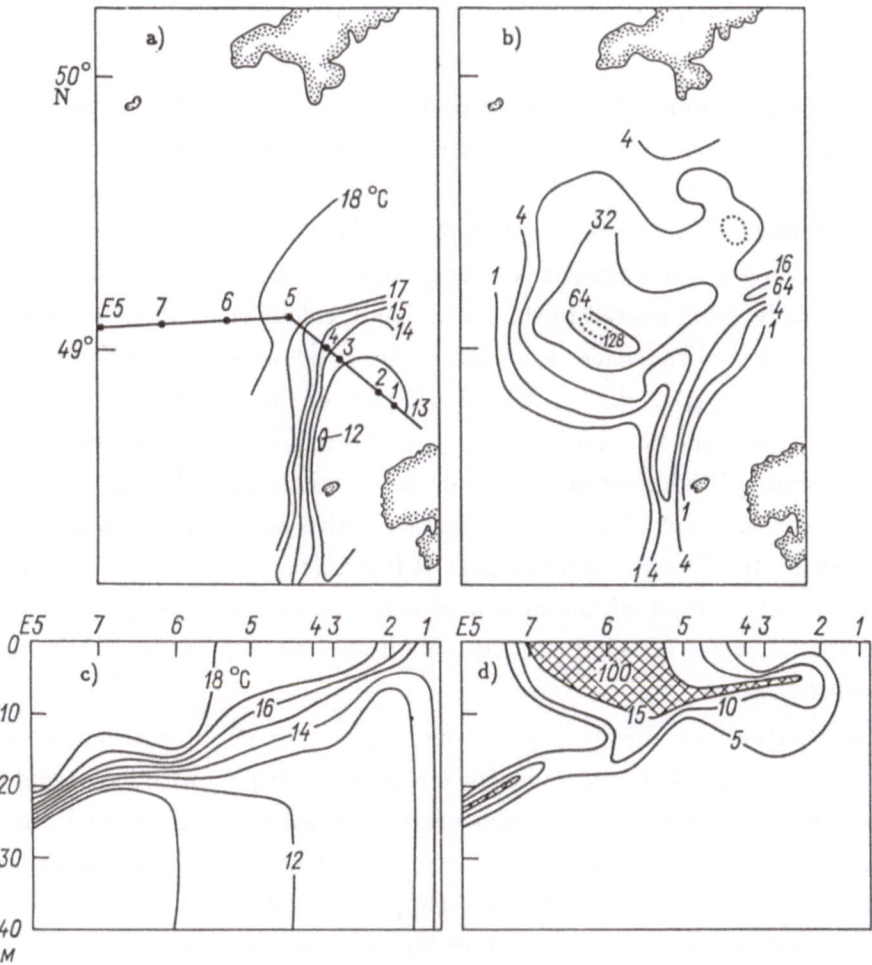

Fig. 3.39. Distribution of temperature at the surface of the ocean (a) and in a section (b), and the distribution of chlorophyll-a (c and d) (mg/m^3) in the shelf frontal zone west of Ushant at the entrance to the English Channel in August 1976, based on Simpson and Pingree's data [229].

Shelf fronts like the ones detected and studied on the shelves of the British Isles and Europe are widespread in the World Ocean. In the literature, there have already been descriptions of such fronts near the coast of New Zealand [92] (where the use of the Simpson-Hunter criterion produces equally good results) and in the Bering Sea [96, 97].

It should be said in conclusion that the Simpson-Hunter criterion is also fully applicable to estuarine fronts in situations where they are associated with tidal mixing in the shallow parts of the estuaries (see subsection 3.3.2).

3.5. Surface phenomena of a frontal nature

In section 1.1, we already spoke of various visible manifestations of fronts, at the free surface of the ocean, which are sometimes accompanied by sound effects as well. So far, the literature has not had a better description of these manifestations than the one published by Uda back in 1938 [249], so those interested can refer to this initial source to their own great advantage. Many of the surface manifestations of fronts, especially those related to the formation of colour or brightness contrasts in reflected or scattered radiation, are what makes it possible to see and photograph oceanic fronts from space [29] in the visible part of the spectrum. However, the convergent flows in the near-surface layer, which modulate the steepness of surface waves and create brightness contrasts or form regions of high concentrations of passive constituents (plankton, trash, surface-active substances, foam, etc.), are characteristic, strictly speaking, not only of fronts (see section 1.2). Alternations, in the near-surface layer of the ocean, of convergence and divergence regions of varying spatial scale, are characteristic of internal waves, Langmuir circulations and convection cells.

The literature contains descriptions and an analysis of observations in the Andaman Sea of gigantic bands of unusually steep waves ("rips") with individual waves reaching a height of up to $1.8\ m$ in calm weather. These are associated with internal solitons [199, 203]. The bands were observed in packets of $5 - 6$ with an interval of $12\ h\ 36\ min$ between packets, which clearly points to the tidal origin of the initial disturbance which generated the train of internal solitons. The bands were about $0.8 - 1.0\ km$ wide, and were spaced $15\ km$ apart within a packet. The length of the bands amounted to $150\ km$. A deep depression (up to $50 - 80\ m$) of the thermocline coincided with each band of current rip, the first soliton in the train always being the longest and of the greatest amplitude. The orbital velocities of particles in the solitons amounted to 2 knots. It is these that interacted resonantly with the surface waves of about $0.6\ m$ in height, generating a steep, high and irregular rip. Each rip band coincided with the zone of convergence at the front of each soliton, i.e. it was located in the place where the orbital motions encountered unperturbed water. Temperature isopleths plotted on the basis of measurements taken during the passage of a soliton [199] have shown that the deep depression of the thermocline which corresponded to a soliton formed unusually sharp horizontal temperature gradients at the front and in the rear of the wave, amounting to $5 - 6°C/km$ in a $100 - 160\ m$ layer. In the upper quasi-homogeneous $0 - 60\ m$ layer, no temperature changes were observed during the passage of internal solitons.

The dark bands on the surface of the Andaman Sea, which fit the above-described phenomenon, have been observed and photographed from the Soviet-American space station "Soyuz-Apollo" (fig. 3.40).

Fig. 3.40. Photographic depiction of solitons in the Andaman Sea, obtained in the Soviet-American "Soyuz-Apollo" experiment [200].

Like observations from research vessels, these photographs reveal numerous characteristic features which are typical of oceanic fronts, whereas we are actually dealing with a wave phenomenon.

Our recent observations during the 34th cruise of the research vessel "Akademik Kurchatov" showed that trains of large internal waves which modulated the state of the ocean surface (including its thermal state) were very often observed near sharp, sloped frontal interfaces which did not emerge at the surface. In this case, the alternating bands of rips and slicks were the only surface manifestation of subsurface fronts.

The sharp modulations of the temperature field near the ocean surface during calm or low-wind (up to 2–3 m/s) weather with intensive solar heating are also most probably associated with internal waves. The thermal inhomogeneities that develop in this case (as already mentioned in sections 1.2 and 2.1) are referred to by us as "calm weather thermal inhomogeneities". The mechanism of their formation in the case of a strictly periodic ("monochromatic") internal wave can easily be visualized (fig. 3.41). Orbital motions, which become horizontal alternating convergent-divergent flows as they near the surface, form accumulations of passive constituents, solar-heated water and surface-active material in the zones of convergence. One would expect the pattern of disturbances to be periodic, but because of the stochastic nature of the internal waves, no periodicity was actually ever observed. This is why it took so long to discern the true nature of the frequently observed patchiness of the temperature field and passive constituents (e.g. plankton) in the near-surface layer of the ocean. The thermal inhomogeneities appeared to be especially intriguing; let us examine them in greater detail.

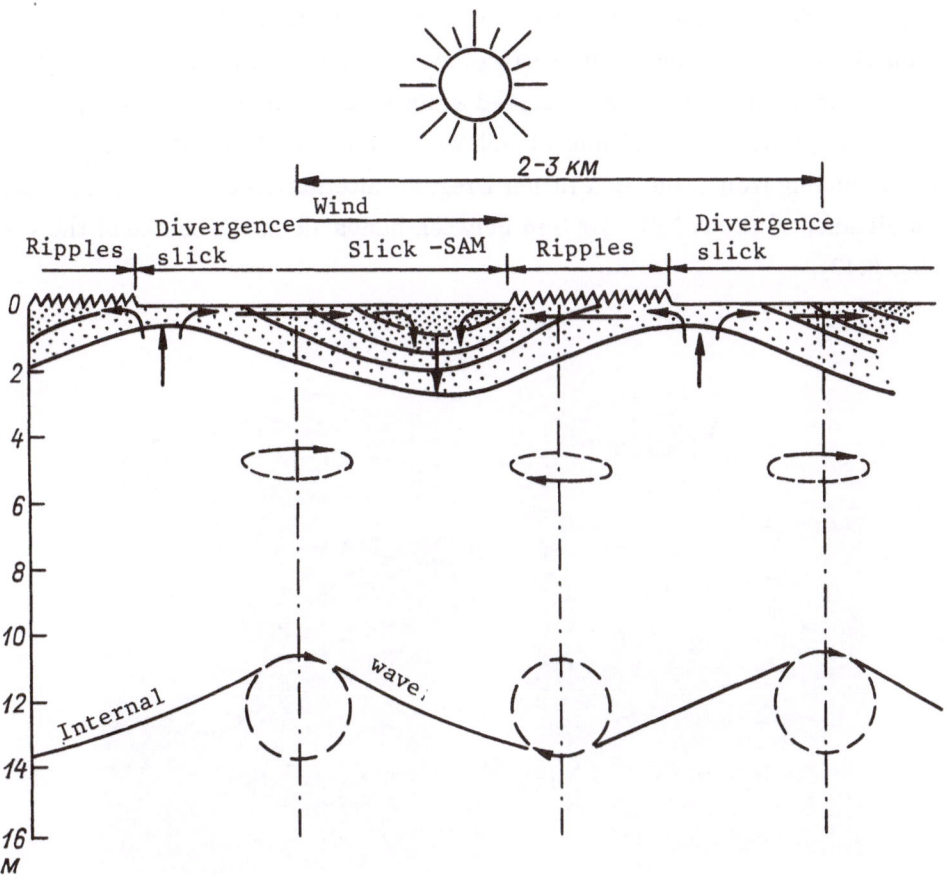

Fig. 3.41. Diagram of modulation of the thickness of a heated near-surface layer and the state of the sea surface by a periodic internal wave. Dashed lines mark the orbits of particles at different levels. Bold lines – isotherms. Dots – suspended particles. Arrows indicate direction of motion. SAM – surface active material.

A number of researchers [31, 34, 69, etc.)] have noted that the temperature distribution near the surface of the ocean acquires a complex patchy (sometimes quasi-periodic) character during intensive solar heating in calm weather. Typical dimensions of the alternating cold and warm patches lie within the range of $1 - 10\ km$. The temperature between the patches may differ by $1 - 2°C$, while the temperature gradients at the boundaries of the patches can reach $0.3 - 0.5°C/km$, and sometimes even $1 - 2°C/km$ [69], as is typical, generally speaking, of sharp oceanic fronts. From the literature, we know that the patchiness of the temperature field under such conditions sometimes correlates quite well with the patchiness of phytoplankton distribution in the surface layer of the ocean. In the Skagerrak at the beginning of July 1977, we observed calm weather thermal inhomogeneities, simultaneously with accumulations of yellowish brown dinoflagellates, drawn out in bands resembling fronts, but in a rather irregular alternation of random orientation with a distance of about $500 - 1000 \, m$ between bands along the course of the vessel (see fig. 3.42).

Fig. 3.42. Accumulation of dinoflagellates in convergence zones at the sea surface in the Skagerrak in July 1977 (author's photo).

Calm weather thermal inhomogeneities (fig. 3.43) appear at approximately $10.00 - 11.00 \, h$ local solar time. As a rule, they attain their maximum amplitude by $14.00 - 15.00 \, h$, i.e. at the time of maximum solar heating of the surface layer. At the same time, the temperature differences between the level measured by a towed sensor closest to the surface ($z = 0.15 \, m$) and a level of several metres reach their maximum values. During intensive heating, inhomogeneities of a kilometre scale are also recorded at levels of $3 - 4 \, m$ by about $14.00 \, h$ (fig. 3.43 a). As measurements with a profiling instrument have shown, horizontal temperature differences at points close to each other are observed down to the $8 \, m$ level at this time [69].

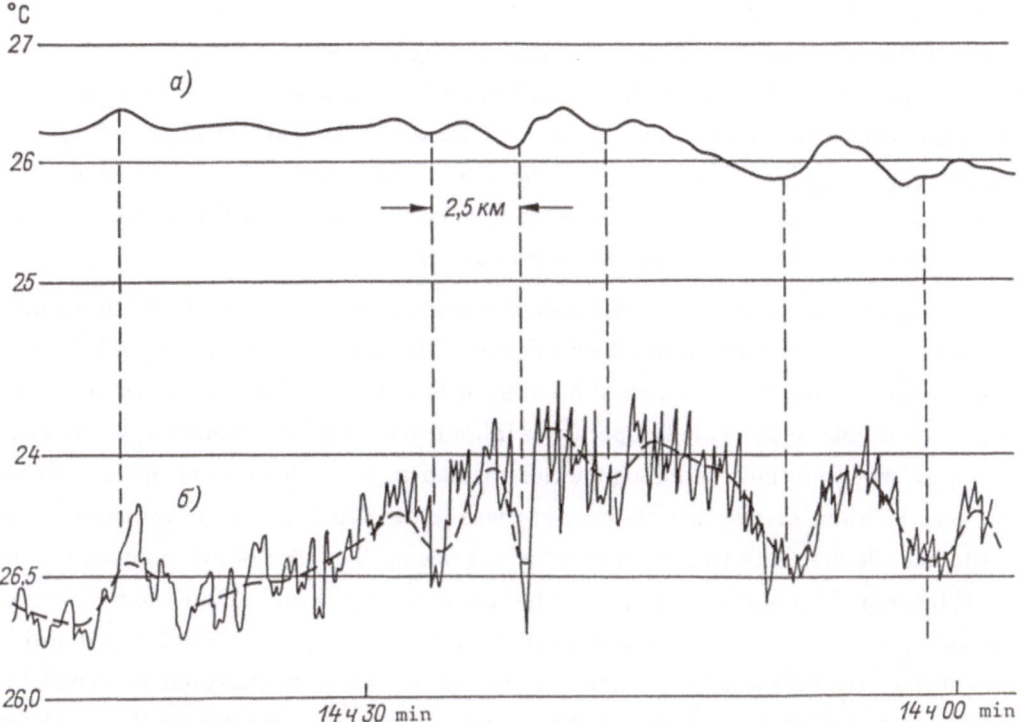

Fig. 3.43. Simultaneous recordings of calm weather thermal inhomogeneities from a moving vessel. a – at a depth of 3.8 m with the help of a sensor placed in the ship's water intake; b – at a depth of 0.15 m with the help of a towed temperature sensor. Sloped dashed lines indicate phase shift for the same disturbances due to the inertia of the temperature sensor in the water intake and a distance of about 120 m between sensors.

Before $14-15\ h$, measurements near the surface ($z = 0.15\ m$) also clearly show small-scale temperature fluctuations with an amplitude of several tenths of a degree, and with a characteristic horizontal scale of the order of 100 m (fig. 3.43 b). As a rule, they disappear between 14.00 and 15.00 h local time. They persist for a longer time only occasionally, under conditions of extreme solar heating in absolutely calm weather. After $14 - 15\ h$, inhomogeneities of a kilometre scale usually begin to decrease in amplitude, but there have been cases where they developed during the second half of the day together with the onset of a calm. The calm weather temperature inhomogeneities at the very surface of the ocean usually disappear completely by 21.00 h. In the deeper layers and especially after highly intensive solar heating during the day, kilometre-scale inhomogeneities continue to exist up to the early hours of the morning. This has been confirmed by temperature measurements with an AIST probe in a flow-through system [68] with water intake at the 3-metre level. With the onset of even a comparatively light wind $(2-3\ m/s)$, accompanied by the appearance of ripples on the surface of the ocean, the calm weather inhomogeneities disappear within an hour at any time of the day. In any case, the lifetime of the calm weather temperature inhomogeneities never exceeds 24 h.

The physical nature of the horizontal calm weather temperature inhomogeneities has not yet been established for certain. The literature has presented discussions of the possible local causes of heating or heat losses which could result in the formation of kilometre-scale temperature inhomogeneities [34]. However, it was still not clear whether these inhomogeneities formed as a result of local differences in the heat balance, or whether their origin was related to nonuniform redistribution of the heat flowing into the upper layer of the ocean. The lack of information about the thickness of the layer encompassed by inhomogeneities and about the horizontal variability of the vertical thermal structure made it impossible to reach any definite conclusion. We also had to be sure that the temperature fluctuations recorded by towed temperature sensors were not the result of the sensor altering its depth in the near-surface layer which has a high vertical temperature gradient.

We believe that the extensive measurements carried out by us in the Atlantic Ocean under conditions of calm and light winds during the 27th cruise of the research vessel "Akademik Kurchatov" in 1978 provided a key to understanding the causes of calm weather inhomogeneities [69]. First of all, the measurements of the calm weather thermal inhomogeneities were subjected to standard statistical processing [69]. For each of the six temperature measurements from a moving vessel $T(x)$, we calculated the mean $\overline{T}°C$, the standard deviation of the mean $\sigma°C$, the skewness As and kurtosis Ex, and then plotted a histogram and spectrum $E_T(k)$ (table 3.6). Examples of the histogram are given in fig. 3.44, and examples of the spectra in fig. 3.45. Of the six plotted histograms, we present one with a quasi-normal probability distribution (fig. 3.44 a), and another one with a well-defined negative skewness (fig. 3.44 b). Of the three spectra presented in fig. 3.45, one diminishes ideally according to the k^{-2} relationship (fig. 3.45 b), which apparently is an indication of very sharp boundaries of thermal inhomogeneities, the second has a noticeable peak (approximating the expected one) at a wavelength of about 1.5 km (fig. 3.45 a), while the third is distinguished by a very high spectral level at all wavelengths from 1 to 4 km (fig. 3.45 c). From these characteristic examples and from the figures given in the table, we can conclude that the characteristics of calm weather inhomogeneities are quite diverse. In some cases, these inhomogeneities come close to having a random character, and in other cases contain a more or less clearly defined periodic component. It does not appear possible to reach any conclusions regarding the physical nature of the calm weather thermal inhomogeneities on the basis of a statistical analysis. A comparison of the results of measurements carried out by different methods gives better grounds for such conclusions.

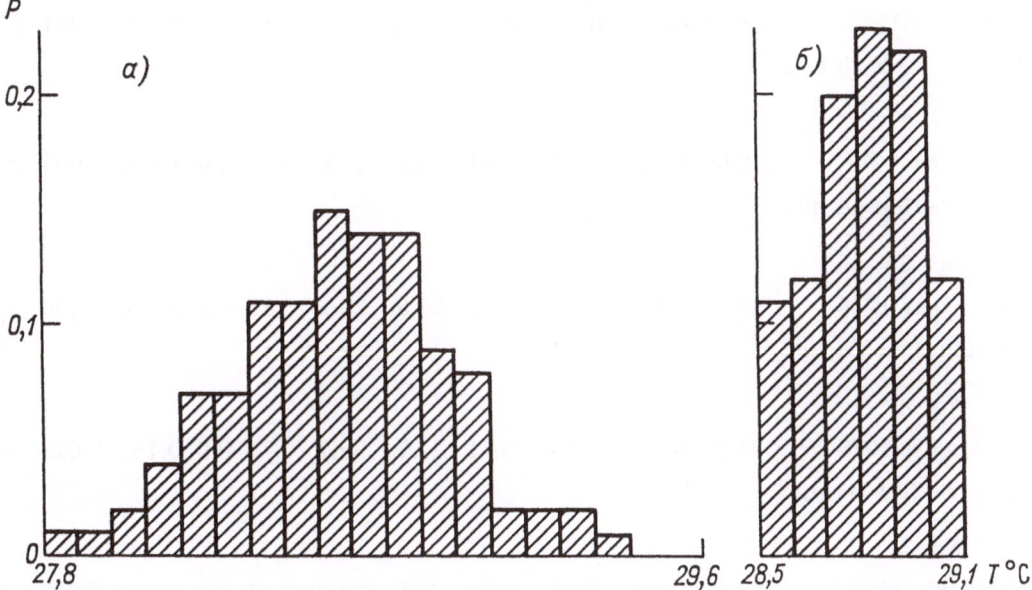

Fig. 3.44. Examples of histograms (with a 0.1°C digitizing interval) for the recorded temperature inhomogeneities. a – entry No. 1 (see table 3.6), b – entry No. 5 (see table 3.6)

Table 3.6

Entry No., date and time	$\overline{T}$ °C	σ °C	As	Ex	T_{min} °C	T_{max} °C	K_{min} km^{-1}	K_{max} km^{-1}	$E_T(k_{min}) \cdot 10$ $(°C)^2 \cdot km$	$E_T(k_{max}) \cdot 10$ $(°C)^2 \cdot km$	L km
No. 1 8 August 1978 12 h 39 min–16 h 25 min	25.03	0.27	−0.50	0.27	24.15	25.75	0.10	0.9	0.852	0.014	102
No. 2 9 August 1978 12 h 52 min–16 h 48 min	26.88	0.30	0.30	−0.57	26.3	27.6	0.10	0.9	1.346	0.007	106
No. 3 24 August 1978 13 h 10 min–16 h 08 min	28.80	0.16	−0.67	1.40	28.5	29.1	0.15	0.9	0.175	0.006	67
No. 4 24 August 1978 22 h 25 min–23 h 54 min	28.20	0.20	−0.18	−0.11	27.7	28.7	0.26	0.9	0.162	0.011	40
No. 5 20 Sept. 1978 16 h 21 min–19 h 41 min	28.60	0.29	−0.05	0.32	27.8	29.6	0.11	0.9	0.572	0.016	90
No. 6 23 Sept. 1978 13 h 32 min–16 h 35 min	28.10	0.25	−0.40	−0.20	27.4	28.6	0.12	0.9	0.690	0.009	82

Note: The maximum wave number k_{max} for all entries is the same and equal to $\frac{1}{4}\Delta x$, where $\Delta x = 0.28$ km is the digitizing interval from analog records (we note that the Nyquist frequency is equal to $\frac{1}{2}\Delta x$). The minimum wave number k_{min} was selected as approximately equal to $10/\ell$ (where ℓ is the full record length) in order to ensure a relative accuracy of determination of the spectrum of the order of 0.1.

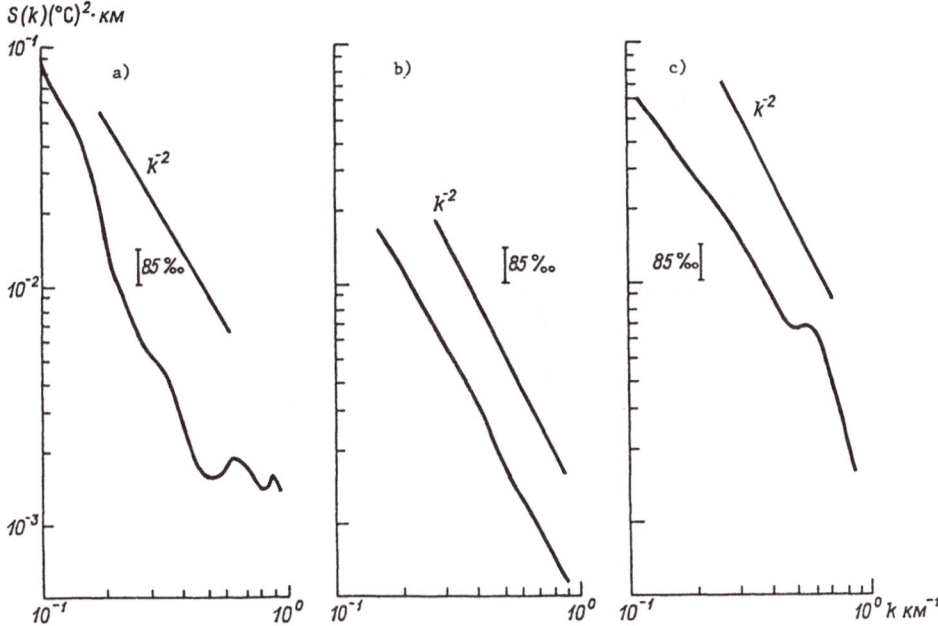

Fig. 3.45. Examples of spectra of temperature fluctuations for entry No. 1 (a), No. 3 (b) and No. 5 (c) (see table 3.6). The vertical segment corresponds to an 85% confidence interval.

The measurements, which were carried out with a profiling instrument under the conditions of an absolute calm and intensive solar heating in an area of water unperturbed by the drift of the vessel (in the vicinity of the bow $3 - 4\ m$ from the stem of the vessel), revealed significant variability in the vertical thermal structure of the surface layer [69]. Fig. 3.46 depicts four vertical temperature profiles from a series of several tens of profiles obtained on 25 September at station No. 2762. These measurements were preceded by continuous recording of temperature near the surface ($z = 0.15\ m$) with a towed temperature sensor from a moving vessel; temperature recording was continued after the completion of the station, an hour after the measurements with a profiling instrument. Comparison of the data from the continuous temperature measurement with the vertical profiles (fig. 3.46) obtained at the station showed that the maximum and minimum temperatures in the kilometre–scale inhomogeneities recorded from a moving vessel coincided almost completely with the maximum and minimum temperatures near the surface from the profiles of the profiling instrument (fig. 3.46). Enthalpy fluctuations above the level common to all the profiles ($10\ m$) fell within the range of $672 \times 10^4 - 966 \times 10^4\ J/m^2$, and with the maximum temperatures in small-scale fluctuations measured from a moving vessel taken into account, they reached $1176 \times 10^4\ J/m^2$.

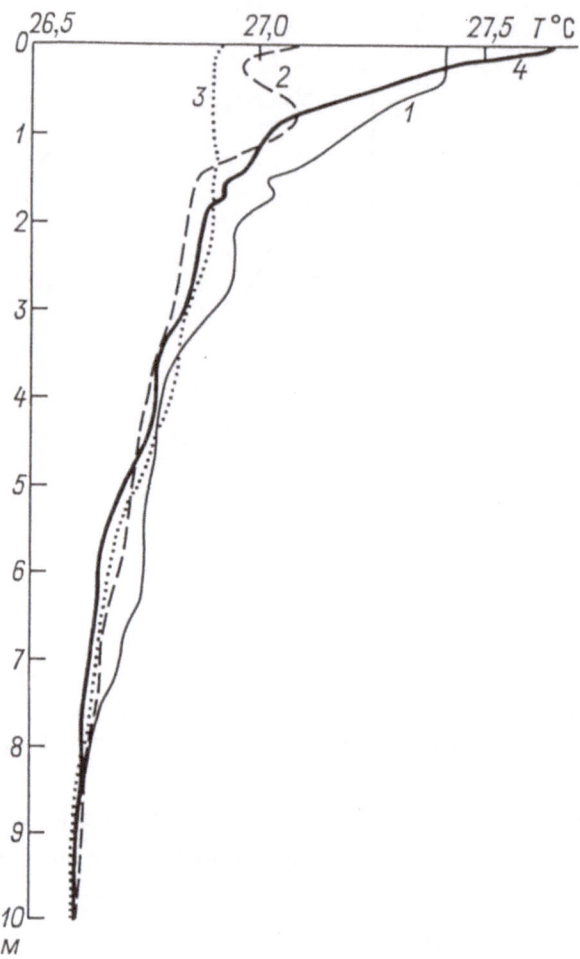

Fig. 3.46. Vertical temperature profiles obtained in calm weather in the Sargasso Sea on 25 September 1978 with the help of a profiling instrument during slight drifting of the survey vessel (not more than 0.5 knot). 1 – at 13 h 11 min., 2 – at 13 h 24 *min.*, 3 – at 13 h 30 *min.*, 4 – at 13 h 32 *min.*

The latter figures and the overall good agreement of the results obtained by two independent methods indicate that the variability recorded by the towed sensor (fig. 3.43) reflects the actual near-surface oceanic kilometre-scale thermal inhomogeneities which can in no way be the result of the vertical "ups and downs" of the sensor. With the actual ±10 *cm* deviations of the sensor from the mean towing level, temperature fluctuations with an amplitude in the neighborhood of ±0.1°C could have resulted, as we can see from the nature of the profiles in fig. 3.46. It cannot be ruled out that the above-mentioned small-scale inhomogeneities with a length in the neighborhood of 100 m did occur precisely in this way, though they could have had a natural origin (e.g. caused by horizontal variability in the depth of convection, see below).

The fact that calm weather inhomogeneities have been observed in different parts of the World Ocean prompts us to search for a universal mechanism of their formation in the near-surface layer of the ocean. From our point of view, this mechanism is the modulation of the near-surface layer by internal waves. We know that the visible manifestations of internal waves can appear at the surface of the ocean through the modulation of ripples or wind waves by the horizontal components of the orbital motions of internal waves. Internal waves can also modulate the fields of a passive scalar (e.g. suspension) within the surface layer of the ocean. The above-mentioned manifestations were noticed primarily because of the clearly defined periodicity of the observed modulations. In the ocean, we generally observe a sporadic field of internal waves, which can be described only by statistical methods. A statistical description of the formation of the patchy structure of the field of passive constituents under the influence of a sporadic field of internal waves is given in ref. [42]. A heated $0.5-1$ m near-surface layer of ocean water can with every reason be regarded as the field of a passive constituent. The temperature patches which form during its modulation by a sporadic field of internal waves should have [42] the characteristic horizontal scale $L \simeq (1.1-1.4)\sigma_u/f$, where f is the Coriolis parameter, and σ_u^2 is the mean square horizontal velocity at the surface of the ocean. The estimation of this horizontal scale, which was carried out in [69], gave a value of L in the neighborhood of one kilometre. Consequently, internal waves in the seasonal thermocline can actually cause the appearance in calm weather of kilometre-scale thermal inhomogeneities of a random, patchy, or quasi-periodic nature, depending on the nature of the field of internal waves. A number of other factors can also affect the thermal state of the near-surface layer of the ocean, intensifying the inhomogeneities observed in it and in some way altering the general characteristic pattern. We primarily have in mind the patchiness of salinity in the near-surface layer of the ocean, which forms as a result of downpours.

The systematic synchronous measurements of vertical temperature and salinity distribution in the upper 30-metre layer of the ocean, which were carried out with an AIST STD-probe in a discrete-depth recording mode with an interval of $0.5-1$ m [18], have shown intensive horizontal variability of the vertical distribution of salinity in the near-surface layer of the ocean. For example, on days with frequent downpours, we recorded vertical salinity profiles which differed by

0.5*ppt* in the upper 2-metre layer at a distance of only 800 *m* between the points of measurement [18]. The differences in the salinity structure determine temperature variability. In fig. 3.47, for example, we can see how a thin rain-freshened layer affects the nature of daytime heating. The upper, extremely heated (to 29.1°*C*) one-metre layer originated as the result of a salinity jump which prevents convective mixing with the layers below. It is not difficult to calculate that only 7 *mm* of rainfall, i.e. not a very heavy downpour, is required to freshen a one-metre layer by 0.25*ppt*. In turn, a freshened layer heated by 0.8°*C* contains an excessive amount of heat, which on a sunny day can be accumulated in only one hour. Obviously, the same amount of heat will result in weaker heating of the overlying layer when the salinity jump occurs at a greater depth. Measurements during the POLYMODE operations in August–September 1978 showed that the near-surface water temperature in freshened patches in calm weather reached 29 − 32°*C*, whereas it did not exceed 27.6 − 28.5°*C* in adjacent points where no rain fell. At the same time, we observed a negative horizontal T − S correlation, which is comparatively rare for the surface layer of the Sargasso Sea. Consequently, rainfall plays an important part in the formation of a patchy temperature distribution near the ocean surface in calm and low-wind weather. We should note that the spatial dimensions of the rain-freshened patches correspond to the dimensions of the cumulonimbi, i.e. only a few kilometres. The differences in thermal structure preserved in the surface layer from previous days also contribute to the development of kilometre-scale temperature inhomogeneities. Superposed on the already horizontally inhomogeneous structure, the daytime heating creates even sharper calm weather inhomogeneities.

The varying intensity and depth of convection at different points may be another factor which contributes to the patchiness of the thermal field at the surface of the ocean. In calm weather, it is precisely convection that serves as the main physical mechanism of heat transfer from the surface to the deeper layers of the ocean. Although convection usually reaches its maximum intensity during the night at moderate and low latitudes in summer, it can also occur during the day when, with the temperature of the air approximately equal to the water temperature, air humidity does not exceed 60 − 70%. Such conditions are very often observed in the vicinity of the Sargasso Sea. A combination of heat transfer due to evaporation and long-wave radiation with direct radiative (solar) heating can result in characteristic vertical temperature profiles with an inversion in the near-surface layer and a temperature maximum slightly below the surface (from 10 *cm* to 1 − 3 *m*).

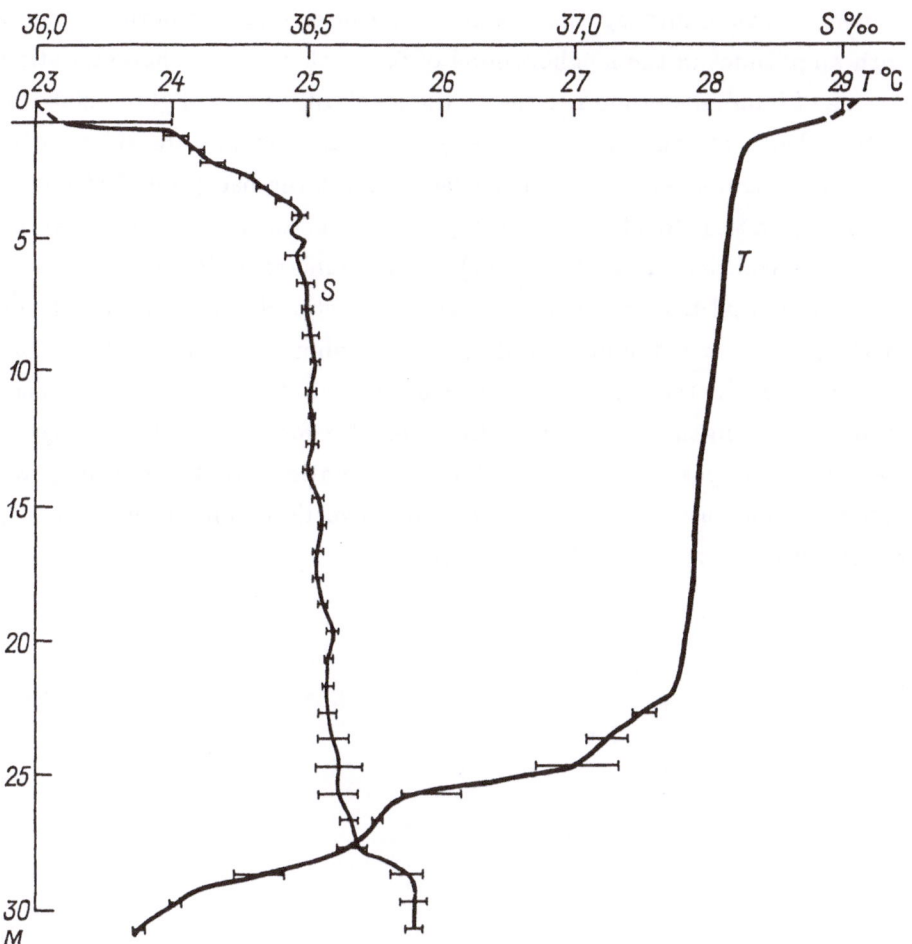

Fig. 3.47. Vertical temperature and salinity profiles recorded with an AIST probe in the Sargasso Sea in the afternoon hours of a calm day with intensive solar heating (15 h 00 min, 25 August 1978) after the freshening effect of rainfall on the preceding days. (Horizontal lines denote the variability range at different levels.)

Such temperature profiles, which have also been observed in the past [34], were repeatedly recorded by us during the 27th cruise of the research vessel "Akademik Kurchatov". According to the investigations of Solov'yov [51] and Woods [264], the depth of daytime convection under the conditions described above shows a clearly defined diurnal variation. Before and around midday, the convective layer is found within the upper 10 *cm*; by 16.00 *h*, its thickness reaches 1 *m*, and then increases very rapidly, exceeding (according to our observations) 10 *m* during the night. On the other hand, the depth of daytime convection can be greatly influenced by patches of films of surface-active substances and by local variations of even the lightest wind. Around midday, these factors can cause variations in the temperature field with amplitudes in the neighborhood of $0.3 - 0.5°C$ and a characteristic horizontal scale of hundreds of metres due to the local differences of $100 - 200\%$ in the penetration depth of convection. These temperature changes are apparently also observable in measurements near the surface against the background of kilometre-scale inhomogeneities (together with temperature fluctuations of "artificial origin" due to the vertical motion of the sensor). Since small-scale (≈ 100 m) patchiness of temperature (real and "artificial") is characteristic of a near-surface layer not greater than 0.5 m, it is the first to disappear (at about 15.00 h) as the convective layer goes deeper. As the thickness of the convective layer continues to increase, we observe a decrease in the amplitude of the kilometre-scale thermal inhomogeneities which at night disappear almost completely. It may also be that after the convective layer begins to increase rapidly, large cells of convectional circulation appear, but the true scales of these are not known as yet.

Chapter 4

FRONTS AND THE STRUCTURE OF OCEAN WATERS

4.1. On the multifrontal structure of frontal zones

In subsection 3.1.3, the example of the frontal zone of a warm Gulf Stream ring already gave the reader some idea of the complex three-dimensional structure of temperature, salinity and density fields, which is formed by the numerous frontal interfaces that occupy the space in a zone measuring only $10 - 20$ km in width. The results of measurements given in figs. 3.12, 3.15 a, b, c and 3.16 give us only the most patchy information about the complexity of this structure, which can be fully established only by organizing highly specialized research and using numerous types of modern specially constructed equipment. However, in order to ascertain that a frontal zone in the ocean is not a gradual transition zone from waters with certain characteristics to waters with other characteristics, it suffices to examine any thermal image of the ocean surface obtained from modern earth-orbiting satellites by means of high-resolution infrared scanners (e.g. see figs. 1.3 and 2.7, as well as [174, 195]). In these images, we are first of all struck by the abundance of local extremes of the horizontal temperature gradient with differing orientations and varying intensity. The frontal zones of such large boundary currents as the Gulf Stream, the Kuroshio, the Agulhas Current, the East Australian and South Australian currents, the zones of subtropical convergence or the frontal zones of coastal upwelling are depicted here as having a complex horizontal structure which in many cases could be called "fine-fibered", and in other cases "rotational". Many earlier detailed ship and aircraft surveys of frontal zones [117, 120, 122, 250, 251] indicated the same, but they were not sufficiently numerous, systematic or descriptive to attract universal attention.

As we now see, the structure of large-scale climatic frontal zones is far more complex than the structure of synoptic-type frontal zones, since the latter are themselves included as elements in the structure of the former. This can be demonstrated using as an example the frontal zone of the Gulf Stream, the structural details of which have been discussed the most in the literature, as in the preceding sections of this book. The structural characteristics of the frontal zone of the Gulf Stream can be arbitrarily divided into at least five groups which combine the following similar structural elements:

1) Gulf Stream rings (with their frontal systems) and meanders which are characterized by transverse dimensions in the neighborhood of $100 - 300\ km$;

2) advective tongues of cold and warm water $10 - 50\ km$ wide and up to $200\ km$ long, which are associated with eddies, rings and meanders;

3) boundary spin-off eddies and eddies within the main stream of this current, measuring $25 - 30\ km$ in diameter (frequently with cores and jets of entrained cold water);

4) isolated cores and bands of cold slope water that have been entrained together with parts of the frontal interfaces into the main stream of the current, or into the frontal zones of anticyclonic rings. These bands are $5 - 10\ km$ in width with a length of $100 - 500\ km$ and form alternating fronts every $5 - 20\ km$;

5) bands of cold and fresh shelf water $100 - 200\ km$ long and $5 - 10\ km$ wide, which spread along the main front of the Gulf Stream from the slope water side and are bounded by sharp frontal interfaces.

It is this presence of numerous frontal interfaces of various scales within the same well-identified (geographically) frontal zones that particularly emphasizes the previously mentioned (in the foreword) parallel between frontal phenomena in the ocean and the vertical thermohaline finestructure of its waters. Just as not all oceanic layers are characterized by an equal "intensity" of the vertical finestructure, the abundance and diversity of frontal interfaces are not typical of all areas of the World Ocean, but only of its frontal zones. If we rightly compare the vertical finestructure with the "signature" of vertical turbulent mixing, then it would be equally fair to interpret the frontal interfaces (including those encountered within the same frontal zone) as the "signature" of horizontal turbulent mixing. At the same time, it would be useful to remember that the concept of horizontal turbulence in the stratified ocean covers a scale spectrum several orders of magnitude wider, and includes a far greater diversity of physical processes and motions than does turbulent vertical mixing. From this stems the diversity and omniscale nature of frontal phenomena in the ocean.

During the 34th cruise of the research vessel "Akademik Kurchatov" in December 1981 – April 1982, continuous recording of temperature near the surface ($z \approx 0.15\ m$) was carried out along the entire route by means of a towed thermistor, and water samples were taken from the surface to determine salinity. During the cruise, a number of large and fairly stable (geographically) climatic frontal zones were traversed, including the zone of convergence of the Kuroshio and Kurile currents (fig. 4.1), as well as the northern and southern subtropical convergences in the Pacific Ocean.

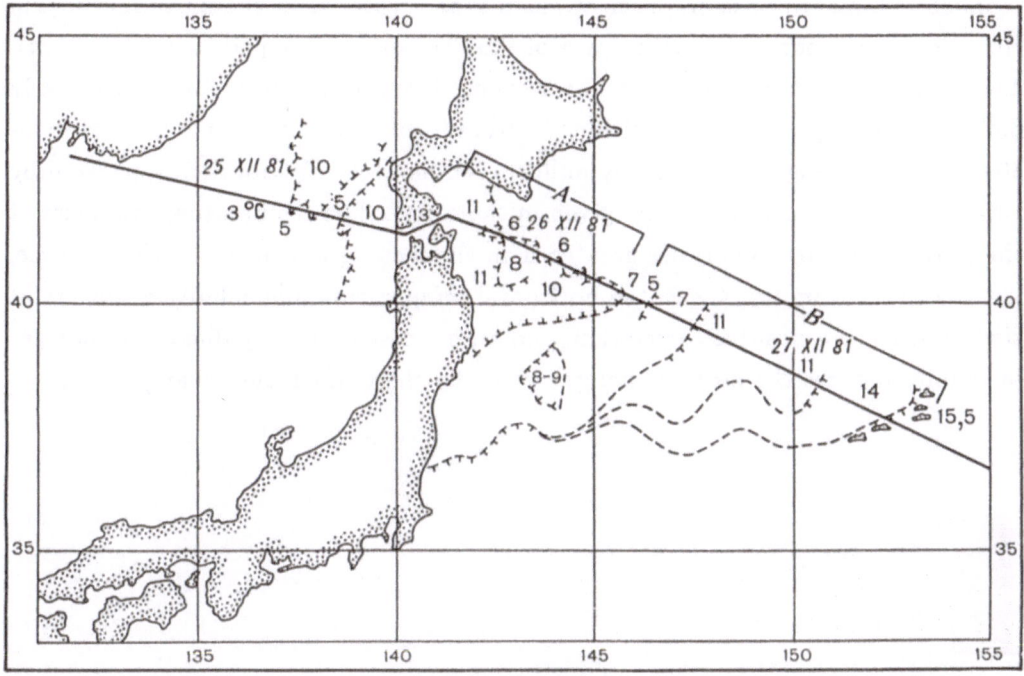

Fig. 4.1. Thermal structure of the ocean surface, reproduced on the basis of temperature measurements and an IR image from a NOAA satellite. A, B - position of profiles depicted in fig. 4.2.

The distribution of SST in the mixing zone of the Kuroshio and Kurile currents east of Honshu is shown in fig. 4.2. We know that this zone has a highly complex structure which is characterized by numerous branchings of the currents, the alternation of warm and cold streams, and intensive eddy formation on various scales [11]. Analysis of temperature measurements together with the available IR image of the Kuroshio area for 21 December 1981 (based on the data of the NOAA-7 satellite) made it possible to compile a chart of the location of thermal fronts (fig. 4.1). Profile A (fig. 4.2) reflects motion along the meandering frontal interface between the region of warm water (a branch of the Tsushima Current flowing through Tsugaru Strait) with a temperature of $10 - 11°C$ and the cold waters of the Kurile Current $(5 - 6°C)$. Consequently, the SST measurements indicate that this frontal interface was crossed many times. The distribution of SST and salinity at the surface in the frontal zone between the waters of the Kurile Current $(5°C)$ and Kuroshio Current $(15.5°C)$ is shown in fig. 4.2 B. The total length of the frontal zone measures approximately 300 km, the temperature change

10.5°C and the salinity change 1.4 *ppt*. The mean temperature gradient is equal to 0.035°C/km, which is an order of magnitude greater than the mean meridional climatic gradient for the northern part of the Pacific Ocean (equal to 0.003°C/km). The internal structure of the temperature and salinity fields in this zone has a clearly defined multistage character. Within the frontal zone, we distinguish four intervals 50 – 150 km in length with a fairly uniform temperature distribution, separated by very sharp fronts (characteristics given in table 4.1). The temperature gradients at these fronts are 10 – 400 times greater than the mean gradient in the frontal zone. As we can see from fig. 4.2 B, the horizontal temperature and salinity gradients at the fronts show a positive correlation, and an increase in density due to an increase in salinity reduces the density change caused by the temperature change.

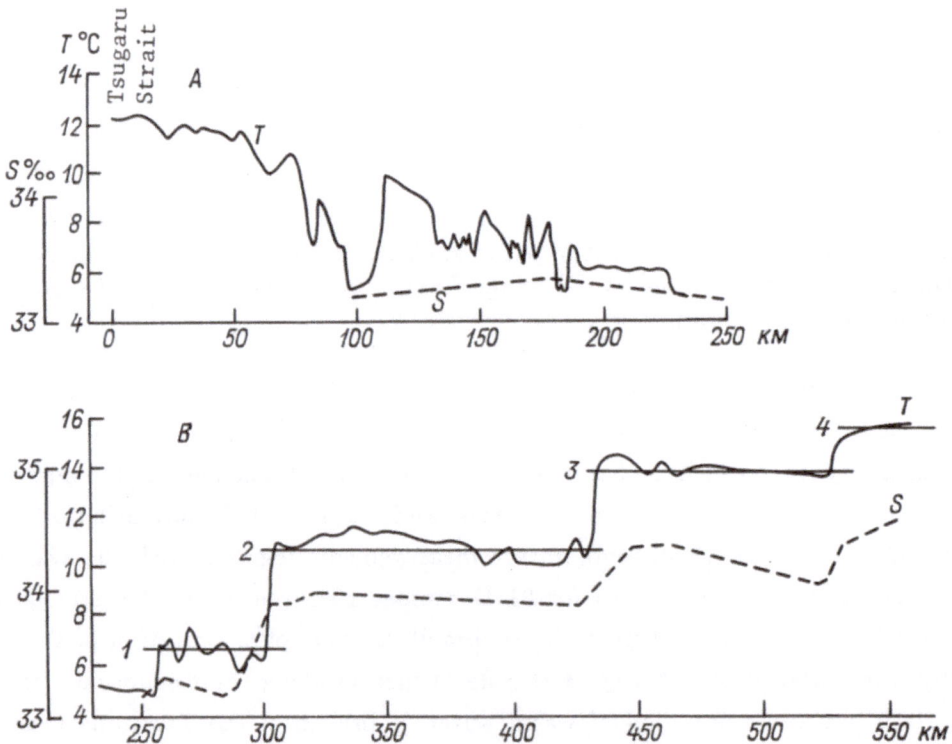

Fig. 4.2. Temperature distribution in an area intersecting the convergence zone of the Kuroshio-Kurile Current.

Table 4.1

Characteristics of frontal interfaces in the frontal zone of the Kuroshio-Kurile Current

No.	ΔT	ΔS	ℓ	$\partial T/\partial x$	$\partial S/\partial x$	$\Delta\sigma_t$	$\partial\sigma_t/\partial x$	$\dfrac{\partial T/\partial x}{\partial T/\partial x}$
	$°C$	ppt	km	$°C/km$	ppt/km	(units)	(units/km)	
1	1.45	0.16	0.3	4.8	0.53	0.054	0.18	137
2	4.50	0.70	0.3	1.5	2.33	0.170	0.57	429
3	3.60	0.45	1.9	1.9	0.24	0.360	0.19	54
4	1.50	0.30	4.8	0.3	0.06	0.092	0.02	9

However, the temperature and salinity changes are far from fully density compensating, and all four fronts are sharply defined in the density field.

It should be said that a similar structure with frontal interfaces extended along the frontal zone has also been observed in the Gulf Stream [250]. It would appear that a similar multi-stage structure with fronts of the same sign should form as a result of the formation of local convergences of the main stream directed along the frontal zone.

The temperature distribution in the zone of the northern sub-tropical convergence (NSC) is of another nature (fig. 4.3). Under usual conditions, the latitudinally extended NSC is located at $31 - 32°N$, and has a temperature change of $4°C$ and salinity change of 0.5 ppt [214]. During our observations, the convergence zone proved to be displaced southward to $27 - 24°N$, and the overall changes in temperature and salinity amounted to $3.5°C$ and 0.4 ppt respectively. Analysis of the meteorological charts received on board the vessel led us to assume that this displacement of the convergence was caused by the fairly strong (up to $20 - 25$ m/s) northerly winds which occurred prior to the crossing of the NSC. The wind-related inflow of cold waters from the north obviously accounts for the fact that the zone as a whole proved to be divided by a region of cold water into two separate parts (fig. 4.3). Unlike the frontal zone between the Kurile Current and the Kuroshio where the thermal structure was of a multistage nature with fronts of the same sign, the zone of the NSC had a temperature distribution which could arbitrarily be called "alternating". It is characterized by the alternation of cold and warm areas of different width ($5 - 20$ miles), separated by fronts of different signs. The mean temperature gradient in areas I and II (fig. 4.3), which measure approximately 200 km in length, is 0.015°C/km. Each area contains up to ten frontal interfaces with local gradients of $0.1 - 2.3°C/km$. The ratio of the gradients at the fronts to the mean gradient in the zone amounts to $5 - 140$. Therefore, the frontal interfaces in the zone of the NSC are relatively less intense than in the convergence zone of the Kurile and Kuroshio currents.

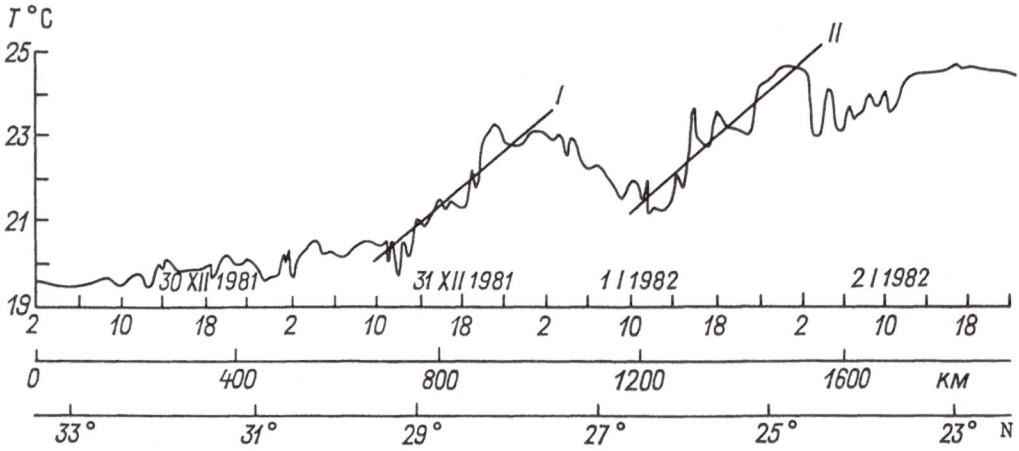

Fig. 4.3. Temperature distribution in the zone of the northern subtropical convergence (NSC).

The alternating structure of the frontal zone can be produced both by advective processes (particularly by the cold waters from the north overriding the warm waters), and by the effect of eddies formed in the frontal zone. Naturally, it is impossible to establish the generation mechanisms of this or that type of structure on the basis of SST measurements along solitary tracks.

Another example of the multifrontal, multistage structure was recorded in the zone of southern subtropical convergence (SSC) in the eastern part of the Pacific Ocean. In this area (fig. 4.4) with a length of about 400 km, the overall temperature change amounted to 3°C (from 20.8 to 17.8°C). The mean gradient was equal to 0.008°C/km. The transitional zone is of a multistage nature, and contains four main, sharpest, frontal interfaces (numbered in fig. 4.4) with gradients of $0.15 - 0.44°C/km$. Their characteristics are given in table 4.2.

Table 4.2

Characteristics of the frontal interfaces
in an area of the southern subtropical convergence

No.	ΔT °C	ΔS ppt	ℓ km	$\partial T/\partial x$ °C/km	$\partial S/\partial x$ ppt/km	$\Delta \sigma_t$ (units)	$\partial \sigma_t/\partial x$ (units/km)	$\dfrac{\partial T/\partial x}{\partial T/\partial x}$
1	0.7	0.5	2.4	0.29	0.21	−0.19	−0.08	36
2	1.0	1.3	2.4	0.42	0.13	0.03	0.01	52
3	0.5	0	3.4	0.15	0	0.12	0.04	19
4	1.1	0.2	2.5	0.44	0.08	0.12	0.05	55

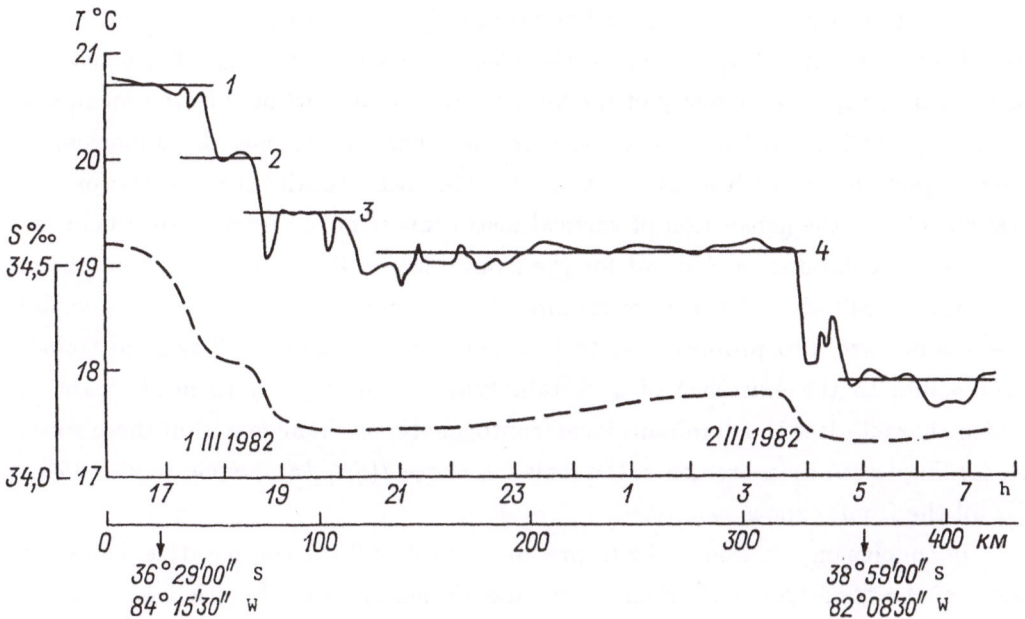

Fig. 4.4. Temperature and salinity distribution in an area of the southern subtropical convergence.

The SST gradients at the main fronts are $20 - 50$ times greater than the mean gradient in the zone. It is interesting to note that, unlike the convergence of the Kuroshio-Kurile Current where all the fronts had the same sign in the density field, the frontal interfaces of the SSC have different signs in the density field. At front 1, for example, the drop in density predominates due to a decrease in salinity, and density diminishes on the whole (regarding the zone as a transition from warm to cold waters). Front 2 is essentially density-compensated, while the density at fronts 3 and 4 increases due to the dominant contribution of the temperature drop.

It would be interesting to check the extent of the earlier mentioned similarity between the horizontal frontal structure of ocean waters and their vertical finestructure. For this, we must establish whether the temperature changes at the frontal interfaces within the frontal zones depend on their mean temperature gradient. By analogy with the vertical finestructure, this can be done by studying the relationship $\sigma'_T / \overline{(dT/dx)}$, where σ'_T is the standard deviation of the temperature values from the average profile with a gradient of $\overline{dT/dx}$. The behavior of the relationship $C_v = \sigma'_T / \overline{(dT/dz)}$ for the vertical finestructure was examined in [72], and it was found that the value of C_v usually varies within a very narrow range. The constancy of C_v means that the intensity of the finestructure is proportional to the mean gradient. C_v itself (sometimes called the Cox number), which has the dimension of length, provides us with a certain value for the mean spatial scale of the process responsible for the generation of vertical disturbances of the temperature field.

The calculations carried out for the frontal zones discussed above have shown that the amplitudes of the temperature changes across frontal interfaces within these zones are also proportional to the mean zonal gradient. This, apparently, also points to the existence of a certain typical, this time horizontal, scale C_h which characterizes the dominant local frontogenetic mechanism within these zones. According to our measurements, the scale $C_h = \sigma_{T''} / \overline{(dT/dx)}$ is equal to $25 - 35\ km$ for all the frontal zones examined.

In conclusion, we would like to present in table 4.3 the comparative characteristics of the "finestructure" of the above-mentioned areas of climatic frontal zones. The table gives the width of the studied area of the frontal zone L, the mean gradient, the type of structure and the number of frontal interfaces, the gradients at the fronts, the ratio of the frontal gradients to the mean zonal gradients, and the values of C_h.

Table 4.3

Comparative characteristics of the thermal structure of frontal zones

No.	Frontal zone	L,km	$\overline{dT/dx}$ $°C/km$	Type of structure, number of frontal interfaces	$\left\lvert\frac{dT}{dx}\right\rvert^*$ $°C/km$	$\frac{dT/dx}{dT/dx}^{**}$	C_h km
1	Convergence zone of the Kuroshio-Kurile Current	300	0.035	Multistage 4	$\frac{0.3-15}{5.5}$	$\frac{9-429}{160}$	27
2	Northern subtropical convergence			Alternating			
	Area I	200	0.015	10	$\frac{0.08-1.4}{0.4}$	$\frac{5-50}{27}$	26
	Area II	220	0.016	8	$\frac{0.07-2.3}{0.8}$	$\frac{5-140}{50}$	28
3	Southern subtropical convergence	400	0.008	Multistage 4	$\frac{0.15-0.44}{0.3}$	$\frac{19-55}{38}$	35

* Frontal gradients: $\frac{min-max}{mean}$ $°C/km$.

** Ratio of gradients: $\frac{min-max}{mean}$.

The results obtained give grounds for the following conclusions:

1. The climatic frontal zones studied include several (up to 10) sharp frontal interfaces with temperature gradients $1-2$ orders of magnitude greater than the mean zonal gradient.

2. Two types of frontal zone thermal structure are distinguished, a multistage structure consisting of homogeneous areas separated by fronts of the same sign, and an "alternating" structure in which warm and cold areas separated by fronts of a different sign alternate.

3. The amplitudes of the thermal inhomogeneities (temperature jumps at the fronts) within the frontal zones are proportional to the mean zonal gradient, which points to the presence of some typical horizontal scale which characterizes the dominant local frontogenetic mechanism within the zone.

4.2. Thermohaline finestructure near oceanic fronts

Over the past decade, it has become apparent that the nature of vertical finestructure is closely related to the degree of horizontal inhomogeneity of ocean waters. As a rule, the frontal zones of the ocean have so many specific forms of thermohaline finestructure (intrusive layers,* temperature inversions) that their appearance on vertical STD and XBT profiles can serve as an indication that we are approaching a frontal interface. The abundance of finestructure characterizes not only large-scale climatic frontal zones, but also synoptic-scale fronts associated with an eddy field. As we have recently seen [66] (see also subsection 3.1.1), frontal systems of different types and thermohaline finestructure of different origins are associated with eddies of a different sign. The available data permit us to assume that oceanic fronts play an important part in the transformation of large-scale and mesoscale horizontal inhomogeneities of the physical fields of the ocean into a thermohaline finestructure, thus serving in a way as suppliers of vertical finestructure of these fields over a wide range of space-time scales [61, 260]. If we look at the variability of the physical fields of the ocean on the scale of a whole oceanic basin, we cannot but admit that the very first sign of the evolution in the vertical, of a horizontally inhomogeneous density field, which is eventually seen as the formation of finestructure, is none other than **f r o n t o g e n e s i s**, the formation of oceanic fronts [62].

It is not by chance that many of the researchers previously occupied with describing and explaining the origin of the thermohaline finestructure of ocean waters later turned to questions related to the structure and dynamics of oceanic fronts. This is the natural course. Obviously, the numerous unsolved problems that have arisen during the study of finestructure can be answered only by studying oceanic frontogenesis, the mechanisms of the self-preservation and evolution of fronts, the processes of cross-frontal heat and mass transfer and, finally, the processes of frontolysis in the ocean.

Research into the structure-forming processes in oceanic frontal zones is now becoming especially important also because of scientists' attempts to establish the processes and mechanisms which ensure sufficiently effective dissipation of the kinetic energy of quasi-geostrophic, oceanic, synoptic-scale eddies. Other than fronts

---

* By "intrusions" or "intrusive layers" (term borrowed from geology) we mean the masses of extraneous water that have intruded into the local water mass at a level of equal density. The historical background of this term is given in section 1.6 of ref. [61].

and internal waves, there are no other physical claimants to the role of "dissipators" of the kinetic energy of eddies in the kinetic energy spectrum for any of the conceivable motions of ocean water between the scales of two-dimensional quasi-geostrophic turbulence (100 km) and the scale of Kelvin-Helmholtz instability (1 m). Fronts ensure the necessary cascade of enstrophy toward smaller scales. Internal fluctuations *per se* are not a dissipative mechanism, but basically they can effectively disperse the excess kinetic energy of any initial disturbance (e.g. a transient eddy or front) in space, transporting this energy in all directions over great distances. For example, Woods [262] believes that only the parallel effect of fronts or internal waves is likely to ensure the necessary flow of kinetic energy from synoptic eddy scales to viscous dissipation scales.* Whereas internal waves can in this case produce the effect "directly", fronts are associated with dissipation scales through the thermohaline finestructure. Therefore, it is hardly possible to discern the roles played by internal waves and finestructure in the destruction of oceanic eddies without taking up the problem of frontogenesis and the structure of fronts in the ocean. At the same time, there has been an increasing number of observations which indicate that internal waves are a particularly powerful source of energy in the frontal zones of the ocean. How can we avoid seeing a genetic relationship between the abundance of finestructure elements and the high intensity of the internal fluctuation field!? However, the time is past when all thoughts concerning the origin of thermohaline finestructure inevitably led to internal waves as the main, if not the only structure-forming process. Today, there is a tendency to regard intrusion as the most widespread mechanism for the formation of structure in the ocean. As it now turns out, internal waves can initiate only a comparatively small-scale intrusive process during their decay. It does not appear possible to attribute the appearance of 10 − 100-metre intrusive layers to internal waves (see subsection 4.2.2). We have to admit that both internal waves and intrusions are in this case by-products of the dynamics of frontal interfaces. They definitely interact with each other, but they also have their own independent physical roles. Because of this, questions related to the origin and dynamics of intrusions are among the most important in the current explosion of interest in oceanic fronts. As a result of this, the latest literature on oceanic fronts and their structure shows a predominance of phenomenological descriptions, as well as studies of individual cases and model problems which describe typical situations, in contrast to the statistical approach which characterized initial interest in finestructure and internal waves. This is precisely why this section is de-

---

* For more details on Woods' concept, see the diagram in ref. [262]

voted mainly to a phenomenological analysis of intrusions, which to a great extent also reflects the author's personal interests.

4.2.1. Characteristics of the thermohaline finestructure of frontal zones

Numerous examples of intrusive finestructure elements have been given in the literature, including papers by the author (see [61, figs. 7, 40, 43] or [62, chapter III, figs. 19-21]). Let us now examine several more examples of intrusive structures which are characteristic of various frontal zones. Fig. 4.5 depicts intrusive laminae of Red Sea waters near a frontal interface associated with their spreading in the Gulf of Aden (observations of A.I. Ginzburg, V.I. Prokhorov and V.Ye. Sklyarov, 22nd cruise of the research vessel, "Akademik Kurchatov", 1976). The nature of the vertical temperature and salinity profiles observed there for a monotonic vertical distribution of density was unusually similar to that observed by the author in 1971 in the Arabian Sea, $150 - 200$ km farther south [61, figs. 7 and 43]. The qualitative identity of profiles occurs both in thin intrusive laminae approximately 5 m in thickness (marked A and B in fig. 4.5), and in larger intrusions with a thickness of up to $30 - 40$ m. In the Gulf of Aden, however, they are all located $200 - 400$ m closer to the surface than in the Arabian Sea, and have a temperature $3 - 4°C$ higher. Here too, the salinity compensation of the warm intrusions with temperature inversions is characterized by an isopycnicity coefficient $R_A > 1$ (see section 4.2 in [61] and the recent more detailed interpretation in [79]).

Fig. 4.6 depicts the isopleths of temperature variation over a 3.5 h period, based on data from repeated STD profiles by the author from a drifting vessel (drift velocity about 1 knot due east) in the Strait of Sicily at about $36°37'N$ and $12°21'E$ during the 18th cruise of the "Akademik Kurchatov" in April 1974. The drift was determined with respect to an anchored buoy. On the isopleth graph, we see a frontal interface (marked F) extending with a slight slope from left to right. It separates the fresher and colder waters of the western basin of the Mediterranean Sea (upper part of the graph) from the warmer and more saline waters of the eastern basin (located below the frontal interface) with a core of Levantine waters with a temperature above $14.5°C$ (dot-shaded area in the graph).

257

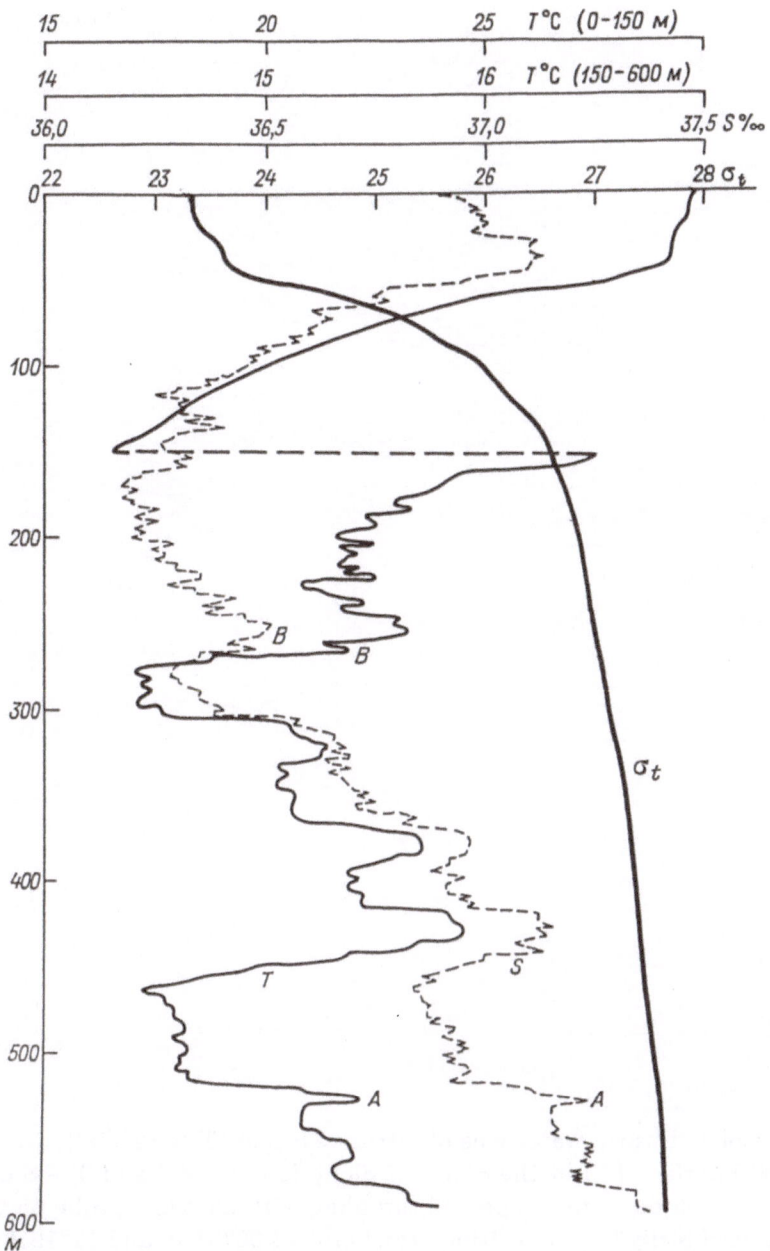

Fig. 4.5. Intrusive thermohaline structure in the 180 − 600 m layer in the Gulf of Aden, associated with the spreading of waters of Red Sea origin, at station No. 1984 of the 22nd expedition of the research vessel "Akademik Kurchatov", 12°09′42″N and 44°24′36″E, 10 June 1976. Thin solid line – temperature in °C; dashed line – salinity in ppt; bold solid line – specific density. A and B mark the thin intrusive laminae similar to those described by the author in [61]. Sampling was carried out with an AIST probe.

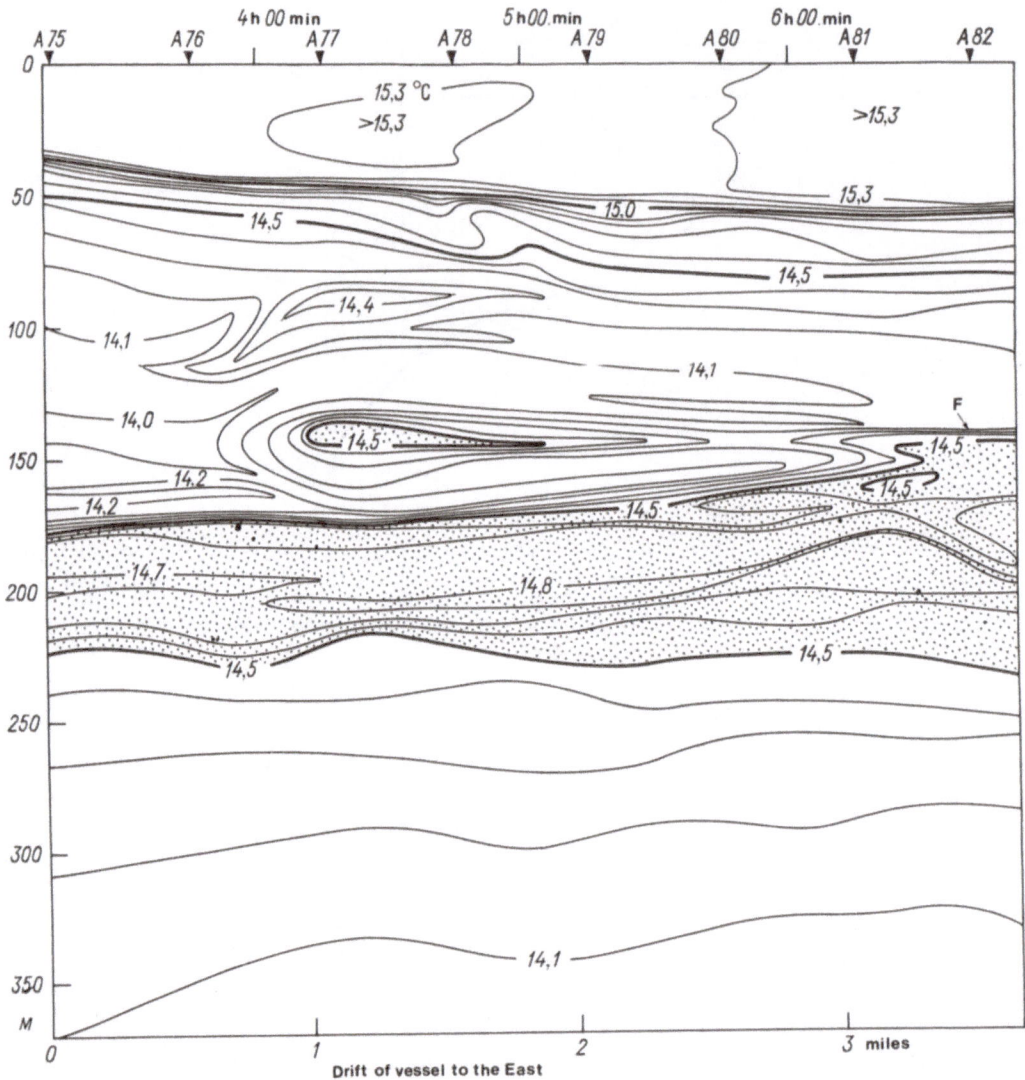

Fig. 4.6. Isolated warm-water lens of intrusive origin $(T > 14.5°C)$, separated from the frontal interface (F) in the Strait of Sicily (see fig. 4.7 and T – S curve in fig. 2.20 d), based on data from repeated sampling with an AIST probe on 2 May 1974 in the Strait of Sicily from a drifting vessel around $36°37'N$ and $12°19'E$ during the 18th expedition of the research vessel "Akademik Kurchatov" (station No. 1569).

The position of the frontal interface is best seen in fig. 4.7 a, b, c, which depicts sections in the temperature, salinity and density fields for a section of the Strait of Sicily closest to the sampling site. The currents to the left and above the frontal interface (but below the surface mixed layer) have a general NW-SE direction, while to the right and below the frontal interface, they flow in the opposite direction (from the eastern basin into the western one). The whole frontal interface is elongated from the northwest to the southeast along the axis of the strait, and is located in the $80 - 250\ m$ layer, i.e. below the seasonal thermocline and surface mixed layer. The frontal interface does not appear at all in the $0 - 80\ m$ layer.

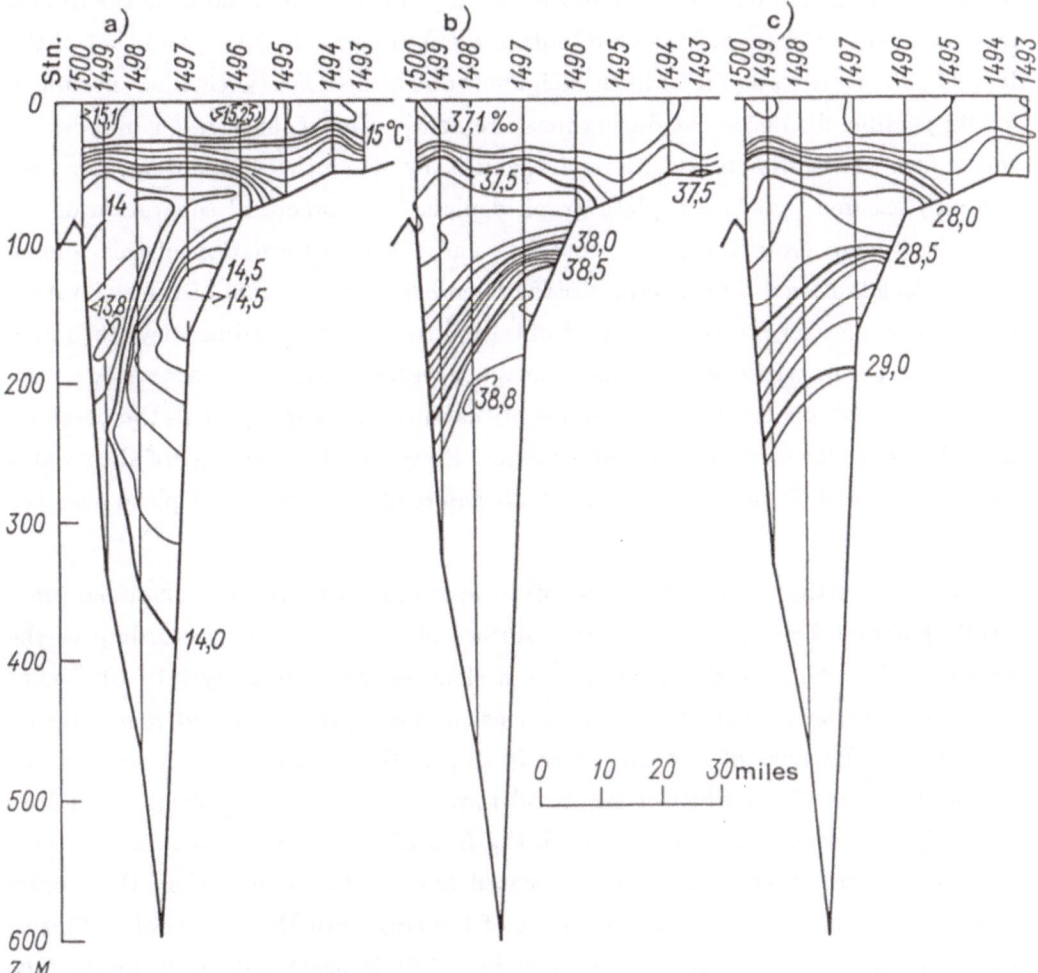

Fig. 4.7. Sections in the field of temperature (a), salinity (b) and specific density (c) across the narrowest part of the Strait of Sicily, executed by the author with the help of an AIST STD probe on 13-14 April 1974 during the 18th cruise of the research vessel "Akademik Kurchatov".

The slope of the frontal interface on the graph of isotherms (fig. 4.6) corresponds to the gradual movement of the vessel eastward in the process of repeated sampling, i.e. at a 45° angle with respect to the direction of the front. A 3.5-mile drift of the vessel eastward corresponds to a 2.5-mile distance along the normal to the frontal interface. Judging by fig. 4.7 a, the isotherms between 14 and $14.5°C$ rise by $38 - 40 \ m$ over this distance. The right end of the frontal interface in the isopleth graph (fig. 4.6) is $36 - 37 \ m$ higher than the left end, which indicates that our reasoning is correct.

The most interesting feature of the isotherm graph (fig. 4.6) is the isolated lens of water, with a core temperature above $14.5°C$, which is separated from the frontal interface. The thickness of the core (dotted area) amounts to $10 \ m$, and that of the lens $20-30 \ m$ with a diameter in the neighborhood of $1-2 \ km$ (with allowance made for its possible displacement during measurement). Unfortunately, it is impossible to confirm the complete isolation of the lens by any other measurements. We can only assume that the isopleth graph depicts the moment of separation of the intrusive tongue from the frontal interface and the transformation of this tongue into an isolated lens. There have already been descriptions [188] of the separation of extraneous water masses across a frontal interface into the adjacent water mass. Applied to the interface between shelf and slope waters in the northwestern Atlantic, this process was given the specific name of "calving", by analogy with the breaking off of blocks of ice from the edge of a glacier. However, the "calving" of shelf-water intrusions and their penetration into the layers of slope water takes place near the bottom [188].

It is interesting to note that the intrusive tongue in this case is extended practically horizontally, instead of along a surface of equal density. According to the section in fig. 4.7 c, such surfaces have a slope of approximately $2.7 \cdot 10^{-3}$ near the 150-metre level, whereas the isotherms in this layer are sloped more steeply $(1.2 \cdot 10^{-2})$. Because of this difference in slopes, the separated warm core of the tongue $(T > 14.5°C)$, which is displaced toward the axis of the strait, should be about $23 - 25 \ m$ in the vertical above the frontal interface where it is observed. On the isotherm graph, this core is located about $30 \ m$ higher than the frontal interface. The line connecting the centre of the core with the tongue of isotherms at the frontal interface (on the right in fig. 4.6) is practically horizontal. The $5-7 \ m$ difference between the isopycnic and actual depths of the core actually does correspond to the observed slope of isopycnals to the horizontal. Therefore, one can assume that the "calving" of the intrusive tongue was caused by some forced motion in the horizontal plane, e.g. an inertial fluctuation. On the other hand, the intrusion could have been isopycnic, in which case the observed insignificant rise of the separated lens in relation to the isopycnic surface could be attributed to the loss of mass and the decrease in water density in the core due to the sinking of salt fingers. Indeed, the isotherms near the lower boundary of the core are much more widely spaced than at the upper boundary. This may be another indirect indication of the presence of salt fingers under the core. We should add that indirect proof of double-diffusive convection was also obtained by us during analysis of the characteristics of the intrusive laminae of cold and freshened shelf waters spreading along the frontal interface on the periphery of a warm Gulf Stream ring [17] (see also figs. 3.18 and 3.19 in subsection 3.1.3).

Detailed data related to the frontal interface in the Strait of Sicily show that this interface is characterized by a high degree of thermoclinicity. As shown in figs. 4.6 and 4.7, the isopycnic temperature gradients at this front amount to $0.1-0.5°C/km$. Even more important is the fact that the separated intrusive lens is characterized by horizontal and isopycnic temperature gradients of the same order. Consequently, the separated intrusive lens is as if it carries off a part of the frontal interface on which it was formed. Therefore, one can say that the intrusive process and the "calving" of intrusions contribute to the fragmentation of frontal interfaces and the development of a "multifrontal character" of frontal zones. One can also picture this process as a continuous transport of frontal-intrusive variability down a cascade of scales from frontal interfaces of a climatic nature to dissipative scales.

The use of the isopycnic and other methods of analysis, which stems from the classical theory of T – S curves, has enabled us to establish that areas close to fronts are on the whole characterized by a high degree of temperature and salinity variability on surfaces of equal density [58, 61, 124, 135, 265]. This variability can be explained only if one accepts the dominant significance of the intrusive process. The author's use of isopycnic analysis for STD data from repeated profiles at a 3-day station executed in parallel with current measurements from a buoy during the "Polygon-70" experiment enabled us to establish that some thermal inhomogeneities of a clearly advective nature and measuring $10-20\ m$ in thickness are observed along isopycnic surfaces for $10-15\ km$. In this case, the isopycnic temperature gradients can amount to $0.8-1.0°C/km$ [61]. This result was recently confirmed by new data in [95] and [265]. Other results from repeated profiles [61] also indicate that well-defined intrusive laminae with an excess of heat and salt in comparison with the ambient water intersect isopycnals as they spread. An analogous and highly visual result was recently obtained for cold laminae by Gregg and McKenzie [136]. These and other results of numerous authors indicate that the intrusive process at fronts is accompanied by double-diffusive convection (e.g. salt fingers) at the boundaries of the intrusive laminae [58, 135, 257]. Direct instrumental proof of this fact is also available [256]. Thus, earlier assumptions and hypotheses regarding this [57, 234] are confirmed. At the present moment, we still require extensive special measurements which would enable us not only to indirectly observe, but also to correctly measure and quantitatively assess the heat-and-mass-transfer effects of double (thermohaline) diffusion in combination with the intrusive process. In turn, this would enable us to pass over to the quantitative assessment of the role of fronts in oceanic heat and mass transfer.

Research into intrusive stratification at fronts and the causes of it are being intensively continued. Each new endeavour in this area is contributing new and interesting facts. It would therefore be useful at this point to briefly sum up the main facts regarding thermohaline finestructure of frontal origin. From our point of view, they boil down to the following.

1. Near-frontal regions are characterized by the dominance of an intrusion-type thermohaline finestructure, the thickness of the intrusive layers (both the "warm", and the "cold") varying mostly between 10 and $30 - 40\ m$, though individual $1 - 5$-metre intrusive layers are easily observed, while in large-scale climatic frontal zones, intrusive layers can reach a thickness of $100 - 250\ m$.

2. Due to causes not quite clear to us, we often observe the "calving" of frontal intrusions and their penetration to the other side of the front in the form of isolated lenses.

3. The temperature inversions accompanied by "cold" and "warm" intrusions of a frontal origin are characterized by a vertical temperature change ΔT varying from several hundredths of a degree to $1 - 3°C$ with quasi-isopycnic compensation by a salinity change ΔS.

4. There are numerous indirect and so far only a few direct field data on double-diffusive convection (in a diffusive regime and in the form of salt fingers) at the upper and lower boundaries of intrusive layers and laminae. Due to the convective heat and mass transfer at the boundaries, the lifetime of intrusions measuring about $10 - 20\ m$ in thickness is most likely limited to a period of several days.

5. Intrusive layers and laminae intersect isopycnic surfaces in a direction corresponding to the presence of a density (mass) excess or deficit in them with respect to isopycnic balance. A deficit or excess of density (mass) in the intrusions is usually

in good agreement with the direction of their evolution under the effect of double-diffusive convection which develops at their upper and lower boundaries. However, the results of measurements of this sort can be interpreted incorrectly if we do not take into account the origin of the intrusions, their relation to the main water mass from which they were formed, or their separation from it, and the presence or absence of a continuous source of fresh intrusive water.

6. The intersection of isopycnic surfaces by intrusions determines the high degree of temperature variability along isopycnic surfaces in frontal zones. The "calving" of intrusions helps to intensify the alternation of frontal interfaces within frontal zones.

7. The only comment we can make about the alternation of the intrusions themselves along the surface of a frontal interface is that it is definitely associated with the diverse signs of frontal instability due to various processes. No studies in this area have been published as yet.

All these phenomenological characteristics of the vertical thermohaline structure of frontal zones also determine the numerous characteristic features of its statistical properties. Let us dwell on some of them.

a) With an increase in the wave number k, we observe a faster decrease in the power spectrum $E_{T'}(k)$ of the finestructure disturbances T' of vertical temperature profiles, than could be expected in the case of a finestructure associated with the kinematic effect of internal waves. In the opinion of some authors (see discussion in [79]), an intensive intrusive finestructure of frontal origin is characterized by spectra which decrease according to the relationship $k^{-2.5}$ both for temperature and salinity, in contrast to the relationship k^{-2} which characterizes the spectrum for the relatively smooth intrusion-free profiles of "calm" areas of the ocean away from fronts.

b) We observe higher levels of the power spectrum of finestructure disturbances of temperature, $E_{T'}(k)$, and salinity, $E_{S'}(k)$, for the same values of k by comparison with analogous spectra for "calm" areas of the ocean [10, 79, 135].

c) We observe particularly high values and a high variability of the finestructure Cox numbers [135, 179], i.e.

$$C_T = \frac{\overline{(\partial T'/\partial z)^2}}{(\partial T/\partial z)^2} \quad \text{and} \quad C_s = \frac{\overline{(\partial S'/\partial z)^2}}{(\partial S/\partial z)^2}, \qquad (4.1a,b)$$

which represent the normalized mean square measure of the variability of the vertical temperature and salinity gradients, dT/dz and dS/dz.

d) The observed values of the spectral stratification parameter $m(k) = ((E_{S'}(k)/E_{T'}(k))^{\frac{1}{2}}$ approximate the ratio α/β, where

$$\alpha = -\frac{1}{\rho}\left.\frac{\partial\rho}{\partial T}\right|_{P,T,S} \quad \text{and} \quad \beta = \frac{1}{\rho}\left.\frac{\partial\rho}{\partial S}\right|_{P,T,S}$$

in the range of wave numbers k characteristic of an intrusive finestructure (from 10^{-2} to 10^{-1} cycles per metre).*

e) The power spectra of microstructure (turbulent) fluctuations of the vertical temperature gradient at the upper and lower boundaries of intrusions are significantly higher than in their cores [135, 179], which is in good agreement with Williams' measurement of velocity fluctuations [242] and is proof that turbulence is generated at the boundary regions of intrusions by double-diffusive convection.

Recently obtained new observational data on the nature of thermohaline finestructure near fronts pose a whole series of problems in reviewing the relationships between the mean fields, stratified finestructure and turbulence within the framework of established concepts of statistical hydrodynamics. The conventional concept of temperature, salinity and velocity components as the sum of the mean and fluctuating components

$$a) T = \overline{T} + T';$$

$$b) S = \overline{S} + S';$$

$$c) u_i = \overline{u}_i + u'_i \tag{4.2}$$

no longer satisfies researchers encountering the more complex structure of hydrophysical fields in the ocean. A terminological distinction had to be drawn between "finestructure" and "microstructure" (see footnote on p. 113 in [62]). It also turned out that the equations for the balance of the intensity of temperature and salinity inhomogeneities, which are based on a division as in (4.2), do not describe the anisotropic intrusive process very well, and that the differences between the behavior of finestructure and turbulent microstructure are more adequately described by a distribution of the following type:

---

* See p. 101-103 in [61] for data on the stratification parameter $m(k)$. A helpful analysis and examples of the applicability of this parameter are given in [79].

$$a) T = \overline{T} + \tilde{T} + T';$$

$$b) S = \overline{S} + \tilde{S} + S';$$

$$c) u_i = \overline{u}_i + \tilde{u}_i + u'_i, \tag{4.3}$$

where the overbar of averaging still belongs to the mean field, the tilde corresponds to the elements of finestructure (especially to anisotropic intrusive stratification), and the accent signifies turbulent microfluctuations (of the microstructure) of a homogeneous and isotropic nature. It is quite clear that the conditions and scales of averaging imposed on each of the three components in (4.3) should differ. The possibility of this type of distribution was discussed by a number of authors [62, 233], while its successful application belongs to Joyce [154] who used the principle on which it is based to measure the horizontal flux of variability contributed near the fronts by finestructure inhomogeneities of an intrusive nature. Basically, Joyce repeated on another level the reasoning of Osborn and Cox [200] who in their time introduced the Cox numbers (4.1); however, this time, the balance was assumed between the horizontal flux of variability due to the smoothing by intrusions of the mean concentration gradient of some passive constituent, and the vertical flux due to a more intensive vertical exchange related to intrusive interleaving. For temperature, this balance takes the form of

$$\overline{\tilde{u}_H \tilde{T}} \frac{\partial \overline{T}}{\partial x_H} = -\tilde{K}_T^V \overline{\left(\frac{\partial \tilde{T}}{\partial z}\right)^2}, \tag{4.4}$$

from which, having designated

$$\tilde{u}_H \tilde{T} = -\overline{K}_T^H \frac{\partial \overline{T}}{\partial x_H}, \tag{4.5}$$

we can obtain

$$\overline{K}_T^H = \tilde{K}_T^V \overline{\left(\frac{\partial \tilde{T}}{\partial z}\right)^2} \Big/ \left(\frac{\partial \overline{T}}{\partial x_H}\right)^2. \tag{4.6}$$

In equations (4.4)–(4.6), K denotes the effective (turbulent or, more correctly, "non-molecular") coefficient of vertical (V index) or horizontal (H index) exchange, and u_H denotes the randomly oriented horizontal component of velocity. The scales of averaging are marked with a tilde or an overbar. The significance of the end result (4.6) lies in the fact that, knowing the mean coefficient of vertical heat transfer K_T^V

across the boundaries of intrusions where the mean vertical temperature gradients reach values of $d\tilde{T}/dz$, and knowing the mean horizontal temperature gradient at the front $d\overline{T}/dx_H$, we can determine the mean coefficient of horizontal cross-frontal heat transfer $\overline{K}_T^H$ due to intrusive stratification. A similar expression can be obtained for salinity or any other constituent.

4.2.2. The formation, evolution and destruction of frontal intrusions

Let us dwell briefly on the present-day understanding of the physical nature of the forces that control the formation, motion and destruction of intrusions. Initially, intrusions were identified mainly with the discharge of extraneous (highly saline and warm, or supercooled and fresher) waters from marginal and inland basins into the ocean. A more general hypothesis on the origin of intrusive interleaving ("lateral convection") can be found in Stommel and Fedorov's paper [238] which discusses systems of lateral exchange between different pairs of stably stratified vertical structures, in close proximity to each other, in the process of their attaining a combined stratification of maximum stability. Basically, they were dealing precisely with conditions in frontal zones where differently stratified thermohaline structures are in close proximity on both sides of a frontal interface. The question of what classes of motion could carry out this type of transfrontal transfer has not yet been discussed to a sufficient extent. The authors suggest [238] that one of the possible mechanisms for intrusive interleaving may be the cross-frontal Ekman flux in upper and lower turbulent boundary layers at the boundaries of homogeneous lenses of water ("laminae"), the cores of which are in geostrophic motion along the front. Using this assumption, the typical value of the flow speed of the laminae proved to be similar to the one in reality, i.e. several centimetres per second.

Later, the author and A.S. Monin [43] suggested that long-period internal waves with a vertical wave vector and particularly inertial fluctuations, could perhaps be the cause of the omnidirectional motions in the vertically adjacent, relatively thin layers of water, and may cause intrusive interleaving in the presence of sharp horizontal temperature and salinity gradients. The role of inertial fluctuations in the advective transfer of thermohaline inhomogeneities was especially emphasized by the author [61, 62] in connection with observational confirmation of the thin-layer nature of the motion of oceanic waters [44], including that on scales of inertial fluctuations.

In general, an awareness of the fact that oceanographic fronts, even the most large-scale ones, are never a strictly geostrophic phenomenon, but display a large number of ageostrophic characteristics in their dynamics, provides the key to understanding the physical nature of intrusions. It permits us to say that all the ageostrophic forms of motion that develop in the frontal zones, i.e.

a) the Ekman drift,

b) tidal and inertial fluctuations,

c) viscous and transient currents, and

d) centrifugal and centripetal accelerations in areas of fronts with a significant curvature (meanders),

can take part in the formation and evolution of an intrusive thermohaline finestructure. All of these motions are "stirring ones" in the Eckartian sense with respect to the sharp horizontal and vertical temperature and salinity gradients that arise in the process of frontogenesis, and the appearance of the first (or initial) intrusions at frontal interfaces can apparently be attributed mainly to them. One can also agree with Woods [263] that initial shallow intrusive temperature inversions can arise by the deflection of the upper part of the thermocline (see section 2.4) under the influence of a current with shear in the process of frontogenesis itself. However, the meandering of a frontal interfaces with the creation of centrifugal and centripetal components of motion along the frontal inerface is apparently a more effective mechanism. On the basis of his own laboratory experiments [221], Turner* recently expressed the opinion that double diffusion could also be the original direct cause of the intrusive stratification of frontal zones. This opinion was disputed by Woods [263] who claimed that "first intrusion, and then double diffusion". We can support this position as to the appearance of the first (initial) intrusions at frontal interfaces. In our opinion, these initial intrusions are needed so that the most favourable conditions could develop for double-diffusive convection. The further evolution of intrusions is most likely related to the fact that the various forms of double-diffusive convection (primarily salt fingers) which develop under favourable conditions at their upper and lower boundaries begin to play a dominant role in the destruction or transformation of the intrusive layers, causing, as our own laboratory experiments have shown [16, 24, 65], the break-up of the initial intrusions into numerous thinner intrusive layers and laminae. The thing is that the uniformly mixed layers that form, for example, by means of the transport of negative buoyancy by salt fingers, cannot remain unchanged for long surrounded by a stratified fluid, and

* Report presented at the 2nd International Symposium on Oceanic Turbulence, Liège (Belgium), 7-18 May 1979.

so they form new intrusions in the process of "collapse" [155, 270]. G.I. Barenblatt helped to relate the ideas of "collapse" to the concept of intrusion in the ocean.

In 1977, G.I. Barenblatt expressed the opinion that intrusions in the ocean can develop locally as a result of the collapse and viscous spreading of a mixed patch of limited dimensions in an ambient stratified fluid. His numerical model of the process [9], which predicted the dependence of the transverse dimension X of the axisymmetric (cylindrical) "patch" on dimensionless time Nt (where N is the Väisälä-Brunt frequency) as $X \sim (Nt)^{\frac{1}{10}}$, was experimentally confirmed that same year by the author of this book and a group of co-workers from his laboratory [23]. The theoretical and laboratory results obtained generated the optimistic hopes that a universal theory for the formation of a thermohaline finestructure (including that of an intrusive nature) would be developed on the basis of local mixing in a stratified ocean due to the energy of breaking and shear instability of internal waves. However, a thorough analysis with the use of energy arguments [22] has shown that the space-time scales of intrusive layers usually detected near fronts do not correspond to the relatively small scales of intrusions which can develop on the basis of the energy generated locally during shearing instability and even the collapse of gravitational internal waves.

We find that only comparatively thin intrusive layers (from 10 to 50 cm in thickness) can form in the ocean by means of the collapse which accompanies shearing instability or the collapse of large internal waves in the presence of a well defined density stratification (N not smaller than $10^{-2}s^{-1}$). At the same time, their scale ratio h_1/X_1 can at best reach $(2-3) \cdot 10^{-3}$, which would take only $0.6-6$ days when $N = 10^{-2}s^{-1}$. Thicker lenses or laminae could form in the same way only in the case where the initial mixed volumes at the initial moment $t = 0$ have a ratio of the characteristic scales $h(0)/X(0)$ of the order of $10^{-1} - 10^{-2}$, with the initial thickness $h(0)$ in the neighborhood of $10 - 30$ m prior to the onset of the first stage of collapse. In any case, the spreading intrusive layers can retain a ratio of scales h_1/X_1 between $3 \cdot 10^{-3}$ and 10^{-3} for a very long time [22], though for the thinnest layers, this time is limited by their destruction due to molecular heat transfer and diffusion.

As A.G. Zatsepin's recent experiments have shown, a more rapid spreading of intrusive layers proportional to $t^{\frac{1}{2}}$ is possible in the case where an intrusion is replenished from some isolated source. If the discharge of fluid Q into an intrusion is constant, then, according to [24], the flow regime is self-similar, and

$$X(t) = A\left(\frac{N^2}{\nu}\right)^{0.1} Q^{0.4} t^{0.5}, \tag{4.7}$$

where ν denotes kinematic viscosity, and $A \simeq 1.02$ is the experimentally determined dimensionless proportionality factor. In this case, the ratio of scales h/X is found to be proportional to $(Nt)^{-0.5}$, which means that the intrusions attain scale proportions of the order of 10^{-3} more quickly (e.g. 1.3 days instead of 6 days). However, spreading with replenishment is more characteristic of intrusions associated with water exchange between basins (e.g. the Mediterranean Sea and the Atlantic Ocean), than of ordinary fronts in the open ocean. During the frequently observed "calving" of an intrusive lens, replenishment ceases, and even if "calving" does not occur, it is not likely that the condition $Q = const$ is fulfilled, due to the fact that the above-mentioned ageostrophic motions are either transient or quasi-periodic in nature.

Ordinary frontal intrusions are apparently characterized more by a viscous flow of a constant volume, beginning with characteristic initial scales such that $h(0)/X(0)$ is of the order of 10^{-2} when $h(0) \geq 10\ldots30\ m$. Initial intrusions should be characterized by such scales. As conjectured earlier, the largest initial intrusions apparently arise as a result of the "mixing" of the frontal gradients by various ageostrophic components of motion characteristic of a near-frontal regime. As our laboratory experiments have demonstrated [16, 65], only one such layer is sufficient under favourable combinations of temperature and salinity gradients at its boundaries for further intrusive stratification to take place as a result of the alternation of double-diffusive convection and collapse. Though they are purely qualitative, the results obtained in these experiments are so interesting that they deserve to be described in greater detail.

In the first experiment, a stable, almost linear stratification was created in a long plexiglass tank $(1.5 \times 0.13 \times 0.3\ m^3)$ by means of the vertical salinity gradient in the water. The value of this gradient was selected from $dS/dz = 0.5 \times 10^2 ppt/m$ to $dS/dz = 1.5 \times 10^2 ppt/m$. The water in the tank had a practically homogeneous vertical temperature $T_w = 18\ldots22°C$. In order to prevent the cooling of the water from the surface due to evaporation, the tank was covered with a heat-insulating and moisture-proof lid.

An intrusion was prepared from a small amount $[(3-5) \times 10^{-3} m^3]$ of coloured water with an intermediate salt concentration (slightly smaller than in the water at the bottom of the tank, but higher than in the water at the surface). The prepared water was heated $20-30°$ above room temperature, and then carefully poured into the tank by means of a sloped trough* placed at the front edge of the tank. The intrusive fluid flowed off evenly along the bottom of the trough to the level of its own density, after which it separated from the sloped bottom and the trough and spread at the same level towards the opposite end of the tank.

The temperature of the water in the intrusion was much higher than that of the rest of the water, and the compensation of the negative temperature contribution to the change in density occurred because the salinity in the intrusion was greater than the salinity of the water in the tank at the level at which the intrusion spread. Thus, the intrusion initially had a higher temperature and salinity as compared with the ambient fluid. This was conducive to the appearance of convection at the boundaries of the intrusion. Indeed, with the spreading of the intrusion (despite the presence of a velocity shear), convection in the form of "salt fingers" quickly developed at the lower boundary of the intrusion (fig. 4.8).

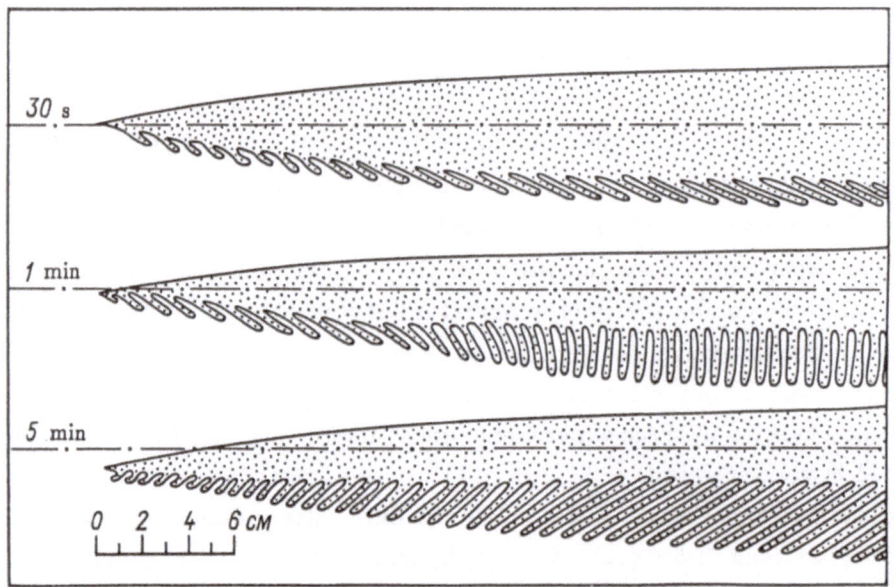

Fig. 4.8. A sketch of the development of convection in the form of salt fingers at the lower boundary of an intrusive layer in the first experiment.

The "salt fingers" displayed a highly uniform (periodic) spatial distribution throughout the lower surface of the intrusion. Each element ("finger") had a transverse size of approximately $4 - 5$ mm and a length of up to $15 - 20$ mm. At first, they were all recurved with respect to the direction in which the intrusion was mov-

---

* The sloped trough made of plexiglass was 0.08 m wide and 0.5 m long, and was fixed in the tank at a 30° angle to the horizontal.

ing. However, after only $3-5\ min$ from the start of the experiment, there appeared a compensating counterflow of water in a thin layer at the bottom of the intrusion, slightly above the bases of the "salt fingers". As a result, the latter straightened out, and then began to slope in the opposite direction. The most intense formation of "salt fingers" was observed at the base of the intrusion (closer to the trough) where the vertical thermohaline gradients were insignificant. Here, because of the transport of negative buoyancy, the "salt fingers" vigorously mixed the layer of water of a certain thickness directly beneath the layer in which they were observed. As a result, a vertically homogeneous thickening, clearly visible (fig. 4.9 a) because of the dye, was formed at the lower boundary of the intrusion. It gradually separated from the initial intrusion, while a wedge of clear water intervened between it and the intrusion, spreading toward the intrusion. Then, the thickening itself collapsed and began to spread as an independent intrusive layer in the form of a wedge below the main intrusion. "Salt fingers" again appeared at its lower boundary, and in this way the intrusion split into several more coloured tongues separated by colourless layers (fig. 4.9 b, c). The splitting of the intrusion into laminae continued for $1-1\frac{1}{2}$ hours until the intrusion reached the opposite end of the tank and stopped.

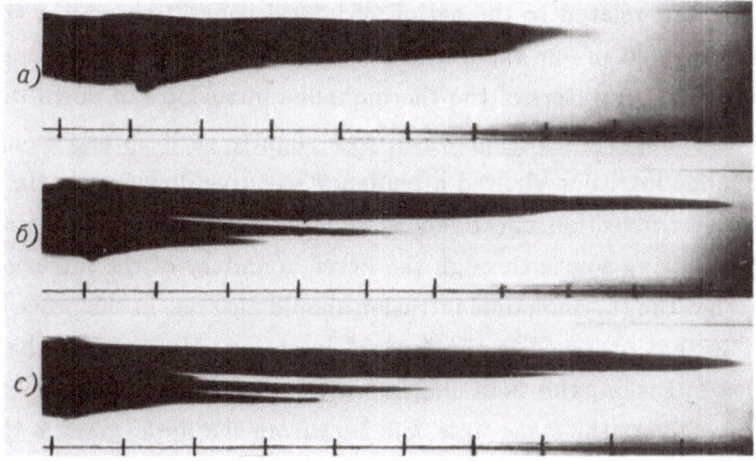

Fig. 4.9. Successive photographic images of the splitting of an intrusive wedge under the influence of salt fingers in the first experiment. $a - t = 370s$, $b - t = 960s$, $c - t = 2500s$.

The temperature profile in the separated part of the intrusion is shown in fig. 4.10. As we can see from the drawing, the region of the intermediate temperature drop is formed by a "cold" layer of colourless water wedged between two coloured "warm" tongues of the intrusion (shaded).

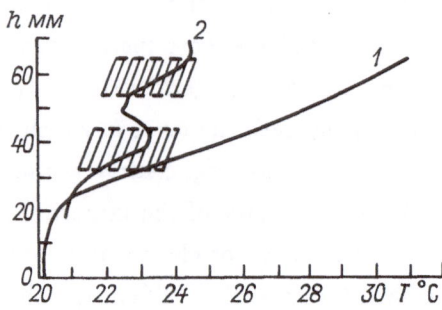

Fig. 4.10. Vertical temperature profile across the lower part of an intrusion in an experimental tank (1st experiment). 1 – before splitting, 2 – after splitting, $t = 960s$.

Similar splitting of an intrusion into laminae was also observed with an intrusion of an aqueous solution of sugar in a stably salinity-stratified medium. This permits us to conclude that the separation of the intrusion into laminae in our experiments was not related to the action of lateral convection which, theoretically, could have developed in our experiment due to the heating of the sloped side wall (trough) as the warm waters of the thermohaline intrusion ran down it.

Unlike the "sugar" intrusion which rose slightly as it spread along the tank, the thermohaline intrusion showed a tendency to curve downwards. In both cases, double-diffusive convection effectively eliminated the excess concentration of the more slowly diffusing solute through the lower boundary of the intrusion. Thus, it would seem that the thermohaline intrusion should also rise in the process of spreading, similar to that observed in the "sugar" intrusion. However, in the case of the thermohaline intrusion, the "salt fingers" carry down not only salt, but the heat (and dye) as well. At the same time, the heat from the upper part of the intrusion is comparatively quickly dispersed upward by molecular heat conductivity. Therefore, the maximum temperature of the intrusion in this case is displaced downward together with the salt transported by the "fingers", and we get the impression that the core of the thermohaline intrusion tends to sink deeper into the ambient fluid as it spreads across the tank. Thus, in the experiments with the thermohaline and the "sugar" intrusions, the spreading was not strictly horizontal, i.e. not quite isopycnic. This non-isopycnic type of spreading was caused by the anisotropic vertical mass flux across the boundaries of the intrusion, associated with both molecular heat and mass transfer, and double-diffusive convection. This experimental result, predicted in [61], was recently confirmed by observations in the ocean [136].

It should be said that even significant changes in the configuration of the experiment did not alter the end result, and in some cases made it even more obvious. Slow pouring of a thermohaline (or sugar) intrusion through a glass tube to a level of equal density, in a tank ($0.75 \times 0.44 \times 0.35$ m^3) containing a fluid with linear salinity stratification, created an initial axisymmetric disk-like intrusive lens, at the lower boundary of which salt (or sugar) fingers developed immediately. This type of lens with salt fingers (lateral view) is shown in fig. 4.11. The scales can be judged by the diameter of the tube, which in reality is equal to 1 cm. The end result of the evolution of this type of intrusion over a $57min$ period is depicted in fig. 4.12. The alternating action of the "salt fingers" and collapse broke up the entire mass of intrusive fluid poured into the tank into a multitude of disk-like layers which spread slowly at various density levels. Judging by the coordinate scale (in the background) with 2×2 cm squares, the maximum thickness of the intrusive laminae in this experiment did not exceed $2 - 3$ cm. Background stratification was characterized by a Väisälä- Brunt frequency of $N \simeq 0.5$ s^{-1}. The excess temperature of the intrusion with respect to the ambient fluid at the start of the experiment amounted to $10°C$.

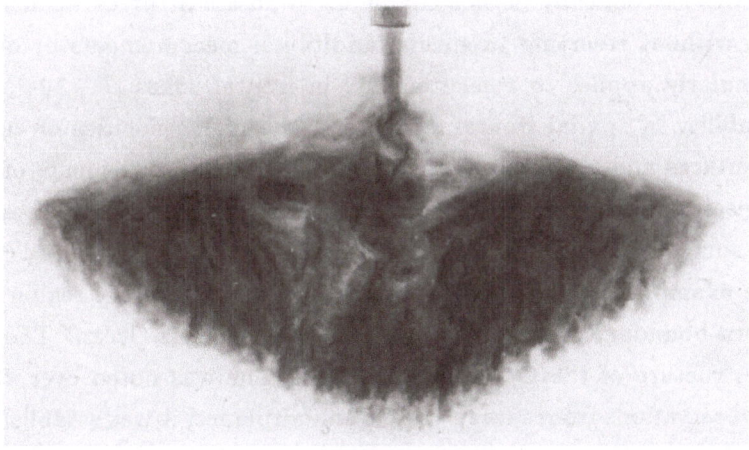

Fig. 4.11. Photograph showing the sinking of salt fingers from an axisymmetric thermohaline intrusion in the initial phase of spreading ($t \approx 450$ s) in the second experiment.

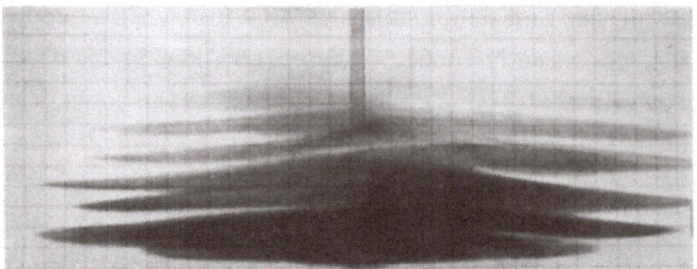

Fig. 4.12. Final stratification of an intrusive mass in an axisymmetric case (photo, lateral view) under the influence of salt fingers ($t \approx 3400\ s$) in the second experiment.

4.3. Characteristic features of the three-dimensional spatial structure of frontal zones (as in the example of the Gulf Stream)

The study of a number of oceanic processes and phenomena would be far more successful and productive if oceanographers had a complete picture of the three-dimensional structure of the objects of study. Measurements from one or even several research vessels in areas with a high space-time variability of hydrographic characteristics theoretically do not enable us to obtain a three-dimensional synoptic picture without resorting to special additional measurements or observations. This particularly applies to measurements in frontal zones [71, 194], due to the rapid variability of spatial frontal structures, the high velocities of currents near frontal interfaces and the unpredictability of the spatial movements of the frontal zones themselves. Measurements from artificial earth-orbiting satellites are a good additional source of information in this case.

A fine example of an unusually variable frontal zone is the region adjacent to the northern boundary of the Gulf Stream in the Atlantic Ocean. The complexity of the finestructure of the Gulf Stream frontal zone was noted over 25 years ago [250]. By observations from survey vessels and airplanes, it was established that the main flow of the Gulf Stream near its northern boundary was divided into separate thin jets which differed greatly in their temperature. As it turned out, the frontal zone itself included a multitude of separate frontal interfaces which were traced over distances of $100 - 500\ km$ and even less. The "multi-frontal" or "multi-step" structure of the northern boundary of the Gulf Stream was also established in our own investigations in the Soviet-American POLYMODE program [17]. In another case, analysis of systematic satellite IR data enabled us to establish certain two-dimensional (horizontal) eddy-type characteristics in this finestructure [173]. As we shall demonstrate below, the temperature patterns of the ocean surface obtained from a satellite with the help of scanning high-resolution IR radiometers ensured a more accurate interpretation of the synchronously obtained data of oceanographic sections across frontal zones, and enabled us to reveal details of the spatial structure of these zones, which would have been impossible to even imagine without the satellite data. For the first time, oceanographers had the opportunity to study the three-dimensional pattern of the finestructure of oceanic frontal zones.

The measurements analyzed in this section were carried out in the frontal zone of the Gulf Stream in March 1978, in the vicinity of $37°N$ and $73°W$ during the Soviet-American POLYMODE experiment. Ship measurements with the help of an AIST probe (at stations) and a towed sensor (from a moving ship) were carried out at the author's request from the research vessel "Akademik Kurchatov" (26th cruise) [50] on 20-21 March along the section depicted schematically in fig. 4.13 $(P - P)$. The temperature patterns of the ocean surface in the IR range were obtained with a scanning IR very-high-resolution radiometer (VHRR) from the NOAA-5 satellite on 20 March at $14\,h\,03\,min$ and 23 March at $14\,h\,19\,min$ GMT. The corresponding numerical data were prepared under the supervision of E.P. McClain (NOAA). In 1979, the experimental data were exchanged between the Soviets and Americans in accordance with an agreement on cooperation in researching the Earth's natural resources from space. Fig. 4.14 shows the results of a comparison of SST data based on the results of measurements with a towed sensor at a depth of about 0.15 m (1) and with an IR radiometer (2) on the basis of numerical data from NOAA.

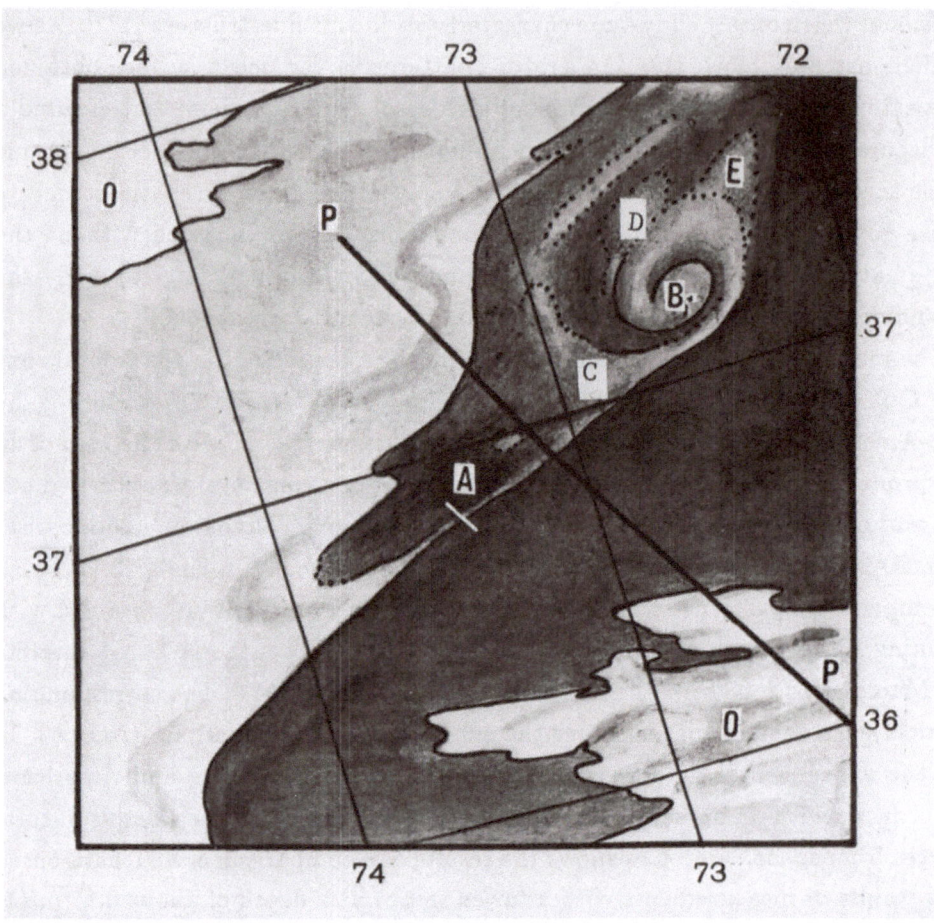

Fig. 4.13. Schematized fragment of an IR image of the investigated section of the Gulf Stream frontal zone for 20 March 1978. $P - P$ – position of oceanographic section based on navigational data; A – region of cold entrainment which, according to calculation, included the oceanographic section executed by the "Akademik Kurchatov" on 21 March 1978; B_1 – centre of cyclonic eddy on 20 March 1978; C, D, E – areas of relatively cold water near the cyclonic eddy; 0 – clouds.

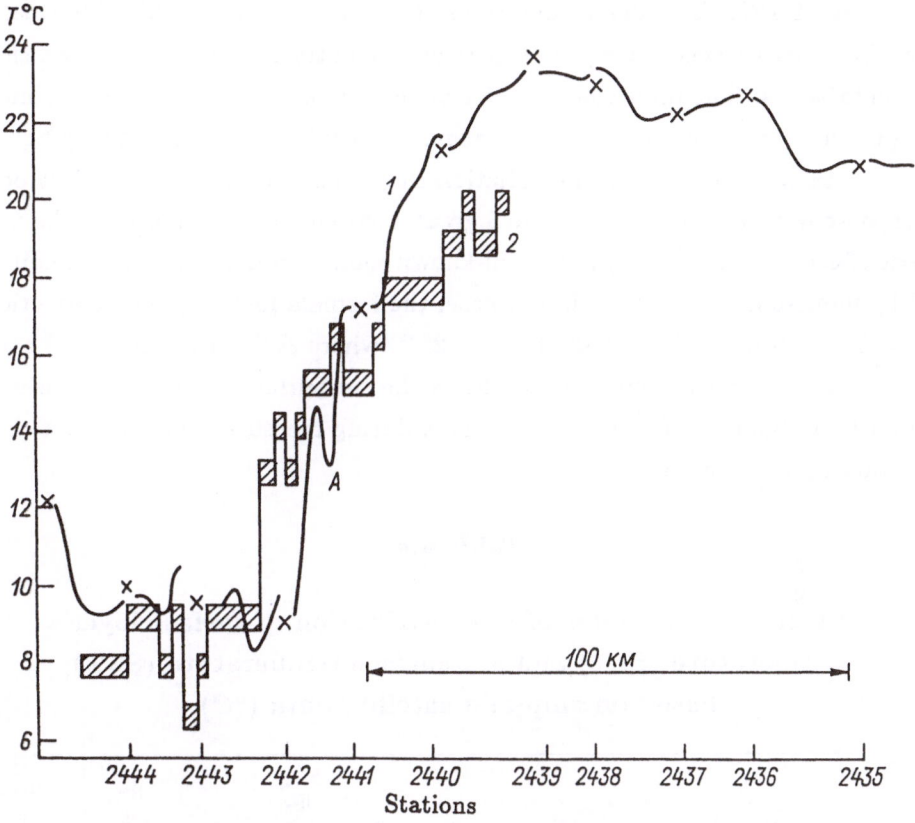

Fig. 4.14. Change in sea-surface temperature (SST) during the crossing of the Gulf Stream frontal zone on 20-21 March. 1 – based on the data of a towed sensor; 2 – based on the data of satellite measurements (NOAA-5); A – intermediate temperature minimum associated with cold entrainment; × – temperature measurements at stations.

It is not difficult to see a good qualitative agreement between the ship and satellite data in the frontal zone of the Gulf Stream. Small details in the SST distribution (based on satellite data) do not always agree with ship data, and the absolute values of radiative temperature are on the whole too low due to the effect of the atmosphere, despite the introduction of a constant correction (approximately $2°C$) in the numerical data during initial processing of the information. The disadvantage of using numerical satellite information is related to its spatial averaging (the element of resolution on the radiation maps obtained by us from NOAA constitutes 2.2×2.6 km) and the approximate quantization of radiative temperature. For instance, one gradation of radiation temperature differs from another by approximately 1.3°C. In order to determine the reliability of satellite information, V.Ye. Sklyarov carried out some statistical calculations, the results of which are given in table 4.4. The table contains SST values and a series of statistical values for data (a) where the atmospheric effect was partially taken into account (top figures), i.e. for data taken directly from radiation maps, and (b) where the effect of the atmosphere was taken into account to a greater extent (bottom figures). The atmospheric effect was calculated by the well-known semi-empirical algorithm commonly used by numerous researchers. In our case, the formula for additional correction is $\Delta T = (2.45 + 0.1072T_c)\exp(0.00012\theta^2) - 2°C$, where ΔT is the additional correction for the atmospheric effect in °C, T_c is the sea-surface temperature measured with an IR radiometer, θ is the zenith angle during measurements, and 2°C is the correction derived earlier.

Table 4.4

Statistical estimates of the distribution of surface layer temperature (SLT) and sea-surface temperature (SST) based on ship and satellite data (°C)

Area	No. of measurements	$\overline{T}_{sh}$	$\overline{T}_{sat}$	$\sigma_{T_{sh}}$	$\sigma_{T_{sat}}$	$\Delta\overline{T} = \overline{T}_{sh} - \overline{T}_{sat}$	r	$\Delta_{sh} = \overline{T}_G - \overline{T}_{sl}$	$\Delta_{sat} = \overline{T}_G - \overline{T}_{sl}$
Entire area of front	41	14.15	13.4 / 15.29	5.26	4.35 / 4.8	0.75 / −1.14	0.93 / 0.91		
Gulf Stream waters	6	22.08	19.28 / 21.88	0.66	0.59 / 0.67	2.8 / 0.2	—	12.58	10.79 / 12.09
Slope waters	16	9.5	8.49 / 9.79	0.48	0.77 / 0.84	1.01 / -0.29	—		

$\overline{T}$ – mean value; σ_T – standard deviation; ΔT – mean difference between ship and satellite data; r – correlation coefficient; Δ_{sh} – mean difference between SLT of the Gulf Stream and slope waters; Δ_{sat} – the same for SST based on satellite data.

The results of the estimates given in the table show that the coefficient of correlation between the ship values of SLT and the radiation values of SST is quite high. The temperature gradient between the slope waters and Gulf Stream waters based on satellite data after a more complete estimate of the atmospheric effect ($12.09°C$) proved to be very similar to the measurement made from the ship ($12.58°C$). Consequently, on the basis of these calculations, we can conclude that the satellite SST data reflect the actual distribution of sea-surface temperature in the frontal zone of the Gulf Stream. Special mention should be made of the difficulty of comparing satellite and ship data in regions of active frontal zones with their complex multi-frontal structure. The velocities of spatial movement of frontal interfaces, eddies and other structural elements of frontal zones are comparable with the speed of the research vessel carrying out the complex measurement program. Because of this, the spatial patterns of frontal structures which are based on ship data always contain distortions (e.g. see [266]). At the same time, the satellite sensors give us quasi-instantaneous images of the thermal field of the ocean surface. The comparison of quasi-instantaneous and distorted patterns of temperature distribution and especially numerical comparison of the temperatures at fixed geographic coordinates can be the source of significant errors of interpretation. In our further interpretations of the results of satellite and ship data correlation, we will have to consider some of the consequences of the displacement of structural elements of the Gulf Stream frontal zone during the experimental period.

Fig. 4.15 depicts the distribution of temperature with depth in the most interesting part of a profile across the frontal zone of the Gulf Stream. The structure of the frontal zone at the profile is greatly complicated by the relatively cold and relatively warm laminae in the layer between the surface and the 200 m level. The frontal zone separates the warm waters of the Gulf Stream (on the right of the profile) from the cold slope waters, the temperature of which drops to $8.5-8.9°C$ at the surface near the frontal interface. The presence of an intermediate SLT minimum (about $13°C$) between stations No. 2441 and No. 2442, recorded by means of a towed temperature sensor, separates the frontal zone in the upper layer of the ocean into two separate frontal interfaces which come together only below 250 m. The interval between the two frontal interfaces is filled with intrusive laminae which, apparently, point to the instability and destruction of this type of "multi-stage" frontal structure. It is practically impossible to imagine the three-dimensional pattern of this structure, its spatial extent, form and origin on the basis of ship data alone.

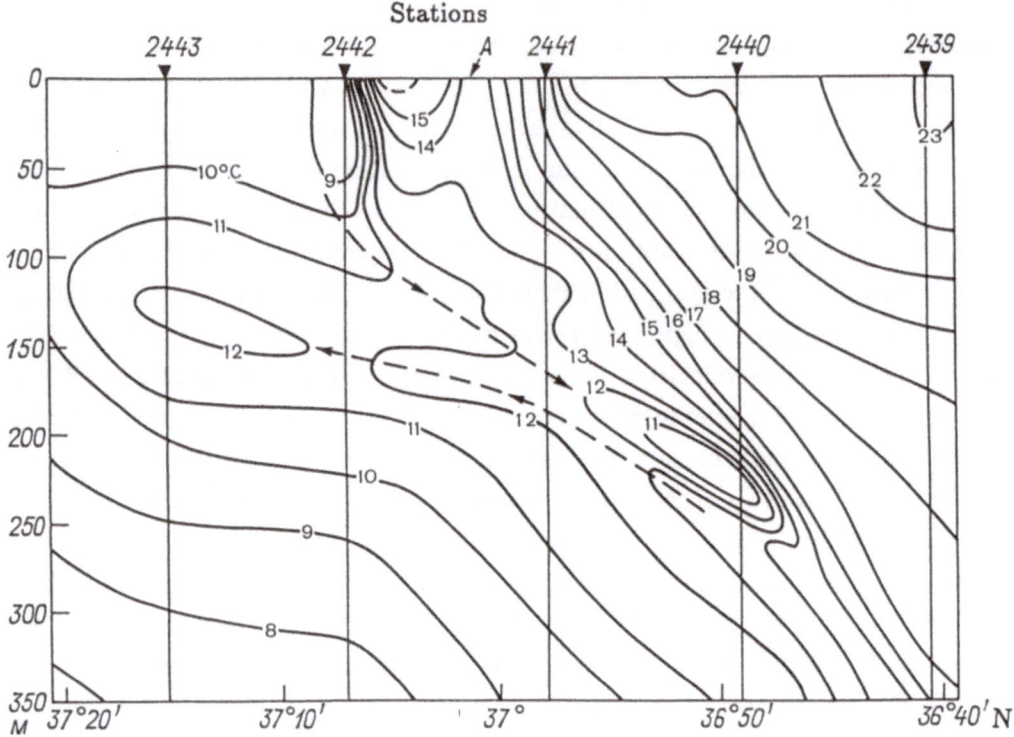

Fig. 4.15. Temperature distribution at a section across the frontal zone of the Gulf Stream. A – region of cold entrainment (intermediate temperature minimum). Arrows indicate the conjectural direction of intrusion.

Two consecutive IR images of the Gulf Stream frontal zone for March 20th and 23rd make it possible to estimate the changes in position and structure of this zone during the execution of the profile (March 20th-21st). A comparison of the two positions of the northern Gulf Stream boundary, determined by the maximum thermal contrast on IR images, shows (fig. 4.16) that in the three days between March 20th and 23rd, the general position and structure of the frontal zone in the vicinity of the profile changed significantly. However, fig. 4.13 clearly shows that, travelling along the profile during the first half of the day on March 21st, the ship actually did traverse two separate frontal interfaces: the first a moderate one, judging by the thermal contrast, between stations No. 2440 and No. 2441 (closer to the latter), and the second a very sharp one in the vicinity of station No. 2442. The results of the measurements taken with a towed temperature sensor (fig. 4.14) make it possible to estimate the horizontal temperature gradient at the fronts. It is approximately $1 - 2°C/m$ for the first front, and $5 - 6°C/km$ for the second one.

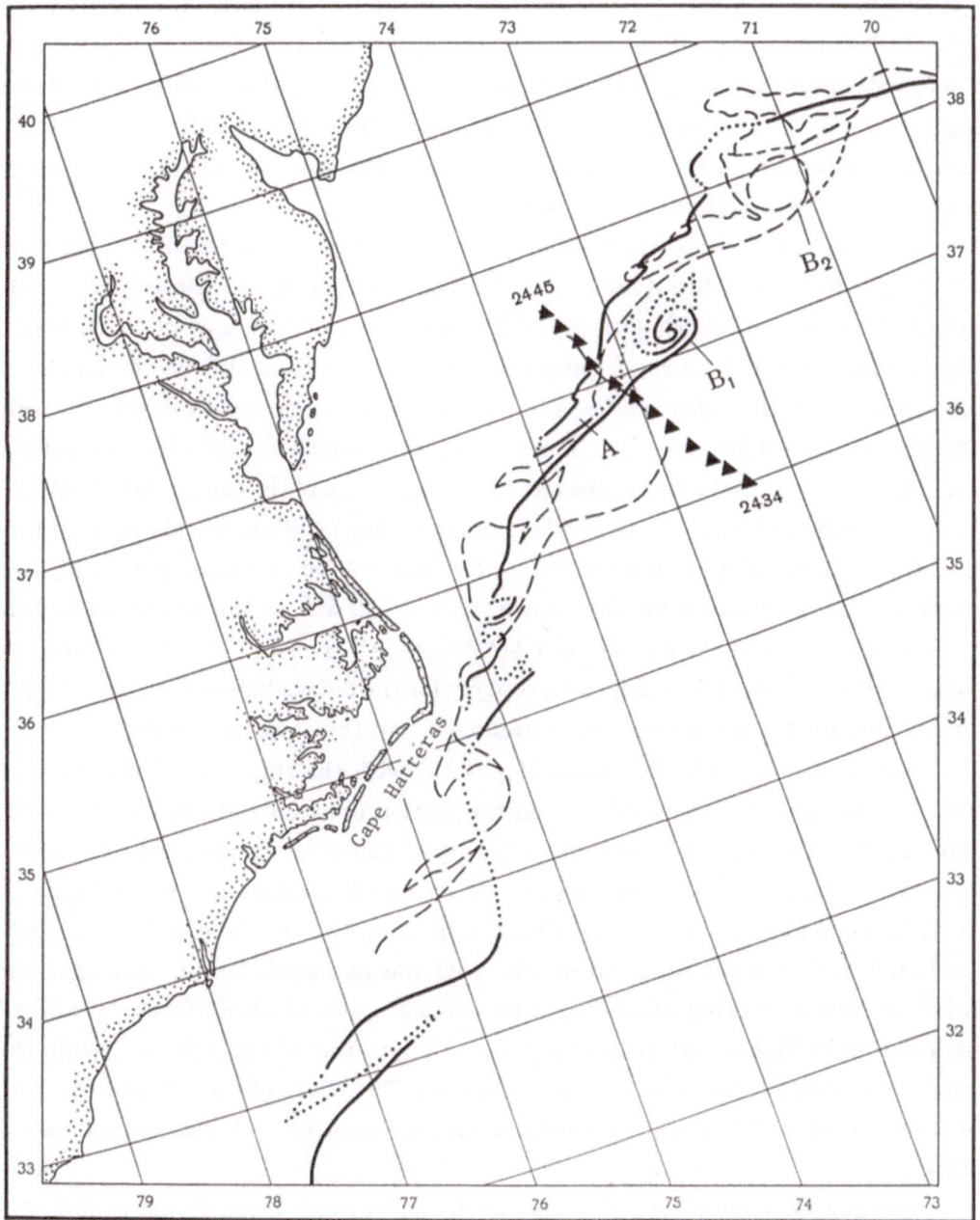

Fig. 4.16. Diagram of the positions of the northern Gulf Stream boundary, based on the data from IR images for 20 and 23 March 1978. March 20th: sharp frontal interface – solid line, diffuse frontal interface – dots. March 23rd: sharp frontal interface – dashed line, diffuse frontal interface – short-dashed line; A – the same as in figs. 4.13–4.15; $\triangle$ – profile stations; B_1 and B_2 – positions of cyclonic eddy on March 20th and 23rd respectively.

The positions of these fronts on the IR image for 20 March 1978 (fig. 4.13) differ little from their positions on the section (fig. 4.15) and in the data obtained by the towed sensor (fig. 4.14). This gives us reason to believe that the structure of the frontal zone in the area where it was intersected by the section hardly changed from March 20th to 21st. The main changes visible in this area in fig. 4.16 apparently occurred during the following two days.

We shall now describe the characteristic spatial features of the frontal zone, to which the above-described details of the temperature profile are related (fig. 4.15). On the IR image for March 20th (fig. 4.13), we can clearly see a jet of colder water (lighter tone on the IR image), entrained into the Gulf Stream from the region of slope waters on the other side of the frontal interface, spreading down the Gulf Stream current. This cold entrainment* has a well-defined cyclonic eddy with a diameter of about 25−30 km on the end of it. The length of the entrainment with the eddy is approximately 150 km. Having compared the positions of this entrainment and the cyclonic eddy on the IR images for March 20th and 23rd (fig. 4.16), we come to the conclusion that the eddy travelled 140 km to the northeast in this time, and the base of the entrainment 115 km, which gives us a mean displacement velocity of about 55 and 45 cm/s respectively for the 3 days between images. At the same time, the total length of the entrainment and the eddy increased to 175 km.

A comparison of the IR images (fig. 4.13) with the ship data shows that the intermediate temperature minimum on the section between stations No. 2441 and No. 2442 is the mark of the ship's crossing of the cold entrainment somewhere between the eddy and the place where it separates from the slope water front. In fact, 17 hours had passed from the time the IR image was received at 14.00 h GMT on March 20th and the time station No. 2441 was executed. During this time, the whole structure, moving northeast at an average speed of about 50 cm/s, altered its position by 30 km with respect to the section, so that the section fell within the cold-jet section marked A in figs. 4.13 and 4.16. The width of the entrainment here did not exceed 2−3 km, while the near-surface temperature in it was approximately 13°C.

The development of cyclonic eddies at the Gulf Stream front ("spin-off eddies"), which are similar to the described cold entrainment, is examined in greater detail in subsection 3.1.2 (e.g. see fig. 3.8).

---

* We shall refer to such structural elements as "entrainments", though they are sometimes called "cold intrusions" in the American literature [195] by analogy with the elements of the vertical structure.

However, the spin-off eddies, described by many authors, which are mostly observed off the coast of Florida, are not accompanied by the appearance of an independent cyclonic eddy on the end of the cold entrainment. Apart from the given case, we know of no earlier satellite observations of well-formed cyclonic eddies of this scale in the main current of the Gulf Stream.

The stability of the front must be disturbed for a cold jet of slope water to be entrained into the flow of the Gulf Stream across the frontal interface. There must be a strong local cyclonic vorticity of the flow near the stable Gulf Stream front, which exceeds the mean equilibrium cyclonic vorticity of the front itself. In Mooers' opinion [185], it is enough for the horizontal transverse velocity shear at the front du/dx to exceed the local value of the Coriolis parameter f for the frontal interface to become unstable. We still do not know the primary factor which destabilizes a potentially unstable frontal interface prior to the formation of an entrainment. Legeckis [173] believes that a wave process develops at the front, while Vukovich [255] surmises that cross-frontal advection of cold water by a drift current takes place. The fact that such entrainments move along the front with speeds of $39 - 75$ cm/s ([173, 195] and our own data), i.e. less than the speed of the Gulf Stream itself (150 cm/s), speaks in favour of the wave hypothesis. However, it cannot be ruled out that the cyclonic eddy itself (visible in fig. 4.13) could also be the initial cause of frontal disturbance. This eddy could have been in the Gulf Stream waters even before the formation of the entrainment. It could have then become "visible" in the IR spectrum due to the cold water entrained into it. On the other hand, this type of eddy could have arisen after the formation of the cold entrainment due to disturbance of the local geostrophic balance near the cold jet. This question cannot be resolved on the basis of the data available to us. However, we are absolutely certain of one thing, that well-formed cyclonic eddies with a diameter of $25 - 30$ km and a lifetime of several days can exist and flow along with the main current of the Gulf Stream.

It would be interesting to attempt to recreate on the basis of the section data (fig. 4.15) and IR images for March 20th (fig. 4.13) a three-dimensional pattern of the structure of the Gulf Stream frontal zone in the area of the entrainment together with the observed cyclonic eddy. Such a pattern is depicted in fig. 4.17. It is hypothetical in many of its details, but it is based on the actual position of the $14°C$ isotherm in the section and on general principles of geometrical plotting, which provide some guarantee against an accidentally absurd result. Among other things, it would be illogical to expect during the plotting of a three-dimensional image that the entrained jet of heavier, cold slope waters would "float" in a thin layer on the surface of the warm and lighter waters of the Gulf Stream. The section (fig. 4.15) shows that it is the waters of the "warm sector" between the entrainment and the main slope water front which lie in a thin layer over the cold extension of the frontal interface, as shown in fig. 4.17. The satellite IR image for March 20th suggests that this layer of warm waters is truly thin. The diffuse light (colder) areas within the "warm sector" and around the eddy (C, D, E in fig. 4.13) are an indication of this. It turns out that the wave-like cold extension, or crest of the main front, which retreats obliquely under the warm waters of the Gulf Stream core, is as if "overridden" from the northeast by a thin layer of slowly moving warm waters from the periphery of the Gulf Stream. It would be even more correct to say that, spreading more quickly than the warm peripheral waters of the Gulf Stream, the wave-like cold crest of the frontal interface flows under their thin layer, spinning them in a vortical motion which forms the "warm sector" and cyclonic eddy. The arrows in fig. 4.17 therefore denote the **relative** motion in coordinates immovably fixed with the structure being studied and moving along with it in the northeastern direction with a speed of about 50 cm/s.

The stratified structure of the frontal interface near the cyclonic eddy, depicted in fig. 4.17), has not been confirmed by any direct ship observations. However, the presence of this type of structure within the $11 - 14°C$ isotherm on the section (fig. 4.15) indicates that even the most stable part of the frontal interface at the base of the entrainment is still not completely stable, and is destroyed by intrusive interleaving. It should be said that an absolutely similar stratified structure of intrusive origin was observed earlier in other sections of the Gulf Stream [117] (see also figs. 38 and 39 in [237]). The frontal boundaries of an eddy, which because of its small dimensions is hardly in geostrophic balance, should be even less stable. The strong centrifugal term should in this case definitely promote intensive intrusive interleaving. Finestructure should also result from the action of double-diffusive convection at the boundaries of initial intrusions which appear at a potentially unstable frontal interface with a significant slope (10^{-2}). It is interesting to note that, by substituting the mean values of the vertical temperature and salinity gradients at the front on the basis of the profile data (fig. 4.15) in Ruddick and Turner's formula [221], we can derive a characteristic scale for the thickness of layers in the neighborhood of $100 - 130$ m, which agrees with observations at the section. Although it is impossible to determine by which process the observed intrusions are formed on the basis of available data, we do have sufficient grounds for depicting intrusive layers around the cyclonic eddy as well in fig. 4.17.

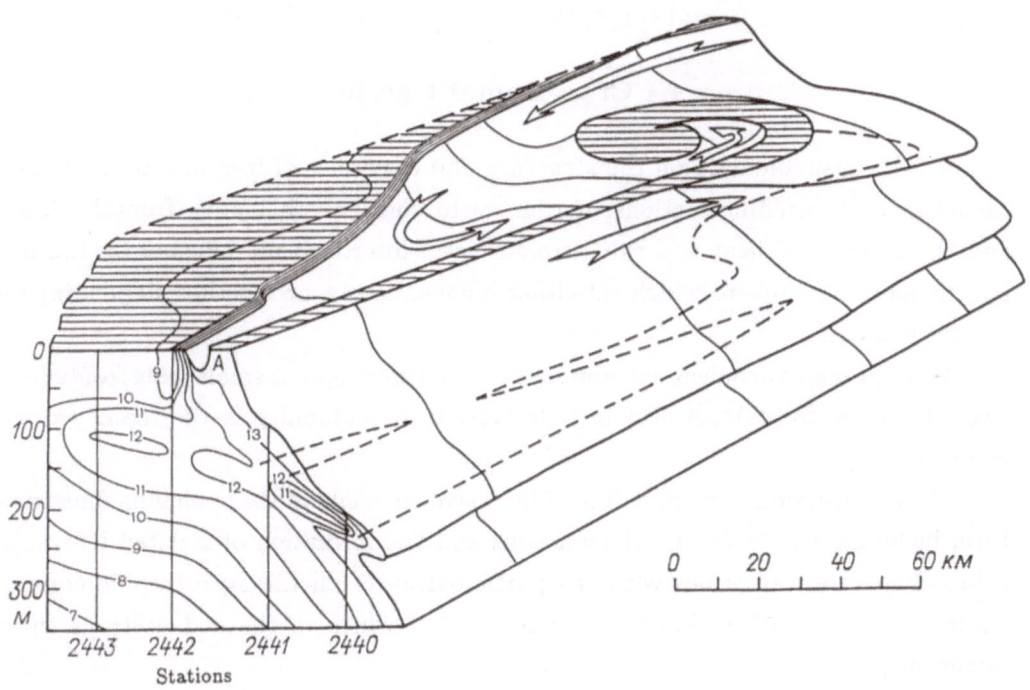

Fig. 4.17. Hypothetical depiction of the three-dimensional finestructure of the frontal zone of the Gulf Stream in the vicinity of a cold entrainment and cyclonic eddy.

In conclusion of this section, we can say that our attempt to compile a three-dimensional image of the finestructure of the Gulf Stream frontal zone on the basis of essentially synchronous ship and satellite data proved to be quite fruitful. We have found that a comparison of a vertical oceanographic section of the front with a detailed thermal image of the ocean surface enables us to better understand some of the characteristics of a three-dimensional frontal structure, and to establish the physical relationship between these characteristics. The planning and execution of these synchronous measurements in the future can greatly speed up progress in research into oceanic frontal zones, their structure and variability.

4.4 Cross-frontal transfer

All the available data on the structure and dynamics of frontal interfaces, discussed in the preceding sections, permit us to conclude that cross-frontal mixing and the transfer of heat and salt across frontal interfaces are effected by the following main mechanisms which act either independently, or together, depending on the specific case:

1) small-scale turbulent entrainment of shearing origin at small-scale fronts (e.g. river discharge fronts). A Kelvin-Helmholtz-type instability is the main process involved;

2) the formation and spreading of intrusive elements of thermohaline finestructure, including the "calving" of intrusions and the formation of isolated lenses, in a broad spectrum of scales with the participation of all the ageostrophic components of motion and double-diffusive convection which in the end destroys these intrusions;

3) the formation of eddies with a vertical axis at frontal interfaces; the advection of water by the orbital motions of eddies; the destruction of eddies with the subsequent formation of a slowly dissipating thermohaline finestructure;

4) the effect of the increase of density of seawater during mixing (cabbeling), which promotes the rapid sinking of the products of mixing along the frontal interface, and the maintenance of the latter in a sharp state.

The main physical concepts of cross-frontal mixing and heat and mass transfer are just beginning to be discussed in the literature. Numerical studies are still very scarce. There are simply not enough adequate observational data for a systematic and complete study of all the problems concerning cross-frontal transfer. Therefore, we shall have to limit ourselves in this section to a fairly brief analysis of estimates related to the mechanisms mentioned in points 2 and 4. Strictly speaking, these data are of the greatest interest to us at present, as they show that frontal interfaces are those regions in the ocean where the previously ignored mechanisms of double diffusion and cabbeling are the most effective during mixing.

4.4.1. Estimates of double-diffusive heat and salt transport

In section 4.2 we dwelt on the results of laboratory experiments with thermo-haline intrusions, during which we inevitably observed the formation of multilayer structures generated by double-diffusive convection, and especially effectively by "salt fingers". The possibility of "salt fingers" forming in the presence of very low destabilizing salinity gradients apparently makes them a common phenomenon in the ocean as well.

The manifestation of double-diffusive convection in the form of multilayer structures enables us to acquire fairly reliable data on the vertical fluxes of heat and salt across the layers where this type of convection is persistent, and particularly across the upper and lower boundaries of intrusive layers and laminae. The first attempts [57] to relate the parameters of a multilayer structure observed in the ocean to vertical heat and salt fluxes on the basis of formulae obtained in laboratory experiments [247] were accompanied by a large number of reservations and doubts regarding the stationarity of the process, the effects of horizontal advection and the possible negative effect of the vertical gradient of currents (shear) on the very development of double-diffusive convection. However, as we have already mentioned in section 4.2, observations now provide us with such conclusive indirect and direct proof of how widespread the effects of double diffusion are in the ocean (especially "salt fingers"), that many of our recent doubts appear to be unfounded. Convective fluxes of heat and salt in the case of double- diffusive convection of any type have the same orientation as the mean gradients of temperature $\partial \tilde{T}/\partial z$ and salinity $\partial \tilde{S}/\partial z$ respectively, and so they can be parametrized with the help of the coefficients of effective vertical "eddy diffusion" $\tilde{K}_T^V$ and $\tilde{K}_S^V$. However, the resulting density flux is directed against its mean gradient. This once again proves that the concept "eddy diffusion of density" is absolutely unsuitable for the ocean.

Two methods [144, 154] have been proposed for determining the contributions of intrusive interleaving to lateral or cross-frontal heat and salt transfer between the water masses. Joyce's method [154], which was mentioned in section 4.2, is a generalization of Osborn and Cox's result [199] for the case where horizontal temperature and salinity gradients and horizontal advection are present. The use of this method requires the *a priori* knowledge of the coefficients of effective v e r t i c a l heat and salt transfer. The second method, proposed by Horne et al. [144], is based on the assumption that intrusions, which develop most frequently at frontal interfaces, are destroyed from the top and bottom because of double-diffusive convection, and so the appropriate experimentally derived formulae [247] can be used to determine the vertical heat and salt fluxes.

Below, we present some of the results of local estimates of $\tilde{K}_T^V$ and $\tilde{K}_S^V$ based on field measurements of the parameters of multi-layer structures and intrusions in various frontal zones of the ocean. As we can see from table 4.5, the actual vertical transfer of heat and salt due to double-diffusive convection even in the weakest case is ten times more effective than molecular transfer. As a rule, it is $10^3 - 10^4$ more intense than molecular transfer for heat, and $10^5 - 10^6$ more intense for salt.

The values of the effective horizontal heat and salt transfer due to intrusive transfer directly at the fronts amount to $10^1 - 10^3 m^2/s$ for the coefficients $\overline{K}_T^H$ and $\overline{K}_S^H$ respectively [96, 144, 154]. The integral values of cross-frontal heat and salt transfer obtained on this basis amount to approximately $10^{14} J/(km \cdot day)$ for heat and $(0.4 - 1.5) \times 10^7 kg/(km \cdot day)$ for salt, calculated per kilometre of the length of the frontal interface [144, 254]. Consequently, we now have the first integral quantitative values of cross-frontal heat and mass transfer, which can serve as reference points in the continuously expanding research on oceanic fronts.

The results of Coachman and Charnell's estimates [96] for fronts in the Bering Sea deserve special mention. These authors used the method of calculation proposed by Joyce [154] [see section 4.2, equation (4.6)]. However, they took the mean vertical coefficient of salt transfer $\overline{K}_S^V$ for the scale of intrusions as being the same, and equal to $10^{-6} m^2/s$, both for fronts and between fronts, which is illogical, considering the greater intensity of the intrusive process at frontal interfaces. As a result, the value of $\overline{K}_S^H$ between fronts $(2 \times 10^3 m^2/s)$ turned out to be 7 times greater than the value of $\overline{K}_S^H$ at an internal front $(2.8 \times 10^2 m^2/s)$ and 20 times greater than the value of $\overline{K}_S^H$ at an external front $(10^2 m^2/s)$. With the horizontal salinity gradients observed at the fronts and between them, the horizontal salt flux turned out to be $3 - 4$ times greater between fronts than at the fronts; on the basis of this unlikely result, the authors came to the conclusion that the fronts served as a barrier with respect to horizontal salt transfer. In the light of all the past discussions, it is quite apparent that the vertical coefficient of salt transfer in the intrusive regime at the front should be at least an order of magnitude greater than between fronts. With $\tilde{K}_S^V$ equal at least to about $10^{-5} m^2/s$ at the fronts, the mean coefficients of horizontal salt transfer $\overline{K}_S^H$ at both fronts would be 2.8×10^3 and $10^3 m^2/s$ respectively. In this case, the horizontal salt fluxes across the fronts would be more intensive than between the fronts, which is the opposite of what the authors of [96] claim. In order to establish the truth of this, we must examine the three-dimensional salt balance

Table 4.5

Author	Location	$\tilde{K}_S^V \times 10^{-4}$ m^2/s	$\tilde{K}_T^V \times 10^{-4}$ m^2/s	Stability parameter R_ρ [61]
		For "salt fingers"		
Zenk [272], 1969	Atlantic	7.0	—	1.2
Turner [246], 1967, based on the data of [241]	Atlantic	5.0	~ 2.5	1.15
Fedorov [61], 1976, based on the data of [147]	Atlantic	12.0	4.0	1.1
Lambert, Sturges [169], 1977	Atlantic, Caribbean Sea	5.7	0.7	1.4-1.7
		For the diffusive regime		
Fedorov [61], 1976, based on the data of [192]	Beaufort Sea	—	$0.25 - 0.01$	$2 - 15$
Middleton and Foster, 1980, [182]	Weddell Sea	0.48	1.01	1.5
		For intrusions (sum of both regimes)		
Voorhis et al. [254], 1976	New England shelf, USA	—	5.0	?
Horne et al. [144], 1977; Horne [143], 1978	Nova Scotia shelf, Canada	3.0	8.0	1.24

in the whole frontal zone. It should be said that the case discussed by Coachman and Charnell can be regarded as special because of the very weak horizontal salinity gradients at the fronts ($\sim 0.009\ ppt/km$). Basically, they were dealing with a fairly diffuse frontal zone with two local peaks of horizontal gradients, too weak to be called fronts. In such a case, one should most probably proceed from the equality of the horizontal salt fluxes in the whole cross section of the zone. Then, the values of $\overline{K}_S^H$ at the "fronts" would be $3 - 4$ times smaller than in the interfrontal region and, therefore, the coefficient of vertical salt transfer on the scale of the intrusions $\overline{K}_S^V$ would be twice as high at "fronts" as in the rest of the zone.

4.4.2. Role of the density increase due to mixing

An increase of density due to mixing is an interesting consequence of the nonlinearity of the equation of state of seawater, as a result of which, for example, a mixture of two water masses of equal density ρ_0, but different temperatures (T_1 and T_2) and salinities (S_1 and S_2) has a new density $\rho_* > \rho_0$ and will have a tendency to sink in the ocean. The phenomenon of water becoming denser during mixing has been described in detail in the Soviet literature by N.N. Zubov, K.D. Sabinin and N.P. Bulgakov [12, 26, 27].

It is quite obvious that the effect of density increase due to mixing should be greater with lower temperatures, at which the nonlinearity of the equation of state of seawater is more strongly pronounced. It is in this connection that Fofonoff's [116] and Foster's [118] assumptions regarding the importance of density increase due to mixing in the process of the formation of Antarctic bottom waters can be considered as being quite realistic. However, we must say that the term "cabbeling", which was borrowed by Foster from the German ("Kabbelungen") with a certain degree of distortion, is not an appropriate name for the process of density increase due to mixing from the physical point of view. This term is the result of inaccurate translation from the German of Witte's paper "Zur Teorie der Stromkabbelungen" [258] which should be translated as "On the Theory of Bands of Rough Water". It was this phenomenon, and not the increase of density due to mixing that was conveyed at that time by the German term "Stromkabbelung".

Despite this fairly old assumption regarding the importance of the effect of cabbeling at the interfacial boundaries between water masses [258], researchers began to seriously discuss its hydrodynamic effects only recently [93, 125, 144]. In this connection, special attention should be paid to the earliest examples of some of the effects related to cabbeling, presented in N.N. Zubov and K.D. Sabinin's paper

[27]. These examples also include estimates of the effects of a density change and level decrease in the frontal zone of the Gulf Stream as a result of the mixing of the warm waters of the current with cold Labrador waters. According to one of the estimates, mixing in the frontal zone of the Gulf Stream of equal volumes of Labrador waters with a near-surface temperature of $10°C$ and Gulf Stream waters with a near-surface temperature of $24°C$ and an equal initial density will lead to a $0.24 \times 10^{-3} m^3/t$ decrease in the specific volume of the mixture. As the difference in temperatures ΔT diminishes from $14°C$ at the surface to $4°C$ at the 800 m level, the mean decrease in specific volume throughout the water column will amount to approximately $0.12 \times 10^{-3} m^3/t$, which will lead to an approximately 0.1 m decrease in the level of the ocean in the zone of contact of the mixing waters. Although this value constitutes only about 3% of the total geostrophic difference in level across the Gulf Stream, its dynamical effect depends on the width of the mixing zone. The mean slope of the surface across the $200 - km$-wide frontal zone of the Gulf Stream can be estimated as 1.75×10^{-5}. If active mixing takes place in a zone measuring only $5 - 10$ km in width, then the surface slope in this zone due to cabbeling can be of the very same order.

With the idealized mixing pattern discussed in [27], the surface slope from both sides in the direction of the frontal interface (fig. 4.18) should result in convergent circulation in a plane perpendicular to the frontal interface which, for the sake of simplicity, is regarded as vertical. At the same time, the speed of sinking of the denser waters of the mixture should be related to the integral (over depth) of the convergent flow in the direction of the front. This flow depends on the surface slope, and for the same decrease in level $\Delta \xi$ on the axis of the mixing zone, it will grow weaker as the mixing zone becomes wider. In turn, the width of the mixing zone is nothing more than the width of the equilibrium front B [see section 2.5, equation (2.13)]. Let us try to relate the expected mean velocity of sinking of the denser product of the mixture of two water masses to the width B_m of the equilibrium front, which in this case remains unchanged by the local deformation rate D_m determined by the above-mentioned convergent circulation.

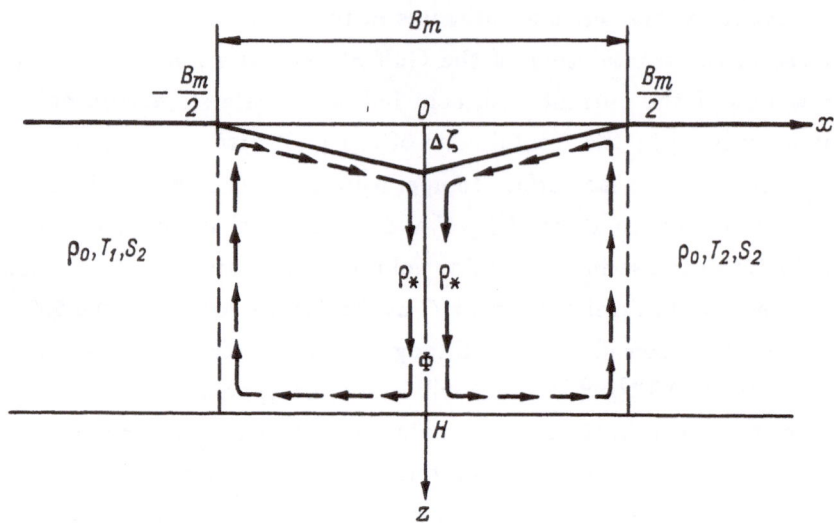

Fig. 4.18. Diagram of the decrease in level and motion in the mixing zone near the frontal interface F. H - thickness of mixing layer, B_m - width of mixing zone.

Let us apply a generalization of the Ekman theory in the form proposed in [56]. Then, the total (integral) horizontal flow S_x in the direction x (in this case, perpendicular to the front) in the near-surface Ekman layer due to the slope of the level $\partial \xi / \partial x$ will be expressed as

$$S_x = \frac{g}{4a^3 K_z} \frac{\partial \xi}{\partial x},\tag{4.8}$$

where K_z is the coefficient of vertical eddy viscosity, $a = \sqrt{f/2K_z}$, and $f = 2\omega \sin \phi$ is the Coriolis parameter.

From the continuity equation, under the condition that the total vertical flow due to the convergence of horizontal flows from both sides takes place in a zone with a width B_m, it follows that

$$S_x = wB_m/2$$

and, accordingly, the mean vertical velocity is

$$\overline{w} = \frac{S_x}{B_m/2},$$ (4.9)

In turn, the slope of the level $\partial\xi/\partial x$ can be expressed as

$$\frac{\partial\xi}{\partial x} = \frac{\Delta\xi}{B_m/2},$$ (4.10)

where $\Delta\xi$ is the decrease in level, on the axis of the mixing zone, due to an increase in the density of the mixture.

From (4.8), with (4.9) and (4.10) taken into account, we get

$$\overline{w} = \frac{gK_z^{\frac{1}{2}}}{(f/2)^{\frac{3}{2}}} \frac{\Delta\xi}{B_m^2}.$$ (4.11)

With $\Delta\xi \simeq 0.1m$, which corresponds to the above-mentioned Zubov and Sabinin estimate [27], $K_z = 5 \cdot 10^{-4}m^2/s, f = 10^{-4}s^{-1}$ and $B_m = 1\ km = 10^3\ m$, we get $\overline{w} = 6\cdot10^{-2}m/s$ or $5184\ m/day$ from (4.11). This sinking rate appears to be extremely high, and is definitely associated with the extremely high temperature difference of the two mixing water masses. However, we should note that the mixing pattern examined by us corresponds to "the unstratified case" in Garrett and Horne's study [125], which, when analyzed on the basis of an analogy with convection near a heated wall, gave a vertical speed in a zone $500\ m$ wide and $100\ m$ deep of approximately $5\ cm/s$ with a smaller temperature gradient. At the same time, Garrett and Horne maintained that horizontal turbulent mixing dominated over all the other types of processes.

It should be said that all the estimates in [125] indicate that the maximum sinking rate due to cabbeling is directly proportional to the coefficient of horizontal turbulent transfer K_ℓ. From the results of the case studied here, it would appear that the vertical velocity cannot be directly proportional to K_ℓ, as the more intensive the horizontal turbulent mixing, the wider the zone of mixing [see equation (2.13)] and the weaker the circulation in the plane perpendicular to the front. However, as we shall see below, this type of dependence can arise if in the frontal zone there is a balance between horizontal turbulent diffusion and convergence at the front due to cabbeling. On the other hand, direct proportionality between $\overline{w}$ and $K_z^{\frac{1}{2}}$ in (4.11) is from our point of view more natural, as the ageostrophic convergent flow from both sides to the front should increase with K_z. With $K_z = 0$, the effect of a level decrease in the zone of mixing will be purely geostrophic, and there will simply be no convergence at the front. It appears that precisely this type of relation of the main dynamical factors at fronts not wider than 1 km, where we observe maximum deviations from the geostrophic regime, is responsible for the significant role of cabbeling in maintaining near-frontal convergence (or a corresponding local deformation rate D_m) against the constant opposition of horizontal diffusion. Approximately the same point of view is expressed by Bowman and Okubo in their analysis [93]. In this case, there should be a relationship between D_m and S_x, which is quite obvious in the diagram of mixing and convergent circulation when $S_{z=0} = 0$ (see fig. 4.18), i.e.

$$D_m = \frac{\partial \overline{u}_c}{\partial x} = \frac{S_x}{HB_m/2} = \frac{gK_z^{\frac{1}{2}}}{H(f/2)^{\frac{3}{2}}} \frac{\Delta \xi}{B_m^2}, \tag{4.12}$$

where $\overline{u}_c$ is the vertical mean of the convergent horizontal velocity along the normal to the front in a mixing layer with thickness H.

Having compared (4.12) and (2.13), we find that in this case, the coefficient of horizontal turbulent transfer K_ℓ is automatically determined by the initial values of K_z, H, $\Delta \xi$ and f, i.e.

$$K_\ell = \frac{2gK_z^{\frac{1}{2}} \Delta \xi}{H(f/2)^{\frac{3}{2}}} \tag{4.13}$$

and, in turn, it follows from (4.11) that

$$\overline{w} = D_m H = \frac{K_\ell H}{2B_m^2}, \tag{4.14}$$

i.e. the greater the deformation rate D_m, or the smaller the equilibrium width of the front (or width of the mixing zone) B_m, the more quickly the waters sink due to cabbeling, all other conditions being equal.

We shall now show that equations (4.12), (4.13) and (4.14) give us fairly reasonable values of D_m, K_ℓ and $\overline{w}$ when the values of B_m, K_z, H and the temperature change across the front ΔT are realistic.

According to [125], the increase in density $\Delta \rho$ during mixing of equal masses of seawater of the same density ρ_0, but different temperatures $(T_2 - T_1 = \Delta T)$ can be calculated approximately by the formula

$$\Delta \rho \approx -\frac{\rho_{TT}}{8} \Delta T^2, \tag{4.15}$$

where $\rho_{TT} = \partial^2 \rho / \partial T^2$.

In turn, the change in the level $\Delta\xi$ can be expressed in terms of $\Delta\rho$ as

$$\Delta\xi = \frac{H\Delta\rho}{\rho_0 + \Delta\rho}. \tag{4.16}$$

Then, the equations (4.13) and (4.12) will take the form of

$$K_\ell = \frac{2gK_z^{\frac{1}{2}}}{(f/2)^{\frac{3}{2}}} \frac{\Delta\rho}{(\rho_0 + \Delta\rho)} \tag{4.17}$$

and

$$D_m = \frac{gK_z^{\frac{1}{2}}}{B_m^2(f/2)^{\frac{3}{2}}} \frac{\Delta\rho}{(\rho_0 + \Delta\rho)}, \tag{4.18}$$

where $\Delta\rho$ is determined from (4.15).

With temperature changes ΔT of the order of $3°C$, which are frequently observed at sharp thermohaline fronts with salt compensation, we get $\Delta\rho \simeq 1.5 \times 10^{-2} kg/m^3$ from (4.15) when $\rho_{TT} = -1.3 \times 10^{-5} (°C)^{-2}$ (for the $5 - 10°C$ temperature range). In turn, the equation (4.16) gives $\Delta\xi = 1.5 \times 10^{-3} m$ when $H = 100\ m$. Therefore, when $B_m = 1\ km = 10^3 m$ and $K_z = 5 \times 10^{-4} m^2/s$ [255], we find that $K_\ell = 1.8 \times 10^1 m^2/s$ from (4.17), $D_m = 9.1 \times 10^{-6} s^{-1}$ from (4.18) and $\overline{w} = 9.1 \times 10^{-4} m/s$ or $79\ m/day$ from (4.14). As far as we can judge from all the available estimates [125, 142, 254], such values are quite realistic for frontal interfaces of the given scale. Therefore, we have reason to believe that cabbeling can play a significant role in maintaining thermohaline fronts in a sharp state. The initial compression of temperature and salinity isolines most probably takes place under the influence of a background deformation field which is characterized by a deformation rate D_0. If the equilibrium width of the frontal zone B_0 is at the same time small enough and is combined with a fairly large temperature change ΔT_0, then the self-sustaining deformation mechanism of cabbeling can become involved, and with the new deformation rate $D = D_0 + D_m$, the frontal interface will sharpen to a new equilibrium width $B_m < B_0$. The question as to which values of B_0 and ΔT_0 can be regarded as sufficient for the "involvement" of the cabbeling mechanism merits special investigation in our opinion.

Chapter 5

PROBLEMS FOR FUTURE RESEARCH AND THE CONCERNS OF ASSOCIATED DISCIPLINES

5.1. Some generalizations

In the preceding chapters of this book, we were hardly able to present systematically enough even a small portion of the data on oceanic fronts that have accumulated so far. In this abundance of information, the unprepared reader can easily stray and miss the main point. The main point lies in the fact that our concepts of the motion and structure of oceanic waters are obviously ready to move on to a new stage which, on the one hand, is more complex than the previous one, and on the other, is characterized by greater physical logic. Back in the 1940-1950's, the view that the ocean constituted a very thin wind-driven shell of our planet led V.B. Shtokman and many of his contemporaries to the theory of total fluxes, which enabled us to take in at a glance the general horizontal (two-dimensional) circulation of all the oceans of the world, and to gain some insight into their physical nature not only as a whole, but in a number of important details. At the same time, the picture of the general circulation was depicted as one gigantic whole which was connected throughout the World Ocean. Certain arguments allowing for the third (vertical) dimension enabled us to realize that the mean value of the vertical velocity in the ocean should be as many times smaller than the mean horizontal velocity, as the mean depth of the ocean is smaller than its cross section, i.e. $10^3 - 10^4$ times. Further details were added to this knowledge through the discovery and description of individual cells, within the general circulation, which could be modelled and studied separately from the rest of the ocean, since they were highly isolated under natural conditions as it was. These primarily include the circulation cells of coastal upwelling, which are associated with the local effect of the wind, as well as the river discharge lenses of large rivers which flow into the ocean. Researchers then discovered and described synoptic-scale oceanic eddies with their characteristic three-dimensional circulation. The results of the present concentration of research at oceanic fronts show the same tendency of adding details to the general circulation pattern of the ocean due to the inclusion of numerous new cells and discontinuities. There is no doubt that the near-surface Ekman layer of the ocean is divided into very small circulation cells under the transient and inhomogeneous influence of the atmosphere (or to be more exact, interaction with the atmosphere), as well as under the influence exerted from below by the internal waves and eddies developing in the thermocline. The near-bottom Ekman layer is apparently also divided into cells, though we still know very little about this. Finally, the thermocline (or to be more exact, the pycnocline) is an area where we observe the action and interaction of eddies and Rossby waves, where the division into cells occurs on larger scales, but nevertheless does take place. The typical ratio of vertical scales to horizontal ones in the majority of natural circulation cells is equal to 10^{-2}, and can even reach 10^{-1} in coastal upwelling. Therefore, one can expect that the vertical velocities in these cells can reach $10^{-1} cm/s$ and sometimes even $1 cm/s$, and should be of the order of $10^{-2} cm/s$ on the average, which is $10 - 100$ times greater than the most optimistic estimates of the past and 1000 times greater than the most conservative concepts of recent years.

Is it possible to sum up in a popular form all the new information on oceanic fronts obtained over the past ten years? Not only is it possible, but it is also necessary to do so.

Though this book was devoted mainly to the physical aspects of the problem, it would be wrong to ignore the most important chemical and biological aspects when summarizing. With the latter taken into account, our summary is as follows:

- The general view on the physical nature of oceanic fronts underwent fundamental changes. The abundance, diversity and omniscale nature of frontal interfaces in the ocean prompt us to renounce the traditional concept of them as boundaries of climatic regions, and permit us to regard them as important elements of the complex three-dimensional structure of ocean waters, associated with the local limits of various elements of the general circulation, the vortical nature of the water motion and the diverse processes of mixing. Oceanic fronts and eddies are inseparably associated with each other over the entire range of spatial scales characteristic of them.

- We have established the frontogenetic role of oceanic eddies, and the complex "multifrontal" structure of oceanic frontal zones which is determined by it and a whole series of other factors.

- Oceanic fronts and the waves, eddies and thermohaline finestructure that develop at them play an important role in the transfer of kinetic energy and enstrophy along a cascade of scales from one region of the spectrum to another, to the scales of dissipation, as well as in the transformation of the available potential energy into kinetic energy, and vice versa.

- We have established a relationship between oceanic fronts and the processes which form the thermohaline finestructure of ocean waters. We have established the importance of intrusive finestructure in cross-frontal heat and mass transfer, and its dissipative nature. We have found that frontal zones represent regions of active vertical and horizontal mixing in the ocean, where, because of an intrusive structure, intensive transfer of heat and salt across a hydrostatically stable pycnocline takes place. Our present knowledge already allows us to carry out realistic quantitative estimates of this transfer.

- The variability in the positions and structure of large-scale oceanic frontal zones, on the one hand greatly affects the variability of the weather and climate in the atmosphere above the ocean and continents, and on the other, can serve as an indicator of significant climatic tendencies of our planet.

- Intensive vertical motions associated with frontal interfaces, as well as the characteristics of water stratification on both sides of a front, create characteristic spatial inhomogeneities in the concentrations of various chemical compounds dissolved in seawater, including biogenic elements, as well as suspensions and surface-active substances. Basically, frontal interfaces can be included among the few dynamical characteristics with which open ocean upwelling may be associated. All this determines many of the characteristic tendencies in the development of ecosystems near frontal interfaces (see section 5.2).

- Frontal zones appear as regions of high biological productivity, and are extremely important commercially. This feature of frontal zones cannot be ignored when evaluating the total biological production of the open ocean, which was most probably underestimated in the past. From this stems the theoretical possibility of conducting fishing in the open ocean where fronts are not as scarce as they once appeared to be.

- Frontal interfaces can serve as natural boundaries of different ecosystems, which can be utilized during the future development of a planned fishing industry and aquaculture in the ocean, especially in its coastal areas.

- On the strength of their convergent nature, frontal zones concentrate various pollutants and toxic substances of anthropogenic origin at the surface of the water and in their ecosystems.

All the above-mentioned characteristics of oceanic frontal phenomena stress the indisputable fact that research on oceanic fronts today should of necessity be based on a multi-disciplinary approach, or involve a number of branches of science. At the same time, it should be based on a physical foundation corresponding to the present level of development of oceanic hydrodynamics. From here stem the imperative tasks in physical research on oceanic fronts, which will be formulated in the final section of this chapter. The relationship between physical research on oceanic fronts and the tasks of related disciplines and branches of science can be traced on the same basis.

5.2. Research on the physics of frontal phenomena in the ocean and associated problems of other disciplines

One has only to look at two IR images of the thermal state of the same frontal zone, obtained at an interval of 2 − 3 days (e.g. see [50, 195] and fig. 4.16) to see the significant changes that took place in the positions, intensity and number of frontal interfaces and eddies within the frontal zone, while the general position and orientation of the zone itself hardly changed at all during this time. A comparison of thermal images recorded over longer intervals shows significant fluctuations in the position of the frontal zones as well. All the changes of this type cannot but affect the integral heat and mass transfer across the frontal zones and the balance of momentum, as well as the nature and intensity of ocean/atmosphere interaction in the vicinity of the frontal zones. This type of variability is of no significance when studying the mean long-term state of water circulation. All of its effects are usually lost during averaging. However, when studying synoptic-scale processes or developing the physical fundamentals of long-range weather forecasting with the thermal effect of the ocean taken into account, the instability of the frontal structure and motions within the frontal zone should be taken into consideration. Mathematical modelling of the ocean has now reached such heights that the question of parameterizing the role of oceanic fronts in investigations involving the use of numerical models of the general circulation of the oceans, or of the large-scale interaction of ocean and atmosphere, can be regarded as quite topical. We also believe that this type of parameterization is basically quite possible if we regard the frontal interfaces as "sub- gridscale" phenomena. The estimates in section 4.4 convince us of the practicability of such parameterization. It stands to reason that the existing oceanic hierarchy of scales of frontal phenomena should be taken into account in this case too. The term "eddy-resolving model of general circulation" has already taken root in the dynamics of ocean currents. These models already demonstrate the transient nature of deformations of the temperature field in the field of motion of a vortex. Only a few more steps are required for these models to become "front-resolving" on the one hand, and for them to take into account the effects of intensification of horizontal and vertical turbulent exchange, which are associated with frontal interfaces beyond the limits of model resolution.

The fundamentals of frontogenesis described in section 2.6 and illustrated with numerical examples in [178] help us to understand the physical nature of the redistribution of scalar constituents in the vicinity of developing fronts. The gases, chemical compounds, pigments, nutrients and fine suspensions dissolved in seawater can be regarded as scalar constituents with far greater reason than, for example, temperature. The same can be said about the free-floating plants, small trash and certain pollutants (e.g. dispersed oil pollutants) found on the surface. The concentrations of all these substances and objects at fronts, or rather near frontal interfaces, can be many times greater than the local intensification of horizontal temperature, salinity or density gradients. Chemical oceanographers, biochemists or geochemists interested in these local concentrations need a maximum knowledge of the physical processes which determine the advection of the objects of their investigations towards the frontal interface. The possibilty of building numerical models for the redistribution of scalars during frontogenesis, which has been demonstrated by MacVean and Woods [178], should be utilized in investigations where all the necessary initial data are available for this.

As we have already mentioned several times (see foreword, sections 3.1.3 and 4.1), frontal zones mostly have a complex horizontal structure which is characterized by numerous transient frontal interfaces and, apparently, numerous closed cells of vertical circulation extended along the frontal zone. Eddy formation at unstable frontal interfaces complicates this picture even more. In some cases, particuarly sharp fronts at the surface are easily detected by accumulations of foam, trash and free-floating plants. In other cases, such concentrations are not observed, temperature contrasts may be weak, while salinity contrasts are difficult to record. We have good reason to assume that analysis of the concentrations of certain dissolved substances and suspensions in such cases could prove very useful when establishing the "invisible" structure of frontal zones. Spectral analysis of the dimensions of suspended particles can basically be interpreted from the point of view of the values of vertical velocities in the water mass. Mapping the spectral characteristics of suspensions can help to establish the downwelling and upwelling zones. Upwelling zones are usually found at some distance from frontal interfaces, and are difficult to detect clearly by physical means, especially if the water rises from not very deep levels. I think that the cooperation of chemists and physicists in studying the structure of frontal zones can prove very effective scientifically. If we added technologically perfected measurements and observations from space on the one hand, and detailed biological investigations on the other, we would have an ideal model of an integrated program of research for frontal phenomena in the ocean.

Acoustic methods can also be used for studying fronts (see section 1.5); however, acoustic experts may be interested in the structure of frontal zones also from the point of view of estimating or predicting the levels of incoming sound signals in cases where a strong frontal zone is found between the source and the receiver. As demonstrated by Fenner [114], in the frontal zone of the Gulf Stream, where not only the main frontal interface, but also the frontal interfaces of a warm anticyclonic ring lay in the path of a sound beam, the range of sound speed variability amounted to about 40 m/s, while the depth of the sound channel axis varied over a 500-metre layer. When the source of sound was outside the warm ring and the receiver within its core, the incoming signal was enhanced by several decibels, which can be attributed to the focusing effect of the frontal interfaces surrounding the core. Additional acoustic effects, mainly dispersive, should be associated with the intrusive-type finestructure which is always encountered in abundance near frontal interfaces.

Fronts should be of special interest to those studying internal gravity waves in the ocean. Frontal interfaces in a transient state should radiate a broad spectrum of internal oscillations in the inertio-gravity range. We have already dwelt on the high intensity of internal oscillations in oceanic frontal zones (see section 4.2). On the other hand, fronts have a significant effect on the propagation of oscillatory motions in the ocean, and on their nature. For example, inertial oscillations can undergo strong refraction and even complete internal reflection at frontal interfaces. The horizontal velocity shear in a stream directed along the front can greatly alter the effective frequency of inertial oscillations. It cannot be ruled out that motions with inertial and near-inertial periods actively participate in the destruction of oceanic fronts through dissipative processes which take place on the finestructure. All of these important aspects of the coexistence and interaction of fronts and internal oscillations are still very poorly researched.

Concentrating suspensions in convergence and downwelling zones, frontal interfaces cannot but affect sedimentation. Under sharp fronts which have retained a quasi-stationary geographic position in the ocean for a least several tens of millennia, not to mention millions of years, there should be easily discernible qualitative and quantitative anomalies in the structure of sedimentary layers at the bottom of the ocean. The interpretation of these anomalies could help to establish the position of paleofronts in the paleo-ocean in various remote geological epochs. This could shed some light on complex questions concerning the evolution of the general circulation of the waters of the World Ocean due to the evolution of various forms of oceanic basins as a result of continental drift. This would be an additional contribution to our knowledge of the history of the Earth's oceans and the planet itself.

An especially close relationship should exist between physical and biological research on frontal zones. Apparently, coordinates associated with the characteristics of the physical structure of frontal zones, instead of geographic coordinates, should determine the network of stations for biological sampling and biochemical measurements. The high spatio-temporal variability of the physical fields in frontal zones makes it very difficult to fulfil this requirement. However, there is reason to assume that biological and biochemical sampling and measurements themselves can serve as a valuable aid to the physicist when determining the boundaries of frontal structures. This has been confirmed by the author's own participation in the 17th biological survey of the research vessel "Akademik Kurchatov" in 1973-1974.

We have already mentioned that frontal regions in which waters with different characteristics come into contact are distinguished by high biological productivity. This was noted several decades ago [25, 232]. However, apart from purely quantitative differences, frontal ecosystems are also characterized by a highly significant qualitative diversity. The quantity of plankton in convergence zones increases either because of the mechanical accumulation of plankton entrained by horizontally convergent currents [15, 137], or because of the complex biological processes that arise from the contact of biological communities of diverse maturity found at different stages of succession (development). At the same time, the energy flux between the communities in contact is from the younger to the more mature community, and the more mature system "exploits" the younger one [14, 119]. At the same time, the "maturation" of the younger system is delayed, and it remains in a state of dynamical stability at a low level of development, continuing to export energy (organic matter) to the more mature system.

In turn, the more mature system develops more quickly than if it were restricted to its own resources, and its productivity appears to be high, especially at the final trophic levels such as plankton-eating fish, macroplankton, etc.

Depending on the micro- and mesoscale physical structure of any definite frontal zone and the specific character of the population of the waters in contact, the biological processes and end results of the interaction of two ecosystems can differ significantly. In some cases, as, for example, on the edge of the Peruvian shelf, we observe the development of mass accumulations of macroplanktonic crustaceans, as well as the concentration of plankton-eating and large carnivorous fish and squid [13, 78]; in other cases, tuna [240] or whales [132] turn out to be the final links of the communities concentrated in a frontal zone.

Neither can we ignore the fact that high-speed downwelling, which can entrain plankton usually found near the surface to depths of $100 - 200 \ m$ and more, takes place at the frontal interfaces themselves. For example, the downwelling in the frontal zone of a coastal upwelling is particularly intensive from the seaward side of the frontal interface. On the seaward side of the frontal interface, the pycnocline is close to the surface itself, as is promoted by the divergence in the Ekman layer observed here (see section 2.9). All this gives rise to a sharp qualitative and quantitative asymmetry in the development of communities of living organisms with respect to the frontal interface.

The degree of maturity of the interacting communities, the productivity of the younger one, the vertical structure and intensity of mixing and vertical water motions in the zone of contact, the velocity of the horizontally convergent flow in the direction of the frontal interface, the appearance and lifetime of local eddies and probably many other factors not yet known to us should have the most significant effect on the formation and development of mixed frontal zone communities, and on the production of their final trophic links (fish, squid). All of these questions are still rather poorly researched, and require urgent and profound study both from the point of view of collecting new facts by specialized observations and measurements, and by the method of numerical modelling for a better insight into the functioning of all the elements of frontal ecosystems.

Another poorly researched problem is the role of frontal interfaces and frontal zones (especially quasi-stationary zones of climatic origin) in the behaviour of fish and other marine animals which migrate over long distances. R.M. Laurs' paper presented at the international discussion on oceanic fronts, organized by SCOR in Brest (France) in 1978, contained information about the transoceanic migrations of tuna populations along fronts in the Pacific Ocean, with the direction of the migrations always determined by the sensitivity to temperature and its contrasts.

5.3. Future research tasks

A number of concrete and urgent problems of modern frontological research are formulated in ref. [82]. In the light of the material presented in this book, I would like to review the prospects of future investigations in this field, proceeding from the most general fundamental and practical problems and then gradually passing on to the most specific and concrete scientific problems.

First of all, it seems that researchers today are clearly in need of large-scale systematic observations of oceanic fronts, which could provide answers to the vast number of questions that have already accumulated as a result of earlier field work and past attempts to develop the theory and numerical modelling of oceanic fronts. Therefore, the organization of an extensive program of such observations is a primary task for the whole complex of physical oceanic sciences in cooperation with other branches of science. An effective combination of traditional ship surveys with remote sensing measurements and observations, involving the popular practice of directing vessels at the objects of study with the help of rapid remote sensing information should be one of the important methods of planned expeditions.

Only after the complete assimilation of the results of field measurements and observations carried out within the framework of this type of program, which would take the next 5 − 10 years, would we be able to expect substantial progress in the theoretical field and in numerical modelling. Approximately the same or an even smaller period of time is required for the numerical models of oceanic water circulation to become "front-resolving" on the one hand, and to take into account the integral effect of small-scale fronts on the other.

Numerous practical problems and economic goals today serve as important incentives for detailed field investigations of oceanic fronts. These include:

1) the urgent need for further development of efficient fisheries on a scientific basis in the open ocean, due to the universal introduction of 200-mile economic zones in coastal waters, where the right to fish and develop other resources belongs exclusively to the coastal states;

2) the necessity to gradually progress from fishing to modern aquaculture which includes not only the raising of "sessile" forms of commercial animals and plants, but also the breeding of commercial species of fish and crustaceans, at first in coastal waters, and then in the open ocean;

3) the development of a safe and profitable technology for extracting mineral resources from the bottom of the World Ocean;

4) the growing threat of anthropogenic pollution of ocean waters, resulting in dangerous concentrations of toxic products primarily in frontal zones where accumulations of commercial species are also observed;

5) the urgent need to improve the methods and the quality of long-range forecasting of weather and climatic changes, including the prediction of catastrophic

weather and climatic phenomena such as hurricane, prolonged droughts, floods, extremely severe winter conditions, etc.

It follows from this that even fundamentally-oriented field investigations of oceanic fronts should at the same time provide us with practical characteristics and estimates, such as the coefficients of heat and mass transfer and the coefficient of momentum exchange at frontal interfaces, the frequency of occurrence of fronts in different areas of the World Ocean, the lifetimes of frontal interfaces of various scales, the ranges of variability of the main characteristics of individual fronts, etc. From here stem the main and most general fundamental goals of physical research on oceanic fronts for the next 5 − 10 years. They are:

1) detailed investigations of the **three-dimensional** structure of the temperature, salinity, density and velocity fields in frontal zones, especially in direct proximity to frontal interfaces; determination of the positions and boundaries of near-frontal zones of convergences and divergences (upwelling) in typical situations; investigation of the thermohaline finestructure near fronts;

2) investigation of the fundamental dynamics of quasi-stationary frontal interfaces, with special emphasis on the physical mechanisms that maintain a transverse convergent circulation which keeps the frontal interfaces sharp;

3) investigation of the dynamics of instability of frontal interfaces and the manifestations of this instability by the formation of meanders, eddies and an intrusive structure, as well as the characteristics of the variability of physical fields in frontal zones due to meanders, eddies and intrusive structure.

4) investigation of the mechanisms of heat and salt transfer and momentum exchange across frontal interfaces; estimations of the intensity of vertical and horizontal cross-frontal transfer under typical conditions;

5) investigation of the local mechanisms of frontogenesis within frontal zones; values of the intensity of the deformation field typical of the ocean;

6) investigation of the processes participating in the destruction of frontal interfaces (frontolysis), and determination of the characteristic periods of frontolysis for different scales of the phenomenon;

7) investigation of the interaction of different types of waves, tides and atmospheric factors with frontal interfaces and the jet stream near fronts;

8) numerical modelling of frontal phenomena in order to study the effect of various factors on the dynamics of frontal zones and the prediction of their evolution;

9) investigation of the interaction of the ocean and atmosphere in the region of the most developed quasi-stationary frontal zones of climatic origin, in connection

with the variability of the intensity and positions of the main front in these zones;

10) improvement of the methods and means of physical investigations of frontal phenomena in the ocean, including the methods of remote sensing measurements, in order to obtain prompt three-dimensional synoptic images of the structure of frontal interfaces and trace their evolution in time.

Considering the great importance of biocommercial investigations in connection with oceanic fronts, we can also outline the most important biological problems which can be solved in parallel and in close relation to the solution of the main physical tasks:

1) investigation of the structural and functional characteristics of plankton communities and their dynamics in various parts of frontal zones;

2) study of the contacts, interaction and trophic relationship of communities varying in maturity in the zone of frontal contact; determination of the role of exploitation of young communities by more mature ones;

3) study of the characteristics of time variability (succession) of mixed communities in frontal zones, and the processes that lead to high productivity of the higher trophic links of the communities (fish, benthos) near fronts;

4) quantitative investigation of the effect of the variability of the structure of frontal zones on the evolution of frontal ecosystems;

5) development of forecasting methods of ecosystem modelling for communities of frontal zones, including the higher trophic levels of the communities (fish), in order to predict the distribution of their concentrations in frontal zones;

6) evaluation of frontal zones and interfaces as biogeographic and ecological boundaries for population distribution in the pelagic ocean;

7) development of recommendations for the detection of commercially promising frontal zones in the open ocean.

In the context of this book, it would not be relevant to give too many details about the biological goals that have been set. A special biological study should be devoted to this. However, we should point out some of the specific physical problems (apart from those indicated above) which must be solved before we can make any significant advance in the field of marine biology. One of these tasks is reliable evaluation of the vertical velocities near frontal interfaces. According to Pingree, who dared to dive into the water close to the line of a front at the surface of the ocean, the water sinking along the interface is capable of entraining even large logs of driftwood to considerable depths. It is important to establish whether such high velocities of downwelling are an exception or a rule, and also the width of the near-frontal zones in which such velocities can be observed. It is also important to have a reliable quantitative idea of the circulation in the plane normal to the frontal interface from the ocean surface and to depths where the frontal interface is no longer apparent, in a band at least several kilometres wide on both sides of the front. Another task is spatial mapping of the frontal interface (or interfaces) within a given frontal zone. In the literature, we have so far dealt mostly with individual profiles across the frontal zone. Such profiles can hardly satisfy biologists. Finally, it would be desirable to establish for certain whether synoptic-scale oceanic fronts behave like "warm" frontal interfaces, or "cold" ones, and what the typical velocities of their spatial migration are in this case.

There are other specific tasks, but they are just as interesting from the point of view of hydrodynamicists:

- It would be very interesting to study the spatial coherence of the intrusive finestructure near frontal interfaces. It is also necessary to present a detailed phenomenological description of the "calving" of intrusions from the frontal interfaces, as well as a physical explanation of this phenomenon.

- The properties of the internal Ekman layers which are found next to frontal interfaces and large intrusions require special investigation.

- It is necessary to establish the conditions of frontogenesis in regions with a general anticyclonic vorticity, e.g. to the right of the axis of large western boundary currents such as the Gulf Stream and Kuroshio.

- It is necessary to study the mechanisms which give rise to the various manifestations of fronts on the surface of the ocean (slicks, bands of trash and foam, rough water and rips, colour and brightness contrasts, etc.).

- It would be very interesting to try to obtain with the help of instruments direct evidence in nature for the functioning of double-diffusive convection combined with intrusive interleaving near frontal interfaces, as well as proof of the effectiveness of cabbeling in maintaining fronts in a sharp state.

- It would be interesting to study the specific forms of cloud cover, which form above particularly sharp oceanic fronts.

- It is necessary to fill in the serious gaps in our knowledge about bottom (benthic) fronts; investigation of their dynamics can help us establish the dynamics of near-surface oceanic fronts.

This far from complete list of tasks, like the contents of the preceding chapters and sections of the book, should undoubtedly help the reader realize the complexity, many-sidedness and fascination of studying the fronts of the World Ocean. The author hopes that the impression left by the book will serve as an incentive towards filling in those obvious gaps in our present knowledge of this topic.

REFERENCES

1. Abramov R.V. Bliznichenko V.I., Bulatov R.P., Kazachkina L.I. Temperature inhomogeneities at the surface of the Atlantic Ocean. *Okeanologicheskiye issledovaniya*, 1975, No. 28, p. 62-72.

2. Antonov V.S. Distribution of fluvial waters in arctic seas. *Trudy ANII*, 1957, v. 208, p. 25-52.

3. Antonov V.S. Anomalies of high river discharge in the Arctic and Subarctic region of Siberia. *Problemy Arktiki i Antarktiki*, 1974, No. 18, p. 24-30.

4. Baranov Ye.I. Short-period oscillations of the Gulf Stream front during the winter-spring season of 1963. *Okeanologiya*, 1966, v. 6, No. 2, p. 228-233.

5. Baranov Ye.I. A study of eddies in the frontal zone of the Gulf Stream. *Okeanologiya*, 1967, v. 7, No. 1, p. 78-83.

6. Baranov Ye.I. Dynamics and structure of the waters in the frontal zone of the Gulf Stream. *Okeanologicheskiye issledovaniya*, 1971, No. 22, p. 94-153.

7. Baranov Ye.I. Mean monthly positions of hydrologic fronts in the northern part of the Atlantic Ocean. *Okeanologiya*, 1972, v. 12, No. 2, p. 217-224.

8. Baranov Ye.I., Shmatko M.A. Investigation of the thermal structure of the frontal zone of the Gulf Stream. *Okeanologiya*, 1966, v. 6, No. 5, p. 630-634.

9. Barenblatt G.I. Dynamics of turbulent patches and an intrusion in stably stratified fluid. *Izv. AN SSSR. Atmosphere and Ocean Physics*, 1978, v. 14, No. 2, p. 195-206.

10. Belyayev V.S., Nozdrin Yu.V. Vertical spectra of the finestructure of the temperature field in a frontal zone. *Okeanologiya*, 1979, v. 19, No. 4, p. 549-557.

11. Bulatov N.V. Eddy structure of a subarctic front in the northwestern part of the Pacific Ocean. *Uch. zap. LGU*, 1980, 27, No. 403, p. 61-72.

12. Bulgakov N.P. The cabbeling phenomenon. *Izv. AN SSSR, ser. geofiz.*, 1960, No. 2, p. 346-352.

13. Vinogradov M.Ye. 17th cruise of the research vessel "Akademik Kurchatov" (investigation of plankton communities in areas of intense upwelling in the eastern equatorial Pacific). *Okeanologiya*, 1974, v. 14, No. 6, p. 1082-1086.

14. Vinogradov M.Ye. A spatial-dynamical aspect of the existence of communities in the ocean. IN: Oceanology, Biology of the Ocean, "Nauka" Publishers, Moscow, 1977, v. 2, p. 34-43.

15. Voronina N.M. Communities of temperate and cold waters of the southern hemisphere. Ibid, p. 68-90.

16. Voropayev S.I., Zatsepin A.G., Pavlov A.M., Fedorov K.N. Laboratory simulation of structure-forming processes in a stably stratified fluid. IN: Investigations of the Variability of Physical Processes in the Ocean. Ed. K.N. Fedorov. Moscow, 1978, p. 112-121.

17. Ginzburg A.I. et al. Characteristics of the thermohaline structure of the frontal interfaces of warm Gulf Stream rings /Ginzburg A.I., Zatsepin A.G., Kuz'mina N.P., Sklyarov V.Ye., Fedorov K.N./. *Okeanologicheskiye issledovaniya*, 1981, No. 34, p. 33-48.

18. Ginzburg A.I., Zatsepin A.G., Sklyarov V.Ye., Fedorov K.N. Effects of precipitation on the near-surface layer of the ocean. *Okeanologiya*, 1980, v. 20, No. 5, p. 828-836.

19. Gruzinov V.M. Frontal Zones of the World Ocean. Moscow, "Gidrometeoizdat" Publishers, 1975, 198 p.

20. Zhurbas V.M., Kuz'mina N.P. On the spreading of a mixed patch in a rotating stably stratified fluid. *Izv. AN SSSR, Atmosphere and Ocean Physics*, 1981, v. 17, No. 3, p. 286-295.

21. Zatsepin A.G., Krasnopevtsev A.Yu., Fedorov K.N. Observations of fronts in the POLYMODE area. *Okeanologicheskiye issledovaniya*, 1979, No. 30, p. 86-88.

22. Zatsepin A.G., Fedorov K.N. Conditions for the formation of finestructure in the ocean by the collapse of mixed patches. *DAN SSSR*, 1980, v. 252, No. 4, p. 989-992.

23. Zatsepin A.G., Fedorov K.N., Voropayev S.I., Pavlov A.M. Experimental study of the spreading of a mixed patch in stratified fluid. *Izv. AN SSSR, Atmosphere and Ocean Physics*, 1978, v. 14, No. 2, p. 234-237.

24. Zatsepin A.G., Shapiro G.I. Investigation of an axi-symmetric intrusion in stratified fluid. *Ibid*, 1982, v. 18, No. 1, p. 101- 105.

25. Zenkevich L.A., Brotskaya V.A. Quantitative estimation of the bottom fauna of the Barents Sea. *Trudy VNIRO*, 1939, v. 4, p. 5-126.

26. Zubov N.N. Cabbeling of seawater of different temperature and salinity. Leningrad, "Gidrometeoizdat" Publishers, 1957, 40 p.

27. Zubov N.N., Sabinin K.D. Calculation of the cabbeling of seawater. Moscow, "Gidrometeoizdat" Publishers, 1958, 37 p.

28. Ivanov Yu.A., Neiman V.G. Frontal zones of the Southern Ocean. IN: Antarctica, Moscow, "Nauka" Publishers, 1964, p. 98-109.

29. Kaz'min A.S., Sklyarov V.Ye. An experiment in utilizing visual information from the earth-orbiting satellite "Meteor" for studying oceanic phenomena. *Issled. Zemli iz kosmosa*, 1981, v. 2, No. 6, p. 48-57.

30. Kamenkovich V.M., Reznik G.M. Rossby waves (Section 6. Baroclinic instability of large-scale currents). Okeanologiya, Ocean Physics, v. 2. Hydrodynamics of the Ocean. Moscow, "Nauka" Publishers, 1978, p. 344-358.

31. Karabasheva E.I., Paka V.T., Fedorov K.N. Are thermal fronts frequently encountered in the ocean? *Okeanologiya*, 1978, v. 18, No. 6, p. 1004-1012.

32. Koshlyakov M.N. et al. Synoptic ocean fronts in the POLYMODE polygon (Feb., March, 1978) /Koshlyakov M.N., Borisenko Yu.D., Brekhovskikh A.A. et al./. *Okeanolog. issled.*, 1980, No. 31, p. 42-55.

33. Koshlyakov M.N., Grachev Yu.M. Mesoscale currents in the hydrophysical polygon in the tropical Atlantic. IN: Atlantic Hydrophysical Polygon-70. Moscow, "Nauka" Publishers, 1974, p. 163-180.

34. Krasnopevtsev A.Yu., Fedorov K.N. Thermohaline inhomogeneities of the spatial structure of the upper layer of the ocean. IN: Investigation of the Variability of Physical Processes in the Ocean. Ed. K.N. Fedorov, 1978, p. 59-73.

35. Krasnopevtsev A.Yu., Fedorov K.N. The nature of the inhomogeneities of horizontal temperature and salinity distribution in the surface layer of the ocean. *Okeanolog. issled.*, 1979, No. 30, p. 82-85.

36. Kuz'mina N.P. Oceanic frontogenesis. *Izv. AN SSSR, Atmosphere and Ocean Physics*, 1980, v. 16, No. 10, p. 1082-1090.

37. Kuz'mina N.P. A nonlinear numerical model of oceanic frontogenesis. *Izv. AN SSSR, Atmosphere and Ocean Physics*, 1981, v. 17, No. 12, p. 1318-1325.

38. Kuz'mina N.P., Kutsenko B.Ya. Some models of oceanic frontogenesis. IN: Investigation of the Variability of Physical Processes in the Ocean. Ed. K.N. Fedorov, Moscow, Publishing House of the Institute of Oceanology of the USSR Academy of Sciences, p. 83-99.

39. Lineikin P.S. On the nonlinear wave disturbances in the main oceanic thermocline. *DAN SSSR*, 1978, v. 241, No. 6, p. 1436-1439.

40. Mederich V.S. Nonlinear evolution of large-scale density anomalies in the ocean. *Izv. AN SSSR, Atmosphere and Ocean Physics*, v. 14, No. 11, 1978, p. 1219-1222.

41. Monin A.S. Weather Forecasting as a Problem of Physics. Moscow, "Nauka" Publishers, 1969, 184 p.

42. Monin A.S., Piterbarg L.I. Statistical description of internal waves. *DAN SSSR*, 1977, v. 234, No. 3, p. 564-567.

43. Monin A.S., Fedorov K.N. On the finestructure of the upper quasi-homogeneous layer in the ocean. *Izv. AN SSSR, Atmosphere and Ocean Physics*, 1973, v. 9, No. 4, p. 442-444.

44. Monin A.S., Fedorov K.N., Shevtsov V.P. On the vertical meso- and microstructure of oceanic currents. *DAN SSSR*, 1973, v. 208, No. 4, p. 833-836.

45. Naumenko M.F., Bikulov B.I., Chigarkov K.I. On the large-scale spatial structure of the temperature field of the surface layer. IN: TROPEX-74, v. 2, 1976, p. 25-31. Leningrad, "Gidrometeoizdat" Publishers.

46. Nezhikhovsky R.A. The Neva River. Leningrad, "Gidrometeoizdat" Publishers, 1955, 94 p.

47. Nelepo B.A., Kuftarkov Yu.M., Kosnyrev V.K. On the effect of mesoscale eddies on the temperature of the ocean surface. *Izv. AN SSSR, Atmosphere and Ocean Physics*, 1978, v. 14, No. 7, p. 768-777.

48. Nits G.S., Polosin A.S. On the definition of the terms front and frontal zone. *Bull. TsNIITEIRKh "Promysl. okeanologiya i podvodn. tekhnika"*, 1972, ser. 9, No. 6, p. 1-12.

49. Seidov D.G. Synoptic oceanic eddies. A numerical experiment. *Izv. AN SSSR, Atmosphere and Ocean Physics*, 1980, v. 16, No. 1, p. 73-87.

50. Sklyarov V.Ye. Fedorov K.N. Three-dimensional structure of the frontal zone of the Gulf Stream based on synchronous satellite and ship data. *Issled. Zemli iz kosmosa*, 1980, v. 1, No. 3, p. 5-13.

51. Solov'yov A.V. Thermal finestructure of the surface layer of the ocean in the vicinity of the POLYMODE-77 polygon. *Izv. AN SSSR, Atmosphere and Ocean Physics*, 1979, v. 15, No. 7, p. 750-757.

52. Stepanov V.N. Main convergences and divergences of the waters of the World Ocean. *Bull. Okeanograf. komissii AN SSSR*, 1960, No. 6, p. 15-22.

53. Stepanov V.N. The World Ocean. Moscow, "Znaniye" Publishers, 1974, pp. 95-99, 255.

54. Stommel H. Horizontal temperature changes in a mixed layer of the southern part of the Pacific Ocean. *Okeanologiya*, 1969, v. 9, No. 1, p. 97-102.

55. Tareyev B.A. Unstable Rossby waves and the non-stationarity of ocean currents. *Izv. AN SSSR, Atmosphere and Ocean Physics*, 1965, v. 1, No. 4, p. 426-438.

56. Fedorov K.N. Wind currents in a sea of variable depth. *Izv. AN SSSR, ser. geol.*, 1955, No. 3, p. 223-233.

57. Fedorov K.N. On the multilayer structure of temperature inversions in the ocean. *Izv. AN SSSR, Atmosphere and Ocean Physics*, 1970, v. 6, No. 11, p. 1178-1188.

58. Fedorov K.N. A case of convection with the formation of temperature inversions due to local instability in the oceanic thermocline. *DAN SSSR*, 1971, v. 198, No. 4, p. 822-825.

59. Fedorov K.N. New evidence for the existence of lateral convection in the ocean. *Okeanologiya*, 1971, v. 11, No. 6, p. 994-998.

60. Fedorov K.N. One of the mechanisms for the formation of mesoscale horizontal temperature inhomogeneities in the upper layer of the ocean. *Okeanologiya*, 1976, v. 16, No. 3, p. 403-407.

61. Fedorov K.N. The Thermohaline Finestructure of the Ocean. Leningrad, "Gidrometeoizdat" Publishers, 1976, 184 p. Pergamon, 1978, 170 p.

62. Fedorov K.N. Finestructure of hydrophysical fields in the ocean. IN: Oceanology, Ocean Physics, v. 1. Hydrophysics of the Ocean. Moscow, "Nauka" Publishers, 1978, p. 113-147.

63. Fedorov K.N. Slow relaxation of the thermal wake of a hurricane in the ocean. *DAN SSSR*, 1979, v. 245, No. 4, p. 960-963.

64. Fedorov K.N. Expectations and realities of cosmic oceanology. *Issled. Zemli iz kosmosa*, 1980, v. 1, No. 1, p. 64-78.

65. Fedorov K.N. Axisymmetric intrusions in linear-stratified fluid. 5th All-Union Conference on Theoretical and Applied Mechanics. Alma-Ata, 27 May–3 June 1981. Annotations of reports. Alma-Ata, "Nauka" Publishers, 1981, p. 342-343.

66. Fedorov K.N., Bubnov V.A., Ginzburg A.I., Paka V.T. On the frontal systems of Gulf Stream rings. *DAN SSSR*, 1979, v. 246, No. 5, p. 1227-1231.

67. Fedorov K.N., Bubnov V.A., Prokhorov V.I., Osadchiy A.S. Hydrophysical conditions in the Peruvian coastal region. *Trudy IOAN*, 1975, v. 102, p. 51-55.

68. Fedorov K.N. et al. An experiment in recording the temperature and salinity of the surface layer of the ocean with an AIST probe. /Fedorov K.N., Ginzburg A.I., Zatsepin A.T. et al./. *Okeanologiya*, 1979, v. 19, No. 1, p. 156-163.

69. Fedorov K.N., Ginzburg A.I., Peterbarg L.I. On the physical nature of "inhomogeneities of calm weather" in an oceanic temperature field. *Okeanologiya*, 1981, v. 21, No. 2, p. 203-210.

70. Fedorov K.N., Kuz'mina N.P. Fronts in the ocean. IN: Mesoscale Variability of an Oceanic Temperature Field. Moscow, IOAN, 1977, p. 33-53.

71. Fedorov K.N., Kuz'mina N.P. Oceanic fronts. Scientific results. *Okeanologiya*, 1979, v. 5, p. 4-44. Moscow, VINITI.

72. Fedorov K.N., Plakhin Ye.A., Prokhorov V.I., Sedov V.G. Characteristics of thermohaline stratification in the polygon area in the tropical Atlantic. IN: Atlantic Hydrophysical POLYGON-70. Moscow, "Nauka" Publishers, 1974, p. 236-286.

73. Four-language Encyclopaedic Dictionary of Terms in Physical Geography. Moscow, "Sov. entsiklopediya" Publishers, 1980, 703 p.

74. Shapiro G.I. Dynamics of a small-scale oceanic front exposed to the effect of the wind. *Izv. AN SSSR, Atmosphere and Ocean Physics*, 1981, v. 17, No. 4, p. 419-427.

75. Shevtsov V.P. Investigations of the finestructure of hydrophysical fields by the remote acoustic method. IN: Finestructure and Synoptic Variability of Seas. Expanded abstracts of papers presented at the All-Union Seminar-Symposium. Tallin, 1980, p. 185-189.

76. Shtokman V.B. Wind set-up and horizontal circulation in a shallow inland sea. *Izv. AN SSSR, ser. geogr. i geofiz.*, 1941, No. 1, p. 69-87.

77. Shutko A.M. Investigation of the water surface by ultrahigh-frequency radiometry methods (summary). *Radiotekhnika i elektronika*, 1978, v. 23, No. 10, p. 2107-2119.

78. Shushkina E.A. et al. Functional characteristics of plankton communities of the Peru upwelling /Shushkina E.A., Vinogradov M.Ye., Sorokin Yu.I., Lebedeva L.P., Mickheyev V.N. *Okeanologiya*, 1978, v. 18, No. 5, p. 886-902.

79. Van Aken H.M. The thermohaline fine structure in the North Rockall trough. – Doctoral Thesis, Univ. Utrecht, 1981. 160 p.

80. Ambar I., Howe M.R. Observations of the Mediterranean outflow. *Deep Sea Res.*, 1979, vol. 26, No. 5A, p. 535-568.

81. Amos A.F., Langseth M., Markl R. Visible oceanic saline fronts. *Studies in Phys. Oceanogr.*, 1972, vol. 1, p. 49-62.

82. Anonymous. Proceedings of the Workshop on Oceanic Fronts in Coastal Processes, May 25-27, 1977. IN: Oceanic Fronts in Coastal Processes. Ed. M. Bowman, W. Esaias. Springer-Verlag, 1978, p. 6-13.

83. Bang N.D. Characteristics of an intense frontal system in the upwelling regime west of Cape Town. *Tellus*, 1973, vol. 25, No. 3, p. 256-265.

84. Beardsley R.C., Hart J. A simple theoretical model for the flow of an estuary into a continental shelf. *J. Geophys. Res.*, 1978, vol. 83, No. C2, p. 873-884.

85. Becker G.A., Prahm-Rodewald G. Oceanic fronts. Saline fronts in the German Bight. *Seewart*, 1980, v. 41. No. 1, p. 12-21.

86. Beckerle J.C. Prediction of mid-oceanic front passage confirmed in near-surface current measurement. *J. Geophys. Res.*, 1972, vol. 77, No. 9, p. 1637-1646.

87. Bergeron T. Weather analysis in its three-dimensional interrelationships. *Geophys. Publ.*, 1928, v. 5, No. 6.

88. Bernstein R.L., White W.B. Time and length scales of baroclinic eddies in the Central North Pacific Ocean. *J. Phys. Oceanogr.*, 1974, vol. 4, No. 4, p. 613-624.

89. Bondar C. Consideration theorique sur la dispersion d'un courant liquid de densité réduite et à niveau libre, dans un bassin contenant un liquide d'une plus grande densité. IN: Hydrology of Deltas, IASH/Unesco, Paris, 1970, p. 246-257.

90. Bowman M.J. Spreading and Mixing of the Hudson River effluent into the New York Bight. IN: Hydrodynamics of Estuaries and Fjords. Ed. J.C.J. Nihoul. Elsevier Oceanogr. ser., 1978, vol. 23, p. 373-386.

91. Bowman M.J., Iverson R.L. Estuarine and plume fronts. IN: Oceanic Fronts in Coastal Processes. Ed. M.Bowman, W. Esaias. Springer-Verlag, 1978, p. 87-104.

92. Bowman M.J., Kibblewhite A.C., Ash D.E. M_2 tidal effects in greater Cook Strait, New Zealand. *J. Geophys. Res.*, 1980, vol. 85, No. C5, p. 2728-2742.

93. Bowman M.J., Okubo A. Cabbeling at thermohaline fronts. *J. Geophys. Res.*, 1978, vol. 83, No. C12, p. 6173-6178.

94. Briscoe M.G., Johannessen O.M., Vincenzi S. The Maltese oceanic front: a surface description by ship and aircraft. *Deep-Sea Res.*, 1974, vol. 21, No. 4, p. 847-862.

95. Cairns J.L. Variability in the Gulf of Cadiz: internal waves and globs. *J. Phys. Oceanogr.*, 1980, vol. 10, No. 4, p. 579-595.

96. Coachman L.K., Charnell R.L. On lateral water mass interaction – a case study, Bristol Bay, Alaska. *J. Phys. Oceanogr.*, 1979, vol. 9, No. 2, p. 278-297.

97. Coachman L.K., Kinder T.H. Schumacher J.D., Tripp R.B. Frontal systems of the southeastern Bering Sea shelf. Stratified Flows, 2nd IAHR Symp., Trondheim, 1980, vol. 2.

98. Craig H., Chung Y., Fiadeiro M. A benthic front in the South Pacific. *Earth and Planetary Sci. Lett.*, 1972, vol. 16, p. 50-65.

99. Cromwell T., Reid J.L. A study of oceanic fronts. *Tellus*, 1956, vol. 8, No. 1, p. 94-101.

100. Csanady G.T. Wind effects on surface to bottom fronts. *J. Geophys. Res.*, 1978, vol. 83, No. C9, p. 4633-4640.

101. Curtin T.B., Mooers C.N.C. Coastal upwelling experiment I and II. Surface Hydrographic Fields, Data Report. Rosenstiel School Marine and Atm. Sci., Univ. Miami Sci. Rep., 1974, p. 94.

102. Danielsen E.F. Project Springfield Report. Defense Atomic Support Agency, Washington, D.C., 20301, 1964, 97p.

103. Dazzi R., Frassetto R., Mioni F., Tomasino M. The study of a river plume by means of numerical models and remote sensing. *Rapp. et Proc. Verbe. Reun. Comis. Int. Explor. Sci. Mer. Mediterr.*, Monaco, 1979, vol. 25-26, No. 7, p. 127-130.

104. Defant A. Stable stratification in oceans and associated current systems. *Veröff. Inst. Meer. Univ. Berlin*, 1929, Neufolge A, H. 19.

105. Defant A., Rossby C.G. Dynamics of stationary ocean currents in light of experimental study of currents. *Annalen Hydrographic und Marit. Met.*, 1937, v. 65, 2, p. 65.

106. Dessureault J.G. "Batfish" a depth controllable towed body for collecting oceanographic data. *Ocean Eng.*, 1976, vol. 3, p. 99-111.

107. Dickson R.C., Gurbutt P.A., Pillai V.N. Satellite evidence of enhanced upwelling along the European continental slope. *J. Phys. Oceanogr.*, 1980, vol. 10, No. 5, p. 813-819.

108. Eliassen A. On the formation of fronts in the atmosphere. IN: The Atmosphere and the Sea in Motion. Ed. B.Bolin. Rockefeller Inst. Press; Oxford Univ. Press., 1959, p. 277-287.

109. Ewans M.W. Surface salinity and temperature "signatures" in the Northeastern Pacific. *J. Geophys. Res.*, 1971,vol. 76, No. 15, p. 3456-3461.

110. Farmer D., Smith J.D. Nonlinear internal waves in a fjord. IN: Hydrodynamics of Estuaries and Fjords. Ed. J.C.J. Nihoul. Elsevier Oceanogr. Ser., 1978, No. 23, p. 465-493.

111. Farmer D.M., Smith J.D. Tidal interaction of stratified flow with a sill in Knight Inlet. *Deep-Sea Res.*, 1980,vol. 27, No. 3/4A, p. 239-254.

112. Fasham M.G.R., Pugh P.R. Observation on the horizontal coherence of chlorophyll-A and temperature. *Deep-Sea Res.*, 1976, vol. 23, No. 6, p. 527-538.

113. Fearnhead P.G. On the formation of fronts by tidal mixing around the British Isles. *Deep-Sea Res.*, 1975, vol. 22, p. 311-321.

114. Fenner D.F. Sound speed structure across an anticyclonic eddy and the Gulf Stream north wall. *J. Geophys. Res.*, 1978, vol. 83, No. C9, p. 4599-4606.

115. Flagg C.N., Beardsley R.C. On the stability of the shelf water/slope water front south of New England. *J. Geophys. Res.*, 1978, vol. 83, No. C9, p. 4623-4632.

116. Fofonoff N.P. Some properties of sea water influencing the formation of Antarctic bottom water. *Deep-Sea Res.*, 1956, vol. 4, No. 1, p. 32-35.

117. Ford W.L., Longard J.R., Banks R.E. On the nature, occurence and origin of cold low salinity water along the edge of the Gulf Stream. *J. Mar. Res.*, 1952, vol. 11, No. 3, p. 281-293.

118. Foster T.D. An analysis of the cabbeling instability in sea water. *J. Phys. Oceanogr.*, 1972, vol. 2, No. 3, p. 294-301.

119. Frontier S. Interface entre deux écosystemes: exemple dans le domaine pelagique. *Ann. Inst. Oceanogr.*, Paris, 1978, vol. 54, No. 2, p. 95-106.

120. Fuglister F.C. Multiple currents in the Gulf Stream system. *Tellus*, 1951, vol. 3, No. 4, p. 230-233.

121. Fuglister F.C. Cyclonic rings formed by the Gulf Stream 1965-1966. Ed. G. Arnold. Studies in Phys. Oceanogr. A tribute to G. Wüst on his 80th birhday, 1972, vol. 1, p. 137-168.

122. Fuglister F.C., Worthington L.V. Some results of a multiple ship survey of the Gulf Stream. *Tellus*, 1951, vol. 3, No. 1, p. 1-14.

123. Gade H.G. On some oceanographic observations in the southeastern Caribbean Sea and the adjacent Atlantic Ocean with special reference to the influence of the Orinoco River. *Bol. Inst. Oceanogr.*, Cumanà, Venezuela, 1961, vol. 1, No. 2, p. 287-342.

124. Gargett A.E. Microstructure and fine structure in an upper ocean frontal regime. *J. Geophys. Res.*, 1978, vol. 83, No. C10, p. 5123-5134.

125. Garrett C., Horne E. Frontal circulation due to cabbeling and double diffusion. *J. Geophys. Res.*, 1978, vol. 83, No. C9, p. 4651-4656.

126. Garvine R.W. Physical features of the Connecticut River outflow during high discharge. *J.Geophys. Res.*, 1974, vol. 79, No. 6, p. 831-846.

127. Garvine R.W. Dynamics of small-scale oceanic fronts. *J. Phys. Oceanogr.*, 1974, vol. 4, No. 4, p. 557-569.

128. Garvine R.W. Observations of the motion field of the Connecticut River plume. *J. Geophys. Res.*, 1977, vol. 82, No. 3, p. 441-454.

129. Garvine R.W. An integral hydrodynamic model of upper ocean frontal dynamics. Pt. I. Development and analysis. *J. Phys. Oceanogr.*, 1979, vol. 9, No. 1, p. 1-18.

130. Garvine R.W. An integral hydrodynamic model of upper ocean frontal dynamics. Pt. II. Physical characteristics and comparison with observations. *J. Phys. Oceanogr.*, 1979, vol. 9, No. 1, p. 19-36.

131. Garvine R.W., Monk J.D. Frontal structure of a river plume. *J. Geophys. Res.*, 1974, vol. 79, No. 15, p. 2251-2259.

132. Gaskin D.E. The evolution, zoogeography and ecology of Cetacean. *Oceanogr. Mar. Biol. Ann. Rev.*, 1976, vol. 14, p. 247-346.

133. Georgi D.T. Fine structure in the Antarctic polar frontal zone: its characteristics and possible relationship to internal waves. *J. Geophys. Res.*, 1978, vol. 83, No. C9, p. 4579-4588.

134. Gibbs R.J. Circulation in the Amazon River estuary and adjacent Atlantic Ocean. *J. Mar. Res.*, 1970, vol. 28, No. 2, p. 113-123.

135. Gregg M.C. Microstructure and intrusions in the California Current. *J. Phys. Oceanogr.*, 1975, vol. 5, No. 2, p. 253-278.

136. Gregg M.C., McKenzie J.H. Thermohaline intrusions lie across isopycnals. *Nature*, 1979, vol. 280, No. 5720, p. 310-311.

137. Griffiths R.C. A study of ocean fronts off Cape San Lucas, Lower California. Spec. Sci. Rept., 1965, No. 499, U.S. Fish and Wildlife Service. 54 p.

138. Griffiths R.C., Simpson J.G. Upwelling and other oceanographic features of the coastal waters of Northeastern Venezuela. Proyecto de Investigation y Desarrollo Pesquero. PNUD. FAO. Serie Recursos y Explotacion Pesqueros, 1972, vol. 2, No. 4. 72 p.

139. Halliwell G.R., Mooers C.N.K. The space-time structure and variability of the shelf water/slope water and Gulf Stream surface temperature fronts and associated warm-core eddies. *J. Geophys. Res.*, 1979, vol. 84, No. C12, p. 7707-7726.

140. Holbrook J.R., Halpern D. Coastal Upwelling Ecosystems Analysis Data Report No. 12. STD Measurements off the Oregon Coast. July, August, 1973. Reports Ref. M 74-17, PMEL/ NOAA, Univ. of Washington, Seattle, 1974.

141. Holladay C.G., O'Brien J.J. Mesoscale variability of sea surface temperature. *J. Phys. Oceanogr.*, 1975, vol. 5, No. 4, p. 761-772.

142. Horne E.P.W. Physical aspects of the Nova Scotian shelf-break fronts. IN: Oceanic Fronts in Coastal Processes. Ed. M. Bowman, W. Esaias. Springer-Verlag, 1978, p. 59-68.

143. Horne E.P.W. Interleaving at the subsurface front in the slope water off Nova Scotia. *J. Geophys. Res.*, 1978, vol. 83, No. C7, p. 3659-3671.

144. Horne E.P.W., Bowman M.J., Okubo A. Crossfrontal mixing and cabbeling. IN: Oceanic Fronts in Coastal Processes. Ed. M. Bowman, W. Esaias. Springer-Verlag, 1978, p. 105-113.

145. Hoskins B.J. Atmospheric frontogenesis models: some solutions. *Quart. J. Roy. Met. Soc.*, 1971, vol. 97, p. 139-153.

146. Hoskins B.J., Bretherton F.P. Atmospheric frontogenesis models: mathematical formulation and solution. *J. Atm. Sci.*, 1972, vol. 29, No. 1, p. 11-37.

147. Howe M.R., Tait R.I. Further observation of thermohaline stratification in the deep ocean. *Deep-Sea Res.*, 1970, vol. 17, No. 6, p. 963-972.

148. Huang N.E., Leitao C.D., Parra C.G. Large-scale Gulf Stream frontal study using GEOS-3 radar altimeter data. *J. Geophys. Res.*, 1978, vol. 83, No. C9, p. 4673-4682.

149. Hurlburt H.E., Thompson J.D. A numerical study of loop current intrusions and eddy shedding. *J. Phys. Oceanogr.*, 1980, vol. 10, No. 10, p. 1611-1651.

150. Huyer A. A comparison of upwelling events in two locations: Oregon and Northwest Africa. *J. Mar. Res.*, 1976, vol. 34, No. 4, p. 531-546.

151. Ichiye T. On the hydrography near Mississippi Delta. *Oceanogr. Mag.*, 1960, vol. 11, No. 2, p. 65-78.

152. Ikeda Y., de Miranda L.B., Rock N.J. Observations on stages of upwelling in the region of Cabo Frio (Brasil) as conducted by continuous surface temperature and salinity measurements. *Bol. Inst. Oceanogr.*, San Paulo, 1974, No. 23, p. 33-46.

153. Johannessen O.M., Foster L.A. A note on the topographically controlled oceanic polar front in the Barents Sea. *J. Geophys. Res.*, 1978, vol. 83, No. C9, p. 4567-4571.

154. Joyce T.M. A note on the lateral mixing of water masses. *J. Phys. Oceanogr.*, 1977, vol. 7, No. 4, p. 626-629.

155. Kao T. Principle stage of wake collapse in a stratified fluid: two-dimensional theory. *Phys. Fluids*, 1976, vol. 19, No. 8, p. 1071-1074.

156. Kao T.W. The dynamics of oceanic fronts. Pt. I. The Gulf Stream. *J. Phys. Oceanogr.*, 1980, vol. 10, No. 4, p. 483-492.

157. Kao T.W., Park C., Pao H.P. Buoyant surface discharge and small-scale oceanic fronts: a numerical study. *J. Geophys. Res.*, 1977, vol. 82, No. 2, p. 1747-1766.

158. Kao T.W., Pao H.P., Park C. Surface intrusions, fronts and internal waves: a numerical study. *J. Geophys. Res.*, 1978, vol. 83, No. C9, p. 4641-4650.

159. Katz E.J. Further study of a front in the Sargasso Sea. *Tellus*, 1969, vol. 21, p. 259-269.

160. Keunecke K.H., Magaard L. Measurements by means of towed thermistor cables and problems of their interpretation with respect to mesoscale processes. *Mémoires Soc. Roy. des ci. de Liège*, 6 sér., 1975, t. 7, p. 147-160.

161. Kitano K. Some properties of the warm eddies generated in the confluence zone of the Kuroshio and Oyashio Currents. *J. Phys. Oceanogr.*, 1975, vol. 5, No. 2, p. 245-252.

162. Klemas V. Remote sensing of coastal fronts and their effects on oil dispersion. *Int. J. Remote Sensing*, 1980, vol. 1, No. 1, p. 11-28.:

163. Knauss J.A. An observation of an oceanic front. *Tellus*, 1957, vol. 9, No. 2, p. 234-237.

164. Koshlyakov M.N., Monin A.S. Synoptic eddies in the ocean. *Ann. Rev. Earth Planet. Sci.*, 1978, vol. 6, p. 495-523.

165. Kuo H. Dynamic instability of two-dimensional nondivergent flow in a barotropic atmosphere. *J. Met.*, 1949, vol. 6, No. 2, p. 105-122.

166. Kupferman S.L., Garfield N. Transport of low-salinity water at the slope water – Gulf Stream boundary. *J. Geophys. Res.*, 1977, vol. 82, No. 24, p. 3481-3486.

167. Kupferman S.L., Klemas V., Polis D.F. Szekielda K.-H. Dynamics of aquatic frontal systems in Delaware Bay (abstract). *EOS, Trans. AGU.*, 1973, vol. 54, p. 302.

168. Lai D.Y., Richardson P.L. Distribution and movements of Gulf Stream Rings. *J. Phys. Oceanogr.*, 1977, vol. 7, No. 5, p. 670-683.

169. Lambert R.B., Sturges W. A thermohaline staircase and vertical mixing in the thermocline. *Deep-Sea Res.*, 1977, vol. 24, No. 3, p. 211-222.

170. Lee T.N. Florida Current spin-off eddies. *Deep-Sea Res.*, 1975, vol. 22, No. 11, p. 753-765.

171. Lee T.N., Mayer D.A. Low frequency current variability and spin-off eddies along the shelf off southeast Florida. *J. Mar. Res.*, 1977, vol. 35, No. 1, p. 193-220.

172. Leetma A., Voorhis A.D. Scales of motion in the Subtropical Convergence Zone. *J. Geophys. Res.*, 1978, vol. 83, No. C9, p. 4589-4592.

173. Legeckis R.V. Application of synchronous meteorological satellite data to the study of time-dependent sea surface temperature changes along the boundary of the Gulf Stream. *Geophys. Res. Lett.* 1975, vol. 2, No. 10, p. 435-438.

174. Legeckis R.V. A survey of worldwide sea surface temperature fronts detected by environmental satellites. *J. Geophys. Res.* 1978, vol. 83, No. C9, p. 4501-4522.

175. Legeckis R.V. Satellite observations of the influence of bottom topography on the seaward deflection of the Gulf Stream off Charleston, South Carolina. *J. Phys. Oceanogr.*, 1979, vol. 9, No. 3, p. 483-497.

176. Levine E.R., White W.B. Thermal frontal zones in the Eastern Mediterranean Sea. *J. Geophys. Res.*, 1972, vol. 77, No. 6, p. 1081-1086.

177. Lonsdale P. Inflow of bottom water to the Panama Basin. *Deep-Sea Res.*, 1977, vol. 24, No. 12, p. 1061-1101.

178. MacVean M.K., Woods J.D. Redistribution of scalars during upper ocean frontogenesis. *Quart. J. Roy. Met. Soc.*, 1980, vol. 106, No. 448, p. 293-311.

179. Marmorino O., Caldwell D.R. Temperature fine structure and microstructure observations in a coastal upwelling region during a period of variable winds (Oregon, summer 1974). *Deep-Sea Res.*, 1978, vol. 25, No. 11, p. 1073-1106.

180. Mascarenhas A.C., de Miranda L.B., Rock N.J. A study of the oceanographic conditions in the region of Cabo Frio. IN: Fertility of the Sea, N.Y.; Lond.; Paris, 1971, vol. 1, p. 285-308.

181. McLeish W. Spatial spectra of ocean surface temperature. *J. Geophys. Res.*, 1975, vol. 75, No. 33, p. 6872-6877.

182. Middleton J.H., Foster T.D. Fine structure measurements in a temperature-compensated halocline. *J. Geophys. Res.*, 1980, vol. 85, No. C2, p. 1107-1122.

183. Miranda (de) L.B., Ikeda Y., de Castro B.M.F. Pereira N.F. Note on the occurence of saline fronts in the Ilha Grande (R.J.) Region. *Bol. Inst. Oceanogr.*, San Paulo, 1977, No. 26, p. 249-256.

184. Mollo-Christensen E., Mascarenhas A.S. Heat storage in the ocean upper mixed layer inferred from Landsat data. *Science*, 1979, vol. 203, No. 4381, p. 653-654.

185. Mooers C.N.K. Frontal dynamics and frontogenesis. IN: Oceanic Fronts in Coastal Processes. Ed. M. Bowman, W. Esaias. Springer-Verlag, 1978, p. 16-22.

186. Mooers C.N.K., Collins C.A., Smith R.L. The dynamic structure of the frontal zone in the coastal upwelling region off Oregon. *J. Phys. Oceanogr.*, 1976, vol. 6, No. 1, p. 3-21.

187. Mooers C.N.K., Flagg C.N., Boicourt W.C. Prograde and retrograde fronts. IN: Oceanic Fronts in Coastal Processes. Ed. M. Bowman, W. Esaias. Springer-Verlag, 1978, p. 43-58.

188. Mooers C.N.K., Garvine R.W., Martin W.W. Summertime synoptic variability of the Middle Atlantic Shelf water/ slope water front. *J. Geophys. Res.*, 1979, vol. 84, No. C8, p. 4837-4854.

189. Morgan C.W., Bishop J.M. An example of Gulf Stream eddy-induced exchange in the Mid-Atlantic Bight. *J. Phys. Oceanogr.*, 1977, vol. 7, No. 3, p. 472-479.

190. Moseley W.B., Del Balzo D.R. Horizontal random temperature structure of the ocean. *J. Phys. Oceanogr.*, 1976, vol. 6, No. 3, p. 267-280.

191. Mysak L.A. On the stability of the California Undercurrent off Vancouver Island. *J. Phys. Oceanogr.*, 1977, vol. 7, No. 6, p. 904-917.

192. Neshyba S., Neal V.T., Denner W.W. Spectra of internal waves: in situ measurements in a multiple-layered structure. *J. Geophys. Res.*, 1971, vol. 76, No. 33, p. 8107-8119.

193. Newton C.W. Synoptic comparisons of jet stream and Gulf Stream systems. IN: The Atmosphere and the Sea in Motion. Ed. B. Bolin. Rockefeller Inst. Press, 1959, p. 288-304.

194. Newton C.W. Fronts and wave disturbances in Gulf Stream and atmospheric jet stream. *J. Geophys. Res.*, 1978, vol. 83, No. C9, p. 4697-4706.

195. NOAA, US Department of Commerce. Oceanic and related atmospheric phenomena as viewed from environmental satellites. Illinois: Walter A. Bohan, 1979.

196. Officer C.B. Physical Oceanography of Estuaries (and Associated Coastal Waters). N.Y.: John Wiley, 1976. 465 p.

197. Okubo A. Advection-diffusion in the presence of surface convergence. IN: Oceanic Fronts in Coastal Processes. Ed. M. Bowman and W. Esaias. Springer-Verlag, 1978, p. 23-28.

198. Orlanski I. Instability of frontal waves. *J. Atm. Sci.*, 1968, vol. 25, No. 2, p. 178-200.

199. Osborn A.R., Burch T.L. Internal solitons in the Andaman Sea. *Science*, 1980, vol. 208, No. 4443, p. 451-460.

200. Osborn T.E., Cox C.S. Oceanic finestructure. *Geophys. Fluid Dyn.*, 1972, vol. 3, p. 321-346.

201. Owen R.W. Fronts and eddies in the sea: mechanisms, interactions and biological effects. IN: Analysis of Marine Ecosystems. Ed. A.R. Longhurst, 1981, Lond.; N.Y., Academic Press, p. 197-233.

202. Pearson C.E., Winter D.F. Two-layer analysis of steady circulation in stratified fjords. IN: Hydrodynamics of Estuaries and Fjords. Ed. J.C.J. Nihoul. Elsevier Oceanogr. Ser., 1975, No. 23, p. 495-514.

203. Perry R.B., Schimke G.R. Large-amplitude internal waves observed off the northwest coast of Sumatra. *J. Geophys. Res.*, 1965, vol. 70, No. 10, p. 2319-2324.

204. Petterssen S. Contribution to the theory of frontogenesis. *Geophys. Publ.*, 1936, vol. 11, No. 6.

205. Petterssen S. Weather analysis and forecasting. 2nd Ed. N.Y.: McGraw Hill, 1956, p. 189-213.

206. Petterssen S., Austin J.M. Fronts and frontogenesis in relation to vorticity. *Papers in Phys. Oceanogr. and Met.*, 1942, vol. 7, No. 2, p. 5-37.

207. Pingree R.D. Effects of mixing and stabilization on phytoplankton distributions on the north-west European continental shelf. NATO School on "Spatial pattern in Plankton Communities" Erice, Sicily, 1977 (preprint).

208. Pingree R.D. Cyclonic eddies and cross frontal mixing. *J. Mar. Biol. Assoc. U.K.*, 1978, vol. 58, p. 955-963.

209. Pingree R.D., Bowman M.J., Esaias W.E. Headland fronts. IN: Oceanic Fronts in Coastal Processes. Ed. M. Bowman, W. Esaias. Springer-Verlag, 1978, p. 78-86.

210. Pingree R.D., Griffiths D.K. Tidal fronts on the shelf seas around the British Isles. *J. Geophys. Res.*, 1978, vol. 83, No. C9, p. 4615-4622.

211. Pingree R.D., Holligan P.M., Mardell G.T. The effects of vertical stability on phytoplankton distribution in the summer on the northwestern European shelf. *Deep-Sea Res.*, 1978, vol. 25, No. 11, p. 1011-1028.

212. Richardson P.L., Cheney R.E., Worthington L.V. A census of Gulf Stream rings, spring 1975. *J. Geophys. Res.*, 1978, vol. 83, No. C12, p. 6136-6144.

213. Roden G. Thermohaline structure, fronts and sea-air energy exchange of the trade wind region east of Hawaii. *J. Phys. Oceanogr.*, 1974, vol. 4, No. 2, p. 168-182.

214. Roden G.I. On North Pacific temperature, salinity, sound velocity and density fronts and their relation to the wind and energy flux field. *J. Phys. Oceanogr.*, 1975, vol. 5, No. 4, p. 557-571.

215. Roden G.I. On the variability of surface temperature fronts in the Western Pacific, as detected by satellite. *J. Geophys. Res.*, 1980, vol. 85, No. C5, p. 2704-2710.

216. Römer E. Stationary current cabbeling in West African waters. *Ann. Hydr.*, 1935, 63, p. 10-17.

217. Römer E. Locally and periodically occurring cabbeling along the Mexican and Central American west coast. *Ann. Hydr. Zweiter Koppen*, Bd. 1936, p. 55-65.

218. Rossby C.G. On the vertical and horizontal concentration of momentum in air and ocean currents. *Tellus*, 1951, vol. 3, No. 1, p. 15-27.

219. Rossby C.G. A comparison of current patterns in the atmosphere and in the ocean basins. *IUGG General Assembly*, Brussels, 1951, p. 9-31.

220. Rossby C.G. On the vertical and horizontal concentration of momentum in air and ocean currents. *Tellus*, 1951, vol. 3, No. 1, p. 15-27.

221. Ruddick B.R., Turner J.S. The vertical scale of double-diffusive intrusions. *Deep-Sea Res.*, 1979, vol. 26, No. 8A, p. 903-914.

222. Ryther J.H., Menzel D.W., Corwin N. Influence of Amazon River outflow on the ecology of the western tropical Atlantic (I). *J. Mar. Res.*, 1967, vol. 25, No. 1, p. 69-83.

223. Saltzman B., Tang C.M. Formation of meanders, fronts and cutoff thermal pools in a baroclinic ocean current. *J. Phys. Oceanogr.*, 1975, vol. 5, No. 1, p. 86-92.

224. Saunders P.M. Anticyclonic eddies formed from shoreward meanders of the Gulf Stream. *Deep-Sea Res.*, 1971, vol. 18, No. 12, p. 1207-1219.

225. Saunders P.M. Space and time variability of temperature in the upper ocean. *Deep-Sea Res.*, 1972, vol. 19, No. 7, p. 467-480.

226. Schumacher A. Current cabbeling, particularly in the Guinea Current and its vicinity. *Ann. Hydr.*, 1935, Bd. 63, Hf. 10, p. 373-382.

227. Simpson J.H. Density stratification and microstructure in the western Irish Sea. *Deep-Sea Res.*, 1971, vol. 18, No. 3, p. 309-319.

228. Simpson J.H., Hunter J.R. Fronts in the Irish Sea. *Nature*, 1974, vol. 250, p. 404-406.

229. Simpson J.H., Pingree R.D. Shallow sea front produced by tidal stirring. IN: Oceanic Fronts in Coastal Processes. Ed. M. Bowman, W. Esaias. Springer-Verlag, 1978, p. 29-42.

230. Smith R.L., Mooers C.N.K., Enfield D.B. Mesoscale studies of the physical oceanography in two coastal upwelling regions: Oregon and Peru. IN: Fertility of the Sea. N.Y.; Lond.; Paris, vol. 2, 1971, p. 513-526.

231. Spilhaus A.F. A detailed study of the surface layers of the ocean in the neighbourhood of the Gulf Stream. *J. Mar. Res.*, 1940, vol. 3, No. 1, p. 51-75.

232. Steeman-Nielsen E. The relationship between phytoplankton and zooplankton in the sea. *Rap. Proc.–Verb. Cons. Intern. Explor. Mer.*, 1962, No. 153, p. 178-182.

233. Stern M.E. Lateral mixing of water masses. *Deep-Sea Res.*, 1967, vol. 14, No. 7, p. 747-753.

234. Stern M.E. Salt fingers convection and the energetics of the general circulation. *Deep-Sea Res.*, Suppl. to vol. 16, 1969, p. 263-267.

235. Stern M.E. Ocean Circulation Physics. N.Y.: Acad. Press., 1975. 246 p.

236. Stommel H. Varieties of oceanographic experience. *Science*, 1963, vol. 139, p. 572-575.

237. Stommel H. The Gulf Stream. 2nd ed. Univ. California Press., 1966. 248 p.

238. Stommel H., Fedorov K.N. Small-scale structure in temperature and salinity near Timor and Mindanao. *Tellus*, 1967, vol. 19, No. 1, p. 88-97.

239. Stommel H., Kozyol P. Adjustment of sea surface temperature to underlying advective fields – an elementary model. *POLYMODE News*, 1976, No. 17.

240. Stretta J.M. Characterisation des situations hydrobiologiques et potentialites de peche thonière au Cap Lopez en juin et juillet 1972 et 1974. *Doc. Sci. Centre Rech. Océanogr. Abidjan*, 1975, vol. 6, No. 2, p. 59-74.

241. Tait R.I., Howe M.R. Some observations of thermohaline stratification in the deep ocean. *Deep-Sea Res.*, 1968, vol. 15, No. 3, p. 275-281.

242. Takano K. A complementary note on the diffusion of seawater river flow off the mouth. *J. Oceanogr. Soc. Japan*, 1955, vol. 11, No. 4, p. 147-149.

243. Tang C.L. Mixing and circulation in the Northwestern Gulf of St. Lawrence: a study of a buoyancy-driven current system. *J. Geophys. Res.*, 1980, vol. 85, No. C5, p. 2787-2796.

244. Tang C.L. Observation of wavelike motion of the Gaspe Current. *J. Phys. Oceanogr.* 1980, vol. 10, No. 6, p. 853-860.

245. Tomczak M. Jr. Continuous measurement of near surface temperature and salinity in the NW African upwelling region between Canary Island and Cap Vert during the winter of 1971-72. *Deep-Sea Res.*, 1977, vol. 24, No. 12, p. 1103-1119.

246. Turner J.S. Salt fingers across a density interface. *Deep-Sea Res.*, 1967, vol. 14, No. 5, p. 599-611.

247. Turner J.S. Buoyancy Effects in Fluids. Cambridge Univ. Press, 1973, 367 p.

248. Turner J.S. Double-diffusive intrusions into a density gradient. *J. Geophys. Res.*, 1978, vol. 83, No. C6, p. 2887-2901.

249. Uda M. Researches on "siome" or current rip in the seas and oceans. *Geophys. Mag.*, 1938, vol. 11, No. 4, p. 306-372.

250. Von Arx W.S., Bumpus D.F., Richardson W.S. On the fine structure of the Gulf Stream front. *Deep-Sea Res.*, 1955, vol. 3, No. 1, p. 46-65.

251. Voorhis A.D. The horizontal extent and persistence of thermal fronts in the Sargasso Sea. *Deep-Sea Res.*, Suppl. to vol. 16, 1969, p. 331-335.

252. Voorhis A.D., Hersey I.B. Oceanic thermal fronts in the Sargasso Sea. *J. Geophys. Res.*, 1964, vol. 69, No. 18, p. 3809-3814.

253. Voorhis A.D., Schroeder E.H., Leetmaa A. The influence of deep mesoscale eddies on the surface temperature in the North Atlantic Subtropical Convergence. *J. Phys. Oceanogr.*, 1976, vol. 6, No. 6, p. 953-961.

254. Voorhis A.D., Webb D.C., Millard R.C. Current structure and mixing in the shelf/slope water front south of New England. *J. Geophys. Res.*, 1976, vol. 81, No. 21, p. 3695-3708.

255. Vukovich F.M. An investigation of a cold eddy on the eastern side of the Gulf Stream using NOAA-2 and NOAA-3 satellite data and ship data. *J. Phys. Oceanogr.*, 1976, vol. 6, No. 4, p. 605-612.

256. Williams A.J. Images of ocean microstructure. *Deep-Sea Res.*, 1975, vol. 22, No. 12, p. 811-829.

257. Williams G.O. Repeated profiling of microstructure lenses with a midwater float. *J. Phys. Oceanogr.*, 1976, vol. 6, No. 3, p. 281-292.

258. Witte E. Theory of current cabbeling. *Gaea*, 1902, Bd. 38, p. 484-487.

259. Woods J.D. The structure of fronts in the seasonal thermocline. Proc. Conf. "Strait of Sicily", Saclancent, 1972, vol. 7, p. 144-151.

260. Woods J.D. Space-time characteristics of turbulence in the seasonal thermocline. *Mém. Soc. Roy. Sci. Liège*, 1973, 6e ser., t. 6, p. 109-130.

261. Woods J.D. Diffusion due to fronts in the rotational subrange of turbulence in the seasonal thermocline. *La Houille Blanche*, 1974, vol. 29, No. 7/8, p. 589-597.

262. Woods J.D. Do waves limit turbulent diffusion in the ocean? *Nature*, 1980, vol. 288, No. 5788, p. 219-224.

263. Woods J.D. The generation of thermohaline finestructure at fronts in the ocean. *Ocean Modelling*, 1980, No. 32, p. 1-4.

264. Woods J.D. Diurnal and seasonal variation of convection in the wind-mixed layer of the ocean. *Quart. J. Roy. Met. Soc.*, 1980, vol. 106, No. 449, p. 379-394.

265. Woods J.D., Minnett P.G. Analysis of mesoscale thermoclinicity with an example from the tropical thermocline during GATE. *Deep-Sea Res.*, 1979, vol. 26, No. 1, p. 85-96.

266. Woods J.D., Wiley R.L., Briscoe M.G. Vertical circulation at fronts in the upper ocean. *Deep-Sea Res.*, 1977, vol. 24 (Suppl.), p. 253-275.

267. Wright D.G. On the stability of a fluid with specialized density stratification. Pt. I. Baroclinic instability and constant bottom slope. *J. Phys. Oceanogr.*, 1980, vol. 10, No. 5, p. 639-666.

268. Wright D.G. On the stability of a fluid with specialized density stratification. Pt. II. Mixed baroclinic-barotropic instability with application to the Northeast Pacific. *J. Phys. Oceanogr.*, 1980, vol. 10, No. 9, p. 1307-1322.

269. Wright W.R. The limits of shelf water south of Cape Cod. *J. Mar. Res.*, 1976, vol. 34, No. 1, p. 1-14.

270. Wu J. Mixed region collapse with internal wave generation in density-stratified medium. *J. Fluid Mech.*, 1969, vol. 35, pt 3, p. 531-544.

271. Wüst G. Stratification and circulation in the Antillean-Caribbean Basins. N.Y.: Lond.: Columbia Univ. Press, 1964.

272. Zenk W. Stratification of water in the Mediterranean, west of Gibraltar. Results of the "Meteor" voyage from January to March 1967. Dissertation, Inst. Meer., Kiel, 1969, 83 p.

273. Zenk W., Katz E.J. On the stationarity of temperature spectra at high horizontal wave numbers. *J. Geophys. Res.*, 1975, vol. 80, No. 27, p. 3885-3891.

SUBJECT INDEX